THE
PROTEIN
KINASE
FactsBook

Protein-Serine Kinases

Other books in the FactsBook Series:

A. Neil Barclay, Albertus D. Beyers, Marian L. Birkeland, Marion H. Brown, Simon J. Davis, Chamorro Somoza and Alan F. Williams
The Leucocyte Antigen FactsBook

Robin E. Callard and Andy J.H. Gearing
The Cytokine FactsBook

Steve Watson and Steve Arkinstall
The G-Protein Linked Receptor FactsBook

Ed Conley
The Ion Channel FactsBook

Rod Pigott and Christine Power
The Adhesion Molecule FactsBook

Shirley Ayad, Ray Boot-Handford, Martin J. Humphries, Karl E. Kadler and C Adrian Shuttleworth
The Extracellular Matrix FactsBook

Robin Hesketh
The Oncogene FactsBook

THE PROTEIN KINASE
FactsBook

Protein-Serine Kinases

Grahame Hardie
Department of Biochemistry
University of Dundee

Steven Hanks
Department of Cell Biology
Vanderbilt University School of Medicine

Academic Press
Harcourt Brace & Company, Publishers
LONDON   SAN DIEGO   NEW YORK   BOSTON
SYDNEY   TOKYO   TORONTO

This book is printed on acid-free paper

ACADEMIC PRESS LIMITED
24–28 Oval Road
LONDON NW1 7DX

United States Edition Published by
ACADEMIC PRESS INC.
San Diego, CA 92101

Copyright © 1995 by
ACADEMIC PRESS LIMITED

A catalogue record for this book is available from the British Library

ISBN 0–12–324720–9 (**Part I**)
ISBN 0–12–324721–7 (**Part II**)
ISBN 0–12–324719–5 (Set)

Typeset by Alden Multimedia, Oxford and Northampton
Printed and bound in Great Britain by
WBC, Bridgend, Mid Glam.

Contents

Contents

Contents

Preface

The genesis of this book was in early 1991 when Susan King, then commissioning editor of the new FactsBook series for Academic Press, approached us individually with the idea of putting together a book on protein kinases. Recognizing the enormity of the task, we each declined her initial invitation to take sole responsibility for the project, but eventually agreed to serve as co-editors. D.G.H.'s acquiescence probably arose from his boredom due to an enforced restriction to an orthopaedic ward, while recovering from a skiing accident in which both legs were broken!

Since 1987, S.K.H. has maintained a database of protein kinase catalytic domain sequences which has proved useful for identifying conserved features of primary structure, and as a tool for classification purposes. This database served as the starting point for the book. Most of the information in other volumes of the FactsBook series has been entirely compiled by the academic editors, which has many advantages in terms of uniformity of style. However, the protein kinase field is simply too immense for this to be feasible, and a very early decision was made to enlist the help of contributors who were experts in the study of the individual enzymes. Of course this approach brought its own problems, related to the difficulty of obtaining submissions from some 140 individuals by the agreed deadline. At least one of the potential contributors we contacted did not reply for the very understandable reason that he had died, while others had moved three or four times since publishing their original paper on their protein kinase! Then there was the problem of contributors interpreting the guidelines in different ways, and providing their entries in different styles and formats. This was by no means their fault, as the guidelines themselves evolved to a certain extent as we gained experience.

Despite these inevitable difficulties, the majority of contributors were enthusiastic about the project, and provided their entries on time and in good shape, and we would like to express our gratitude to them all. They made this book possible. Roger Perlmutter and Michael Jaye deserve special thanks for agreeing to compile several important entries at the last minute. Although editing the individual entries took several months, the contributors had the opportunity to update their submissions shortly before the book went to press, so the final versions should be reasonably up to date. Because of restrictions on overall length necessary to maintain the format of the Factsbook series, we very much regret that we were unable to include most of the kinases published after April 1993, although these are included in the Classification Table in the introductory chapter by SKH and Tony Hunter.

Despite these attempts to keep the project down to a manageable size, a rather late decision had to be taken to split the book into two volumes. The only logical manner to do this seemed to be to include the protein-serine/threonine kinases (including some which phosphorylate tyrosine on there target proteins, e.g. Wee1) in one volume, and the "conventional" protein-tyrosine kinase subfamily in the other. To allow each volume to stand alone if necessary, the introductory chapters are reproduced in both, plus a common index covering both volumes.

Finally, it is hard to conceive how this project could have been pulled off in the days before floppy disks, FAX and electronic mail, and we feel the book is a small testament to the power of modern communication. Given the continued expansion of the number of defined protein kinases, electronic methods of publication may also be necessary if future editions are to become a complete resource. Whether the editors can also be replaced by computer programs remains to be seen!

Left: *Grahame Hardie*, Right: *Steven Hanks*

Abbreviations

The list below only includes abbreviations which occur in more than one entry. Abbreviations and acronyms for the protein kinases themselves are not included in the list, but may be found in the index. Standard one- and three-letter abbreviations for amino acids are used throughout the book.

A. thaliana	*Arabidopsis thaliana*
bAR	b-adrenergic receptor
BDNF	brain derived neurotrophic factor
C subunit	catalytic subunit
c-	cellular (protooncogene) form of
C-terminal	carboxyl-terminal
C. elegans	*Caenorhabditis elegans*
CaM	calmodulin
cAMP, cyclic AMP	cyclic adenosine-3′,5′-monophosphate
Cdk	cyclin-dependent kinase
cDNA	complementary DNA
cGMP, cyclic GMP	cyclic guanosine-3′,5′-monophosphate
CNS	central nervous system
CREB	cAMP response element binding protein
CSF	cytostatic factor
CSF1	colony stimulating factor-1
D. discoideum	*Dictyostelium discoideum*
D. melanogaster	*Drosophila melanogaster*
DM	myotonic dystrophy
dsRNA	double stranded RNA
EGF	epidermal growth factor
eIF2, eIF-2B	eukaryotic initiation factor-2, -2B
EMS	ethyl methane sulphonate
FGF-1, -n	fibroblast growth factor-1, -n
G protein	GTP-binding protein
G-CSF	granulocyte-colony stimulating factor
GAP	GTPase activating protein
GM-CSF	granulocyte/macrophage-colony stimulating factor
HGF	hepatocyte growth factor
HSV-1	herpes simplex virus type-1
IG$_{50}$	concentration giving 50% inhibition
IFN-a, -b, -g	interferon-a, -b, -g
Ig, IgM	immunoglobulin, immunoglobulin-M
IGF-1, -2	insulin like growth factor-1, -2
IL-1, IL-n	interleukin-1, interleukin-n
IRS-1	insulin receptor substrate-1
JH1, JHn	Jak homology-1, -n
kb	kilobases
kDa	kilodaltons
KL	kit ligand

MAP kinase	mitogen-activated protein kinase
MCGF	mast cell growth factor
MOL. WT	molecular weight
N-terminal	amino-terminal
NGF	nerve growth factor
NLS	nuclear localization signal
NT-1, *-n*	neurotrophin-1, *-n*
P-Tyr	phosphotyrosine
P. falciparum	*Plasmodium falciparum*
PCR	polymerase chain reaction
PDGF	platelet-derived growth factor
PDH	pyruvate dehydrogenase
PH	pleckstrin homology
pI	isoelectric point
PI	phosphatidyl inositol
PK	protein kinase
PKI	heat-stable protein inhibitor of cAMP-dependent protein kinase
PLC, PLC-g	phospholipase C, phospholipase C-g
PNS	peripheral nervous system
PS	pseudosubstrate site
PTK	protein-tyrosine kinase
R subunit	regulatory subunit
RFLP	restriction fragment length polymorphism
S. cerevisiae	*Saccharomyces cerevisiae*
S. pombe	*Schizosaccharomyces pombe*
SCF	stem cell factor
SDS-PAGE	sodium dodecyl sulphate-polyacrylamide gel electro-phoresis
SF	scatter factor
SH1, SH2, SH3	Src homology-1, -2, -3
SLF	steel factor
T. parva	*Theileria parva*
TCR	T cell receptor
TGF-a, -b	transforming growth factor-a, -b
TNF-a	tumour necrosis factor-a
TPA	12-O-tetradecanoylphorbol 13-acetate
ts	temperature sensitive
v-	viral (oncogene) form of
X. laevis	*Xenopus laevis*

THE INTRODUCTORY CHAPTERS

1 Introduction

AIMS OF THE BOOK

Our aim in compiling this book was to provide a handy reference for people working in the cell signalling field, so that basic information about a particular protein kinase could be retrieved rapidly and with the minimum of searching in libraries and databases. To keep the final purchase price reasonable, and to ensure a size and general format in line with other volumes in the FactsBook series, the publishers imposed a limit on overall length. This created particular difficulties for this volume, since the protein kinases are now the largest known protein superfamily, with the list of members still expanding at a remarkable rate.

The protein kinases we have included in the book were based on literature searches compiled up to March 1993. We decided to restrict our list to those which had been defined by sequencing of the catalytic subunit, with the exception of a small number which were well defined biochemically but not cloned. In the end, sequences appeared for most of the latter during production of the book, and the only entry which does not have a sequence is **EF2K**. The March 1993 cutoff was arbitrary, and we are well aware that a large number of new protein kinases have appeared since then. It was unfortunately not possible to include them because of the retrictions on overall length, and this is therefore one book which is out of date even before it is published. We have tried to minimize this problem in two ways. First, Table 1, Chapter 2 includes all protein kinases of which we were aware at the time of going to press, although not all have entries in the book. Secondly, although assembling and editing the entries took almost a year, all contributors were given the opportunity to update their entries shortly before the book went to press.

Having assembled a list in March 1993, the decision was taken to include only one representative of each protein kinase from vertebrates, rather than having separate entries for homologues from different vertebrate species. However due to the importance of *D. melanogaster, S. cerevisiae, S. pombe, C. elegans* and *D. discoideum* for genetic and developmental studies, we have included homologues from those species even where the kinase was originally characterized in vertebrates. We have also included, where available, one example from higher plants, and a few protein kinases of general interest from other species (e.g. **NimA** from *Aspergillus nidulans*). Applying these criteria was not always straightforward, and with hindsight we know that one or two kinases that should have been included were left out, while in one or two cases we have included as separate entries what have turned out to be mere homologues. The contributor we consulted for each entry was usually, but not always, the senior author from the original sequence paper, and in the case of vertebrates the choice of which species to discuss in detail was left up to him or her, although we had a slight preference for the human sequence. From the references and database accession numbers given, the reader should be able to find information on homologues in other species.

NOMENCLATURE

Nomenclature of protein kinases is a thorny and confused topic, with different members of the superfamily being named either after their substrate(s), after their regulatory molecules, after some aspect of the phenotype of a mutant, or by some arbitrary name or number. We hope that this book will be helpful to the uninitiated in sorting out the morass of alternative names and acronyms. The conventions we have used in naming the protein kinases are discussed below.

Acronyms

All protein kinases have been given an acronym, often derived from the gene name. These acronyms are used to cross-reference protein kinases throughout the book, using bold type. Thus **"Xyz"** in the text means that there is an entry for the protein kinase "Xyz" elsewhere in the book. We have tried to use a consistent style in devising acronyms, which has sometimes meant changing, with their permission, those originally given by contributors. There is a widely accepted convention that acronyms for genes are given in italics, whereas those for gene products are in plain type. However, aside from that there appears to be little conformity in the manner of presenting acronyms, even between researchers working with different yeast species. Wherever possible we adopted the convention now agreed for *S. cerevisiae*, in which gene names (wildtype) are given in upper case and italics, whereas names of the corresponding gene products are given in plain type with only the first letter in upper case. We did not feel that it was necessary to add the suffix p to indicate the gene product. We have retained the use of upper case throughout for some protein kinases which were well characterized biochemically before they were cloned, e.g. **PKA** or **AMPK**. Also, some contributors working with *C. elegans* asked that we stick to the convention in their research community, which is for the gene product to be in upper case throughout, e.g. **LET-23**. Where protein kinases from different species would otherwise have the same acronym, we have used prefixes to avoid confusion. Thus vertebrate examples have no prefix, whereas examples from *D. melanogaster*, *S. cerevisiae*, *S. pombe*, *C. elegans*, *D. discoideum* and higher plants have the prefixes "Dm", "Sc", "Sp", "Ce", "Dd" and "p", respectively. For example, **SpCdc2** refers to the originally described *S. pombe* gene product, whereas **Cdc2** refers to the vertebrate homologue.

Full names

The original name given to a protein or gene at the time it was first isolated is not always, with the benefit of hindsight, the best. Nevertheless researchers often develop a strong emotional attachment to the name they coined, which they see as being associated with the priority of their discovery. Although we tactfully suggested to one or two contributors that they might change the name to one which seemed less confusing to an outsider, in general the full name given is the one chosen by the contributor. However all common alternative names are also given below the full name. In one or two cases we have also given commonly used alternative names when cross-referencing protein kinases, e.g. "MAP kinase/**Erk1/2**".

FINDING PROTEIN KINASE ENTRIES

Although we considered ordering entries alphabetically by acronym, we decided that this would not necessarily be helpful due to the large variety of different names and acronyms in use for a typical protein kinase. The entries have therefore been ordered into subfamilies defined by sequence relationships, using principles discussed in Chapter 2. The quickest way to find the entry for a particular protein kinase is likely to be the index, in which all acronyms and alternative names have been listed. Protein kinases which do not have an entry, but are listed in Chapter 2, also appear in the index, as do other genes or proteins which, although not themselves protein kinases, are mentioned in entries.

ORGANIZATION OF THE DATA WITHIN AN ENTRY

Although the entries are hopefully self-explanatory, the format of a typical entry is explained below. If any section(s) is (are) missing, either the information is not known, or it was not provided by the contributor.

Introductory paragraph

This gives brief details on discovery and known functions.

Subunit structure and isoforms

This information is presented for each subunit and isoform as a table:

SUBUNIT	AMINO ACIDS	MOL. WT	SDS-PAGE
*	**	***	****

*Name of subunit or isoform.
**Number of amino acids in open reading frame.
***Calculated molecular weight of open reading frame (i.e. ignoring processing).
****Apparent molecular weight of mature product on SDS-PAGE.

Information on quarternary structure is also given where known.

Genetics

Chromosomal locations and brief details of mutant forms are given.

Domain structures

With the exception of a few protein kinases which have a catalytic domain only, diagrams of domain structure are included. These have been drawn to a uniform format. With a few exceptions, the same scale was used, but the scale is always given in any case. Although the domains are labelled, a key to the hatch patterns used to denote various types of domain (e.g. kinase, SH2) is shown in Fig. 1. Locations of known physiological phosphorylation sites are also marked.

Database accession numbers

A table of accession numbers for the PIR, SWISSPROT and EMBL/GENBANK databases is given, including those for homologues whose sequence may not be presented in the book.

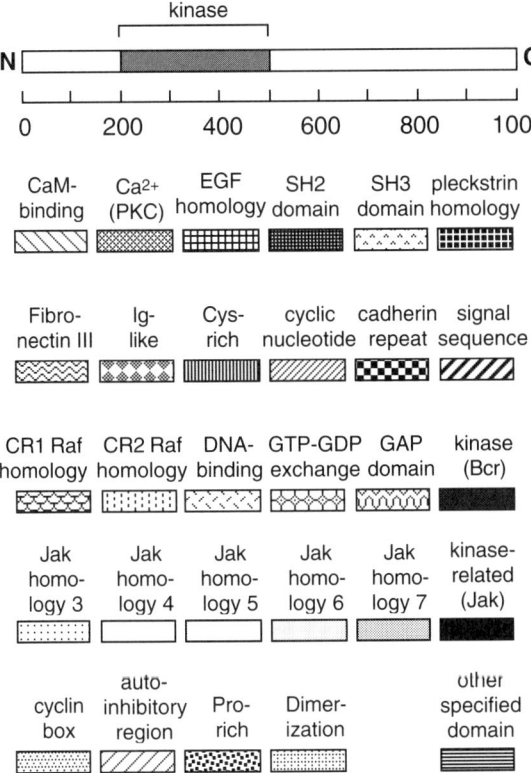

Figure 1: *General format of domain diagram (Top), and key to hatch patterns used to represent different types of domain (Bottom).*

Amino acid sequences

Sequences of catalytic and regulatory subunits, including isoforms, from a single species. Sequences are presented in single letter code with 50 residues per line. Residues are generally numbered from the initiating methionine, but in certain cases where this has been the previous practice, numbering is from the N-terminus of the mature processed protein. Residues of interest (e.g. sites of covalent modification, transmembrane domains) are underlined, and brief details given after the sequence. Phosphorylation sites are double underlined.

Homologues in other species

Brief details of known homologues or related sequences in other species are given.

Physiological substrates and specificity determinants

This gives a description of physiological substrates, where known, or artificial substrates used for assay. Description of consensus recognition sequence and tables of sequences around known sites are also included.

Assay

Brief details of enzymatic kinase assay are given where available; full details are given in references.

Enzyme activators and inhibitors

This gives information on activators and inhibitors which can be used as experimental probes.

Pattern of expression

Expression of mRNA and/or protein, both temporal and spatial (e.g. tissue distribution) is given.

References

A limited set of references is listed, with key references highlighted in bold.

2 The Eukaryotic Protein Kinase Superfamily

Steven K. Hanks and Tony Hunter (Vanderbilt University School of Medicine, Nashville, USA, and The Salk Institute, San Diego, USA)

The largest known protein family is made up of protein kinases identified largely from eukaryotic sources. (The term "superfamily" will be used hereafter to distinguish this broad collection of enzymes from smaller, more closely related subsets which have been commonly referred to as "families".) These enzymes use the gamma phosphate of ATP (or GTP) to generate phosphate monoesters utilizing protein alcohol groups (on serine and threonine) and/or protein phenolic groups (on tyrosine) as phosphate group acceptors. They are related by virtue of their homologous "kinase domains" (also known as "catalytic domains") which consist of ≈250–300 amino acid residues [reviewed in refs 1–3, and see below]. Over the past 15 years or so, previously unrecognized members of the eukaryotic protein kinase superfamily have been uncovered at an exponentially increasing rate, and currently appear in the literature almost weekly. This pace of discovery can be attributed to the development of molecular cloning and sequencing technologies and, more recently, to the advent of the polymerase chain reaction (PCR) which encouraged the use of homology-based cloning strategies. Consequently, at the time of this writing (April 1994), over 175 different superfamily members (products of distinct paralogous genes) had been recognized from mammalian sources alone! The prediction made several years ago[4], that the mammalian genome contains about one thousand protein kinase genes (roughly 1% of all genes) would still appear to be within reason.

In addition to mammals and other vertebrates, eukaryotic protein kinase superfamily members have been identified and characterized from a wide range of other animal phyla, as well as from plants, fungi and microorganisms. Hence, the protein kinase progenitor gene can be traced back to a time prior to the evolutionary separation of the major eukaryotic kingdoms. Studies on the budding and fission yeasts, *Saccharomyces cerevisiae* and *Schizosaccharomyces pombe*, have been particularly fruitful in the recognition of new protein kinases. In these genetically tractable organisms, the powerful approach of mutant isolation and cloning by complementation has netted dozens of protein kinase genes required for numerous aspects of cell function. In many cases now, vertebrate counterparts have been found for these, leading to a growing awareness that protein phosphorylation pathways that regulate basic aspects of cell physiology have been maintained throughout the course of eukaryotic evolution.

While the overwhelming majority of protein kinases identified from eukaryotic sources belong to this superfamily, a small but growing number of such enzymes do not qualify as superfamily members. Most of these are related to the prokaryotic "protein-histidine kinase" family (see below). Included in this category are a putative ethylene receptor encoded by the flowering plant *ETR1* gene[5], the product of the budding yeast *SLN1* gene[6] thought to be involved in relaying nutrient information to elements controlling cell growth and division, the mitochondrial branched-chain α-ketoacid dehydrogenase kinase (**BKDK**), and the mitochondrial pyruvate dehydrogenase kinase (**PDK**). In addition, the **Bcr** protein encoded by the *breakpoint cluster region* gene involved in the Philadelphia chromosome translocation and the

"A6" kinase isolated by expression cloning using an antiphosphotyrosine antibody[7] have kinase domains unrelated to any known eukaryotic or prokaryotic kinase. A single "true" protein-histidine kinase has also been purified from *S. cerevisiae*[8]. Whether this is related to the eukaryotic protein kinase superfamily is not yet known.

What about the prokaryotes? It has been known for years that protein phosphorylation events play key regulatory roles in numerous bacterial cell processes including chemotaxis, bacteriophage infection, nutrient uptake, and gene transcription [reviewed in refs 9 and 10]. The bacterial protein kinases have been divided into three general classes[11]: (1) protein-histidine kinases such as those functioning in two-component sensory regulatory systems (strictly speaking, these are protein-aspartyl kinases as autophosphorylation on histidine is an intermediary step in phosphotransfer to an aspartyl residue in the response-regulator protein); (2) phosphotransferases such as those of the phosphenol pyruvate-dependent phosphotransferase system involved in sugar uptake; and (3) protein-serine/threonine kinases such as isocitrate dehydrogenase kinase/phosphatase. Amino acid sequences have been determined for members of each class, and all were found to be unrelated to the eukaryotic protein kinase superfamily. The bacterial protein kinases belonging to these families are not covered in this book.

Recently, however, clear homologues of the eukaryotic protein kinases have been identified from two species of bacteria, *Yersinia pseudotuberculosis*[12] and *Myxococcus xanthus*[13, 14]. Still, these may be rather special cases. The eukaryotic-like protein kinase "YpkA" from the pathogenic enterobacteria *Y. pseudotuberculosis* is encoded by a plasmid essential for the virulence of this infectious organism. In addition to YpkA, at least two other proteins encoded by genes residing on the virulence plasmid exhibit high similarity to eukaryotic proteins. Thus it seems likely that the virulence plasmid genes were transduced from a eukaryotic host by horizontal transfer. The myxobacterium *M. xanthus* presents a different and perhaps more intriguing picture. Application of the PCR homology-based cloning strategy revealed that at least eight genes encoding members of the eukaryotic protein kinase superfamily are present in the genome of this species[14]. The myxobacteria are unusual prokaryotes in that they undergo a complex developmental cycle upon nutrient depletion, much like that of the eukaryotic slime mould *Dictyostelium*. Given the penchant for protein kinase involvement in regulating growth and differentiation of eukaryotic cells, it is attractive to speculate that the eukaryotic-like protein kinases in *M. xanthus* are specifically involved in regulating their developmental cycle. Indeed, one of these kinases, **Pkn1**, was shown to be required for proper fruiting body formation. Significantly, neither the PCR approach applied to *E. coli*[14], nor extensive sequencing of the *E. coli* genome (now >30% complete), has yielded eukaryotic-like protein kinases. Hence, genes encoding members of the eukaryotic protein kinase superfamily may be present only in bacteria that can undergo a developmental cycle. Potentially, these genes represent the ancestors for the entire eukaryotic protein kinase superfamily.

THE HOMOLOGOUS KINASE DOMAINS

The kinase domains of eukaryotic protein kinases impart the catalytic activity. Three separate roles can be ascribed to the kinase domains: (1) binding and orientation of the ATP(GTP) phosphate donor as a complex with divalent cation (usually Mg^{2+} or Mn^{2+}); (2) binding and orientation of the protein (or peptide) substrate; and (3) transfer of the γ-phosphate from ATP(GTP) to the acceptor hydroxyl residue (serine, threonine or tyrosine) of the protein substrate.

Conserved features of primary structure

The total number of distinct kinase domain amino acid sequences available is now approaching 400 (Table 1). Included in this total are the vertebrate enzymes encoded by distinct paralogous genes, their presumed functional homologues from invertebrates and simpler organisms (encoded by orthologous genes), and those identified from lower organisms and plants for which vertebrate equivalents have not been found. Conserved features of kinase domain primary structure have previously been identified through an inspection of multiple amino acid sequence alignments[1-3]. The large number of sequences now available precludes showing an alignment containing all known kinase domains. Thus in the alignment given in Fig. 1 only 60 different kinase domain sequences are shown. These are drawn, however, from the widest possible sampling of the superfamily and thus provide a good representation of the known primary structures. The kinase domains are further divided into 12 smaller subdomains (indicated by Roman numerals), defined as regions never interrupted by large amino acid insertions and containing characteristic patterns of conserved residues (consensus line in Fig. 1).

Twelve kinase domain residues are recognized as being invariant or nearly invariant throughout the superfamily (conserved in over 95% of 370 sequences), and hence strongly implicated as playing essential roles in enzyme function. Using the type α cyclic AMP-dependent protein kinase catalytic subunit (**PKA**-Cα) as a reference point, these are equivalent to G50 and G52 in subdomain I, K72 in subdomain II, G91 in subdomain III, D166 and N171 in subdomain VIB, D184 and G186 in subdomain VII, E208 in subdomain VIII, D220 and G225 in subdomain IX, and R280 in subdomain XI. The pattern of amino acid residues found within subdomains VIB, VIII and IX have been particularly well-conserved among the individual members of the different protein kinase families, and these motifs have been targeted most frequently in PCR-based homology cloning strategies aimed at identifying new family members.

Relationship between conserved subdomains, higher order structure and catalytic mechanism.

The homologous nature of the kinase domains implies that they all fold into topologically similar three-dimensional core structures, and impart phosphotransfer according to a common mechanism. The larger inserts found within some kinase domains are likely to represent surface elements which do not disrupt the basic core structure. With the solution of the crystal structure of mouse PKA-Cα, in a binary complex with a pseudosubstrate peptide inhibitor (PKI (5–24); TTYAD-FIASGRTGRRNAIHD, the underlined alanine substituting for the serine phospho-acceptor), the general topology of a protein kinase catalytic core structure was revealed for the first time[15, 16]. Later, structures of ternary complexes of PKA-Cα, the pseudosubstrate inhibitor, and either MgATP or MnAMP-PNP (a MgATP analogue) were solved[17, 18]. As a consequence of these studies, precise functional roles for most of the highly conserved kinase domain residues have now been assigned.

```
subdomain            I                                    II
consensus o-----og-G-og-v---------------------oaoK-o-----------
2°struct  < b1>          <- b2->                 <-- b3 -><-aB->
          4 3    50           60                70          80
PKA-Cα    FERIKTLGTGSFGRVMLVKHKE-------------TGNHYAMKILDKQKVVKLK---------
PKG-I     FNIIDTLGVGGFGRVELVQLKSE------------ESKTFAMKILKKRHIVDTR---------
cPKCα     FNFLMVLGKGSFGKVMLADRKG------------TEELYAIKILKKDVVIQDD---------
βARK1     FSVHRIIGRGGFGEVYGCRKAD------------TGKMYAMKCLDKKRIKMKQ---------
S6K       FELLRVLGKGGYGKVFQVRKVTGAN---------TGKIFAMKVLKKAMIVRNAK--------
RSK1(Nt)  FELLKVLGQGSFGKVFLVRKVTRPD---------NGHLYAMKVLKKATLKVR----------
DMPK      FEILKVIGRGAFSEVAVVKMKQ------------TGQVYAMKIMNKWDMLKRG---------
CaMK2α    YQLFEELGKGAFSVVRRCVKVL------------AGQEYAAKIINTKKLSAR----------
skMLCK    MNSKEALGGGKFGAVCTCTEKS------------TGLKLAAKVIKKQTPK------------
Mre4      EITNRIVGNGTFGHVLITHNSKERDEDVCY-----HPENYAVKIIKL----------------
PhKγM     YEPKEILGRGVSSVVRRCIHKP------------TCKEYAVKIIDVTGGGSFSAEEVQE----
Kin1      WEFVETVGAGSMGKVLAKHRY-------------TNEVCAVKIVNRATKAFLHKEQMLP{20}
Snf1      YQIVKTLGEGSFGKVKLAYHTT------------TGQKVALKIINKKVLAKSD---------
Polo      YKRMRFFGKGGFAKCYEIIDVE------------TDTVFAGKIVSKKLMIKHN---------
Cdc5      YHRGHFLGEGGFARCFQIKDD-------------SGEIFAAKTVAKASIKSEK---------
Cdk2      FQKVEKIGEGTYGVVYKARNKL------------TGEVVALKKIRLDTETEG---------
Erk2      YTNLSYIGEGAYGMVCSAYDNL------------NKVRVAIKKISPFEHQT----------
GSK3α     YTDIKVIGNGSFGVVYQARLAE------------TRELVAIKKVLQ--------------
CK2α      YQLVRKLGRGKYSEVFEAINIT------------NNEKVVVKILKPV-------------
Clk       DEIVDTLGEGAFGKVVECADHKA-----------GGRHVAVKIVKNVDR-----------
Ire1      VVSEKILGYGSSGTVVFQGSF-------------QGRPVAVKRMLID-------------
Cdc7      YKLIDKIGEGTFSSVYKAKDITGKITKKFASHFWNYGSNYVALKKIYVTS---------
Cot       NIGSDFIPRGAFGKVYLAQDIK------------TKKRMACKLIPVD-------------
YpkA      VAETDKFAEGESHISIIETK--------------DKQRLVAKIERSIAE-----------
MEK1      FEKISELGAGNGGVVFKVSHKP------------SGLVMARKLIHLEIKPA----------
Ste7      LVQLGKIGAGNSGTVVKALHVP------------DSKIVAKKTIPVEQNNST---------
Ste11     WLKGACIGSGSFGSVYLGMNAH------------TGELMAVKQVEIKNNNIGVPTDNNK{35}
Nek1      YVRLQKIGEGSFGKAVLVKSTE------------DGRHYVIKEINISRMSDK---------
NIMA      YEVLEKIGCGSFGIIRKVKRKS------------DGFILCRKEINYIKMSTK---------
Fused     YAVSSLVGQGSFGCVYKATRKD------------DSKVVAIKVISKRGRATK---------
NinaC     FEIYEEIAQGVNAKVFRAKELD------------NDRIVALKIQHYD--------------
Ste20     YANLVKIGQGASGGVYTAYEIG------------TNVSVAIKQMNLEKQP-----------
Cdc15     YHLKQVIGRGSYGVVYKAINKH------------TDQVVAIKEVVYENDE-----------
Npr1      IKTGADLGAGAGGSVKLAQRIS------------DNKIFAVKEFRTKFENESKRD-------
Pim1      YQVGPLLGSGGFGSVYSGIRVS------------DNLPVAIKHVEKDRISDWGELP------
Ran1      LRFVSIIGAGAYGVVYKAEDIY------------DGTLYAVKALCKDGLNEK---------
Esk       YSILKQIGSGGSSKVFQVLNE-------------KKQIYAIKYVNLEEADNQ---------
Elm1      YTLGVSAGSGQFGYVRKAYSST------------LGKVVAVKIIPKKPWNAQQYSVNQV{13}
SpWee1    FRNVTLLGSGEFSEVFQVEDPVE-----------KTLKYAVKKLKVKFSGPK---------
Wee1(Hs)  FHELEKIGSGEFGSVFKCVKRL------------DGCIYAIKRSKKPLAGSV---------
PKR       FKEIELIGSGGFGQVFKAKHRI------------DGKTYVIKRVKY--------------
Gcn2      LKRLNFSGQGAFGKVARNAL--------------DSRYVAIKKIRNTEE-----------
CK1α      YKLVRKIGSGSFGDIYLAINIT------------NGEEVAVKLESQKA------------
Pkn1      FRLVRRLGRGGMGAVYLGEHVS------------IGSRVAVKVLHAHLTMYPE--------
Ykl516    WKKVRPIGSGNFSTVLLYELMDQSNP--------KLKQVAVKRLKYPEELSNVEQINTSL{8}
Mos       VCLLQRLGAGGFGSVYKATYR-------------GVPVAIKQVNKCTKNRLA---------
ZmPK1     RKFKVELGRGESGTVYKGVLE-------------DDRHVAVKKLENVRQ-----------
Pelle     WSPDNRLGQGGFGDVYRGKWK-------------QLDVAIKVMNYRSPNIDQKMV------
TGFβRII   IELDTLVGKGRFAEVYKAKLKQNTSE--------QFETVAVKIFPYDHY-----------
ActRII    LQLLEVKARGRFGQVWKAQLLN------------EYVAVKIFPIQDK-------------
Raf-1     VMLSTRIGSGSFGTVYKGKWH-------------GDVAVKILKVVDPTPE-----------
Sp1A      LEFGQTIGKGFFGEVYRGTWRE------------TDVAIKIIYRDQFKTKS---------
Src       LRLEVKLGQGCFGEVWMGTWNG------------TTRVAIKTLKPGTM-----------
EGFR      FKKIKVLGSGAFGTVYKGLWIPEGEK--------VKIPVAIKELREATSPK---------
PDGFRβ    LVLGRTLGSGAFGQVVEATAHGLSHSQ-------ATMKVAVKMLKSTARSS---------
```

```
      III                          IV                          V
-------E--oo-------------------h--oo-o---o--------------ooooo*oo------o---o--------
<---- aC ---->                    <-b4->              <-b5->           <- aD ->
      90              100         110                 120              130
QIEHTLNEKRILQAV---------NFPFLVKLEFSFKDN---------SNLYMVMEYVPG--GEMFSHLRRIGR---
QQEHIRSEKQIMQGA---------HSDFIVRLYRTFKDS---------KYLYMLMEACLG--GELWTILRDRGS---
DVECTMVEKRVLALLD--------KPPFLTQLHSCFQTV---------DRLYFVMEYVNG--GDLMYHIQQVGK---
GETLALNERIMLSLVSTG------DCPFIVCMSYAFHTP---------DKLSFILDLMNG--GDLHYHLSQHGV---
DTAHTKAERNILEEV---------KHPFIVDLIYAFQTG---------GKLYLILEYLSG--GELFMQLEREGI---
DRVRTKMERDILADV---------NHPFVVKLHYAFQTE---------GKLYLILDFLRG--GDLFTRLSKEVM---
EVSCFREERDVLVNG---------DRRWITQLHFAFQDE---------NYLYLVMEYYVG--GDLLTLLSKFGER---
DHQKLEREARICRLL---------KHPNIVRLHDSISEE---------GHHYLIFDLVTG--GELFEDIVAREY---
DKEMVMLEIEVMNQL---------NHRNLIQLYAAIETP---------HEIVLFMEYIEG--GELFERIVDEDYH--
KPNKFDKEARILLRL---------DHPNIIKVYHTFCDRN--------NHLYIFQDLIPG--GDLFSYLAKGDCLTS
LREATLKEVDILRKVS--------GHPNIQLKDTYETN----------TFFFLVFDLMKK--GELFDYLTEKVT---
RDKRTIREASLGQIL---------YHPHICRLFEMCTLS---------NHFYMLFEYVSG--GQLLDYIIQHGS---
MQGRIEREISYLRLL---------RHPHIIKLYDVIKSK---------DEIIMVIEYIAG--NELFDYIVQRDK---
QKEKTAQEITIHRSL---------NHPNIVKFHNYFEDS---------QNIYIVLELCKK--RSMMELHKRRKS---
TRKKLLSEIQIHKSM---------SHPNIVQFIDCFEDD---------SNVVYILLEICPN--GSLMELLKRRKV---
VPSTAIREISLLKEL---------NHPNIVKLLDVIHTE---------NKLYLVFEFLHQ---DLKKFMDASALTG-
YCQRTLREIKILLRF---------RHENIIGINDIIRAPTIEQM----KDVYIVQDLMET---DLYKLLKTQH----
DKRFKNRELQIMRKL---------DHCNIVRLRYFFYSSGEKKDE---LYLNLVLEYVPE---TVYRVARHFTKAKL
KKKKIKREIKILENLR--------GGPNIITLADIVKDPVS-------RTPALVFEHVNN--TDFKQLYQT------
YCEAARSEIQVLEHLNT-------TDPNSTFRCVQMLEWFEHH-----GHICIVFELLGL---STYDFIKENGFLP-
FCDIALMEIKLLTESD--------DHPNVIRYYCSETTD---------RFLYIALELCNL---NLQDLVESKNVSDE
SPQRIYNELNLLYIMT--------GSSRVAPLCDAKRVR---------DQVIAVLPYYPH--EEFRTFYRD-----
QFKPSDVEIQACF-----------RHENIAELYGAVLWG---------ETVHLFMEAGEG--GSVLEKLESCGP---
GHLFAELEAYKHIYKTAG------KHPNLANVHGMAVVPYGNRKEEALLMDEVDGWRCSDTLRTLADSWKQGKINSE
IRNQIIRELQVLIIEC         NEPYIVCFYCAFYSD---------GEISICMEHMDG--GSLDQVLKKAGR---
IINQLVRELSIVKNVK--------PHENIITFYGAYYNQHIN------NEIIILMEYSDC--GSLDKILSVYKRFVQ
MVDALQHEMNLLKEL---------HHENIVTYYGASQEG---------GNLNIFLEYVPG--GSVSSMLNNYGP---
ERQESRREVAVLANM---------KHPNIVQYKESFEEN---------GSLYIVMDYCEG--GDLFKRINAQKGAL-
EREQLTAEFNILSSL---------RHPNIVAYYHREHLKAS-------QDLYLYMEYCGG--GDLSMVIKNLKRTNK
ELKNLRRECDIQARL---------KHPHVIEMIESFESK---------TDLFVVTEFALM---DLHRYLSYNGA---
EEHQVSIEEEYRTLRDYC------DHPNLPEFYGVYKLSKPNGP----DEIWFVMECLAG--GTAVDMVNKLLKLDR
KKELIINEILVMKGS---------KHPNIVNFIDSYVLK---------GDLWVIMEYMEG--GSLTDVVTHCI----
ELNDIMAEISLLKNL---------NHNNIVKYHGFIRKS---------YELYILLEYCAN--GSLRRLISRSSTG--
YVKKITSEYCIGTTL---------NHPNIIETIEIVYEN---------DRILQVMEYCEY---DLFAIVMSNK----
NGTRVPMEVVLLKKVSS-------GFSGVIRLLDWFERP---------DSFVLILERPEPV-QDLFDFITERGA---
QKKLQARELALHARVS--------SHPYIITLHRVLETE---------DAIYVVLQCPN--GDLFTYITEKKVYQ-
TLDSYRNEIAYLNKLQ--------QHSDKIIRLYDYEITD--------QYIYMVMECGNI---DLNSWLKKKKS---
TTNMSGNEAMRLMNIEKCRWE{8}NNVHIVRLIECLDSPFS-------ESIWIVTNWCSL--GELQWKRDDDEDILP
ERNRLLQEVSIQRALK--------GHDHIVELMDSWEHG---------GFLYMQVELCEN--GSLDRFLEEQGQLSR
DEQNALREVYAHAVLG--------QHSHVVRYFSAWAED---------DHMLIQNEYCNG--GSLADAI-SENYRIM
NNEKAEREVKALAKL---------DHVNIVHYNGCWDGFDYDPE{24}KCLFIQMEFCDK--GTLEQWIEKRRGEK-
KLSTMISEVMLLASL---------NHQYVVRYYAAWLEEDSMD{112}STLFIQMEYCEN--RTLYDLIHSENLN--
RHPQLLYESKLYKILQ--------GGVGIPHIRWYGQEK---------DYNVLVMDLLG--PSLEDLFNFCSRR--
LVQRFHAEARAVNLI---------GHENIVSIFDMDATP---------PRPYLIMEFLDG--APLSAWVGTP-----
LENSLTRELQVLKSL---------NHPCIVKLLGINNPIFVTSK{14}PPCDMIMSYCPA--GDLLAAVMARNGR--
SRRSFWAELNVARL---------RHDNIVRVVAASTRTPAGSN----SLGTIIMEFSGV--VTLHQVIYGAAGHPE
GKEVFQAELSVIGRI---------NHMNLVRIWGFCSEG---------SHRLLVSEYVEN--GSLANILFSEGGNIL
ELQQSYNELKYLNSI---------RHDNILALYGYSIKG---------GKPCLVYQLMKG--GSLEARLRAHKAQNP
ASWKDRKDIFSDINL---------KHENILQFLTAEEERKTELG----KQYWLITAFHAK--GNLQEYLTRHV----
QSWQNEYEVYSLPGM---------KHENILQFIGAEKRGTSVD-----VDLWLITAFHEK--GSLSDFLKANV----
QFQAFRNEVAVLRKT---------RHVNILLFMGYMTK----------DNLAIVTQWCEG--SSLYKHLHVQETK--
SLVMFQONEVGILSKL--------RHPNVVQFLGACTAGGE-------DHHCIVTEMAGG--GSLRQFLTDHFNLLE
SPEAFLQEAQVMKKL---------RHEKLVQLYAVVSE----------EPIYIVTEYMSK--GSLLDFLKGETGKY-
ANKEILDEAYVMASV---------DNPHVCRLLGICLT----------STVQLITQLMPF--GCLLDYVREHKDN--
EKQALMSELKIMSHLG--------PHLNVVNLLGACTKG---------GPIYIITEYCRY--GDLVDYLHRNKHTFL
```

names used are as defined in Table 1. The single letter amino acid code is used and gaps are indicated by dashes. The entire sequences for the larger

```
subdomain                                           VIA
consensus   -----------------------o-------o--*o--+o-ooh----------------
2°struct                            <------- aE ------->
                                    140       150       160
PKA-Cα      ----------------------FSEPHARFYAAQIVLTFEYLHSL-------------
PKG-I       ----------------------FEDSTTRFYTACVVEAFAYLHSK-------------
cPKCα       ----------------------FKEPQAVFYAAEISIGLFFLHKR--------------
βARK1       ----------------------FSEADMRFYAAEIILGLEHMHNR--------------
S6K         ----------------------FMEDTACFYLAEISMALGHLHQK--------------
RSK1(Nt)    ----------------------FTEEDVKFYLAELALGLDHLHSL--------------
DPKM        ----------------------IPAEMARFYLAEIVMAIDSVHRL--------------
CaMK2α      ----------------------YSEADASHCIQQILEAVLHCHQM--------------
skMLCK      ----------------------LTEVDTMVFVRQICDGILFMHKM--------------
Mre4        ----------------------MSETESLLIVFQILQALNYLHDQ--------------
PhKγM       ----------------------LSEKETRKIMRALLEVICALHKL--------------
Kin1        ----------------------IREHQARKFARGIASALIYLHAN--------------
Snf1        ----------------------MSEQEARRFFQQIISAVEYCHRH--------------
Polo        ----------------------ITEFECRYYIYQIIQGVKYLHDN--------------
Cdc5        ----------------------LTEPEVRFFTTQICGAIKYMHSR--------------
Cdk2        ----------------------IPLPLIKSYLFQLLQGLAFCHSH--------------
Erk2        ----------------------LSNDHICYFLYQILRGLKYIHSA--------------
GSK3α       I---------------------IPIIYVKVYMYQLFRSLAYIHSQ--------------
CK2α        ----------------------LTDYDIRFYMYEILKALDYCHSM--------------
Clk         ----------------------FRLDHIRKMAYQICKSVNFLHSN--------------
Ire1        NLKL------------------QKEYNPISLLRQIASGVAHLHSL--------------
Cdc7        ----------------------LPIKGIKKYIWELLRALKFVHSK--------------
Cot         ----------------------MREFEIIWVTKHVLKGLDFLHSK--------------
YpkA        ----------------------AYWGTIKFIAHRLLDVTNHLAKA--------------
MEK1        ----------------------IPEQILGKVSIAVIKGLTYLREKH-------------
Ste7        RGTVSSKKTW------------FNELTISKIAYGVLNGLDHLYRQY-------------
Ste11       ----------------------FEESLITNFTRQILIGVAYLHKK--------------
Nek1        ----------------------FQEDQILDWFVQICLALKHVHDR--------------
NIMA        Y---------------------AEEDFVWRILSQLVTALYRCHYGTDPAEVGSNLL{16}
Fused       ----------------------MGEEPARRVTGHLVSALYYLHSN--------------
NinaC       R---------------------MREEHIAYIIRETCRAAIELNRN--------------
Ste20       ----------------------LTEGQIGAVCRETLSGLEFLHSK--------------
Cdc15       ----------------------LSENESKTYVTQTLLGLKYLHGE--------------
Npr1        ----------------------MSYEEICCCFKQILTGVQYLHSI--------------
Pim1        ----------------------LQEELARSFFWQVLEAVRHCHNC--------------
Ran1        ----------------------GNSHLIKTVFLQLISAVEHCHSV--------------
Esk         ----------------------IDPWERKSYWKNMLEAVHTIHQH--------------
Elm1        QWKKIVISNC------------SVSTFAKKILEDMTKGLEYLHSQ--------------
Ykl516      ----------------------LEAWLIQRIFTEVVLAVKYLHEN--------------
SpWee1      ----------------------LDEFRVWKILVEVALGLQFIHHK--------------
Wee1(Hs)    SY--------------------FKEAELKDLLLQVGRGLRYIHSM--------------
PKR         ----------------------LDKVLALELFEQITKGVDYIHSK--------------
Gcn2        ----------------------QQRDEYWRLFRQILEALSYIHSQ--------------
CK1α        ----------------------FTMKTVLMLADQMISRIEYVHTK--------------
PKN1        ----------------------LAAGAVVSVLSQVCDALQAAHAR--------------
Mos         GDAGEPHCRTGGQ---------LSLGKCLKYSLDVVNGLLFLHSQ--------------
ZmPK1       ----------------------LDWEGRFNIALGVAKGLAYLHHECLE-----------
Pelle       LPA-------------------LTWQQRFSISLGTARGIYFLHTARGT-----------
TGFβRII     ----------------------ISWEDLRNVGSSLARGLSHLHSDHTPCGRPKM-----
ActRII      ----------------------VSWNELCHIAETMARGLAYLHEDIPGLKDGHKP----
Raf-1       ----------------------FQMFQLIDIARQTAQGMDYLHAK-------------
Sp1A        ----------------------QNPHIRLKLALDIAKGMNYLHGWTP-----------
Src         ----------------------LRLPQLVDMAAQIASGMAYVERM-------------
EGFR        ----------------------IGSQYLLNWCVQIAKGMNYLEDR-------------
PDGFRβ      QHHSDKRRPPSAELYSNALPVG{74}LSYMDLVGFSYQVANGMEFLASK----------
```

inserts are not shown, but excluded residues are indicated as numbers in brackets. Twelve distinct subdomains are indicated by Roman numerals.

```
       VIB                         VII                              VIII
-oohrDok+-Nooo-------------oko+Dfgo+----------------------------g+--o-+pEoo-
<b6>      < b7>        < b8 >    <b9>
       170          180          190                            200       210
DLIYRDLKPENLLIDQQ--------GYIQVTDFGFAKRVKG-----------------RTWTLCGTPEYLAPEII-
GIIYRDLKPENLILDHR--------GYAKLVDFGFAKKIGFGK---------------KTWTFCGTPEYVAPEII-
GIIYRDLKLDNVMLDSE--------GHIKIADFGMCKEHMMDGV-------------TTRTFCGTPDYIAPEII-
FVVYRDLKPANILLDEH--------GHVRISDLGLACDFSKK---------------KPHASVGTHGYMAPEVLQ
GIIYRDLKPENIMLNHQ--------GHVKLTDFGLCKESIHDGT-------------VTHTFCGTIEYMAPEILL
GIIYRDLKPENILLDEE--------GHIKLTDFGLSKEAIDHEK-------------KAYSFCGTVEYMAPEVV-
GYVHRDIKPDNILLDRC--------GHIRLADFGSCLKLRADGTV------------RSLVAVGTPDYLSPEILQ
GVVHRDLKPENLLLASKLKG------AAVKLADFGLAIEVEGEQQ------------AWFGFAGTPGYLSPEVL-
RVLHLDLKPENILCVNTTG------HLVKIIDFGLARRYNPNE--------------KLKVNFGTPEFLSPEVV-
DIVHRDLKLDNILLCTPEPC-----TRIVLADFGIAKDLNSNKE------------RMHTVVGTPEYCAPEVGF
NIVHRDLKPENLLDDD---------MNIKLTDFGFSCQLDPGE-------------KLREVCGTPSYLAPEIIE
NIVHRDLKIENIMISDS--------SEIKIIDFGLSNIYDSRK------------QLHTFCGSLYFAAPELL-
KIVHRDLKPENLLLDEH--------LNVKIADFGLSNIMTDGN------------FLKTSCGSPNYAAPEVI-
RIIHRDLKLGNLFLNDL--------LHVKIGDFGLATRIEYEGE-----------RKKTLCGTANYIAPEIL-
RVIHRDLKLGNIFFDSN--------YNLKIGDFGLAAVLANESE------------RKYTICGTPNYIAPEVLM
RVLHRDLKPQNLLINTE--------GAIKLADFGLARAFGVPVR------------TYTHEVVTLWYRAPEILL
NVLHRDLKPSNSSLNTT--------CDLKICDFGLARVADPDHDHTG---------FLTEYVATRWYRAPEIML
GVCHRDIKPQNLLVDPDT-------AVLKLCDFGSAKQLVRGE------------PNVSYICSRYYRAPELIF
GIMHRDVKPHNVMIDHEH-------RKLRLIDWGLAEFYHPGQ------------EYNVRVASRYFKGPELLV
KLTHTDLKPENILFVQSDYTEA{14}PDIKVVDFGSATYDDE-----------HHSTLVSTRHYRAPEVIL
KIIHRDLKPQNILVSTSSRFTAD{7}LRILISDFGLCKKLDSGQSSFRT------NLNNPSGTSGWRAPELL-
GIIHRDIKPTNFLFNLEL-------GRGVLVDFGLAEAQMDYKSMISSQNDYANDN{72}ANANRAGTRGFRAPEVLM
KVIHHDIKPSNIVFMS---------TKAVLVDFGLSVQMTEDVY-----------FPKDLRGTEIYMSPEVI-
GVVHNDIKPGNVVFDRAS-------GEPVVIDLGLHSRS----------------GEQPKGFTESFKAPELGV
KIMHRDVKPSNILVNSR--------GEIKLCDFGVSGQLIDS------------MANSFVGTRSYMSPERL-
KIIHRDLKPSNVLINSK--------GQIKLCDFGVSKKLINS------------IADTFVGTSTYMSPERI-
NIIHRDIKGANILIDIK--------GCVKITDFGISKKLSPLNKKQN-------KRASLQGSVFWMSPEVV-
KILHRDIKSQNIFLTKD--------GTVQLGDFGIARVLNSTVE----------LARTCIGTPYYLSPEIC-
TILHRDLKPENIFLGSD--------NTVKLGDFGLSKLMHSHD-----------FASTYVGTPFYMSPEIC-
RILHRDLKPQNVLLDKN--------MHAKLCDFGLARNMTLGTH----------VLTSIKGTPLYMAPELL-
HVLHRDIRGDNILLTKN--------GRVKLCDFGLSRQVDSTLG----------KRGTCIGTPSPCWMAPEVVS
GVLHRDIKSDNILLSME--------GDIKLTDFGFCAQINELNL----------KRTTMVGTPYWMAPEVV-
GVIHRDIKAANILLSAD--------NTVKLADFGVSTIVNS-------------SALTLAGTLNWMAPEIL-
GLAHRDLKLDNCVINEK--------GIVKLIDFGAAVVFSYPFSKNLV-------EASGIVGSDPYLAPEVCI
GVLHRDIKDENILIDLNR-------GELKLIDFGSGALLKDT-----------VYTDFDGTRVYSPPEWIR
GIYHRDLKPENIMVGNDG-------NTVYLADFGLATTEPY------------SSDFGCGSLFYMSPECQR
GIVHSDLKPANFLIVD---------GMLKLIDFGIANQMQPDTTSV--------VKDSQVGTVNYMPPEAIK
GCIHRDIKPSNILLDEEE-------KVAKLSDFGSCIFTPQSLPFSDANFEDCFQR----ELNKIVGTPAFIAPELCH
SIIHRDLKLENILLKYSFDDIN{11}NFIELADFGLCKKIENNE---------MCTARCGSEDYVSPEILM
NYVHLDLKPANVMITFE--------GTLKIGDFGMASVWPVP-----------RGMEREGDCEYIAPEVL-
SLVHMMDIKPSNIFISRTSIPNA{14}VMFKIGDLGHVTRIS-----------SPQVEEGDSRFLANEVLQ
KLIHRDLKPSNIFLVDT--------KQVKIGDFGLVTSLKNDG----------KRTRSKGTLRYMSPEQI-
GIIHRDLKPKNIFIDES--------RNVKIGDFGLAKNVHRSLDILKDSQNLPGSSDN--LLTSAIGTAMYVATEVLD
NFIHRDIKPDNFLMGIGRHC-----NKLFLIDFGLAKKYRDNRTRQHIPYR--------EDKNLTGTARYASINAHL
GIVHRDLKPDNIFLVRRNGNA----PFVKVLDFGIAKLADLHMPQT------------HAGIIVGTPEYMAPEQS-
SIVHLDLKPANILISEQ--------DVCKISDFGCSEKLEDLLCFQT----------PSYPLGGTYTHRAPELL-
WVIHCDVKPENILLDQA--------FEPKITDFGLVKLLNRGGSTQ----------NVSHVRGTLGYIAPEWV-
PLIHGDIKPANILLDQC--------LQPKIGDFGLVREGPKSLDAVV----------EVNKVFGTKIYLPPEFR-
PIVHRDLKSSNILVKND--------LTCCLCDFGLSRLGPYSSVDDL---------ANSGQVGTARYMAPEVLE
AISHRDIKSKNVLLKNN--------LTACIADFGLALKFAEGKSAG----------DTHGQVGTRRYMAPEVLE
NIIHRDMKSNNIFLHEG--------LTVKIGDFGLATVKSRWSGSQ----------QVEQPTGSVLWMAPEVIR
PILHRDLSSRNNILLDHNIDPKN{8}IKCKISDFGLSRLKKEQAS----------QMTQSVGCIPYMAPEVF-
NYVHRDLRAANILVGEN--------LVCKVADFGLARLIEDNEYT-----------ARQGAKFPIKWTAPEAA-
RLVHRDLAARNVLVKTP--------QHVKITDFGLAKLLGAEEKEY----------HAEGGKVPIKWMALESI-
NCVHRDLAARNVLICEG--------KLVKICDFGLARDIMRDSNYI----------SKGSTFLPLKWMAPESI-
```

*The consensus line is given according to the following code: uppercase
letters, invariant residues; lowercase, nearly invariant residues; o, positions*

13

```
subdomain                          IX                                        X
consensus  ---------o-----Doo+ogoooo-o------po-------------oo--o---------------
2°struct            <----- aF ----->                     <-- aG ->
                    220        230        240                250        260
PKA-Cα     -----LSKGYNK-AVDWWALGVLIYEMAA-GYPPFFA-------DQPIQIYEKIVSG-KV-RFPSH----
PKG-I      -----LNKGHDI-SADYWSLGILMYELLT-GSPPFSG-------PDPMKTYNIILRGIDMIEFPKK----
cPKCα      -----AYQPYGK-SVDWWAYGVLLYEMLA-GQPPFDG-------EDEDELFQSIMEH-NV-SYPKS----
βARK1      -----KGVAYDS-SADWFSLGCMLFKLLR-GHSPFRQHK-----TKDKHEIDRMTLTMAV-ELPDS----
S6K        -----MRSGHNR-AVDWWSLGALMYDMLT-GAPPFTG-------ENRKKTIDKILKCKL--NLPPY----
RSK1(Nt)   -----NRQGHTH-SADWWSYGVLMFEMLT-GSLPFQG-------KDRKETMTLILKAKL--GMPQF----
DMPK       AV{4}GTGSYGP-ECDWWALGVFAYEMFY-GQTPFYA-------DSTAETYGKIVHYKEHLSLPLVDEG-
CaMK2α     -----RKDPYGK-PVDLWACGVILYILLV-GYPPFWD-------EDQHRLYQQIKAGAYDFPSPEWDT--
Mre4       R{15}EQRGYDS-KCDLWSLGVITHIMLT-GISPFYGD------GSERSIIQNAKIGKLNFKLKQWDI--
PhKγM      CSMNDNHPGYGK-EVDMWSTGVIMYTLLA-GSPPFWH-------RKQMLMLRMIMSGNYQFGSPEWDD--
Kin1       -----KANPYTGPEVDVWSFGVVLFVLVC-GKVPFDD-------ENSSVLHEKIKQGKV--EYPQH----
Snf1       -----SGKLYAGPEVDVWSCGVILYVMLC-RRLPFDD-------ESIPVLFKNISNGVY--TLPKF----
Polo       -----TKKGHSF-EVDIWSIGCVMYTLLV-GQPPFET------KTLKDTYSKIKKCEY--RVPSY----
Cdc5       G----KHSGHSF-EVDIWSLGVMLYALLI-GKPPFQA------RDVNTIYERIKCRDF--SFPRDKP--
Cdk2       -----GCKYYST-AVDIWSLGCIFAEMVT-RRALFPGDSEI---DQLFRIFRTLGTPDEE-VWPGVTSMP
Erk2       -----NSKGYTK-SIDIWSVGCILAEMLS-NRPIFPGKHYL---DQLNHILGILGSPSQE-DLN-CIINL
GSK3α      -----GATDYTS-SIDVWSAGCVLAELLL-GQPIFPGDSGV---DQLVEIIKVLGTPTRE-QIR-EMNPN
CK2α       -----DYQMYDY-SLDMWSLGCMLASMIFRKEPFFHGHDNY---DQLVRIAKVLGTEDLYDYIDKYNIEL
Clk        ------ALGWSQ-PCDVWSIGCILIEYYL-GFTVFPTHDSK---EHLAMMERILGPL-PKHMIQKTRKRK
Ire1       E{24}TKRRLTR-SIDIFSMGCVFYYILSKGKHPFGDKY-----SRESNIIRGIFSLDEMKCLHDRS---
Cdc7       -----KCGAQST-KIDIWSVGVILLSLLG-RRFPMFQSL-----DDADSLLELCTIFGWKELRKCAALHG
Cot        -----LCRGHST-KADIYSLGATLIHMQT-GTPPWVK------RYPRSAYPSYLYIIHKQAPPLEDIAD
YpkA       -----GNLGASE-KSDVFLVVSTLLHCIE-GFEKNPEIKP----NQGLRFITSEPAHVMDENGYPIHRPG
MEK1       -----QGTHYSV-QSDIWSMGLSLVEMAV-GRYPIPPPDA{36}RPPMAIFELLDYIVNEPPPKLPSGV-
Ste7       -----QGNVYSI-KGDVWSLGLMIIELVT-GEFPLGGHN-----DTPDGILDLLQRIPSPRLPKDRI---
Ste11      -----KQTATTA-KADIWSTGCVVIMEFFDPPDF-------SQMQAIFKIGTNTTPEIPSW------
Nek1       -----ENKPYNN-KSDIWALGCVLYELCT-LKHAFEA------GNMKNLVLKIISGSFPPVSPH-----
NIMA       -----AAEKYTL-RSDIWAVGCIMYELCQ-REPPFNA-------RTHIQLVQKIREGKFAPLPDF-----
Fused      -----ADEPYDH-HADMWSLGCIAYESMA-GQPPFCA-------SSILHLVKMIKHEDVKW-PST-----
NinaC      AME--SEPDITV-RADVWALGITTIELAD-GKPPFADMHPT---RAMFQIIRNPPPTLMRPTN------
Ste20      -----SRKEYGP-KVDIWSLGIMIIEMIE-GEPPYLNETPL---RALYLIATNGTPKLKEPEN------
Cdc15      -----GNRGAST-LSDIWSLGATVVEMLT-KNPPYHN------LTDANITTAVENDTYYPPSS------
Npr1       -----FAKYDPR-PVDIWSSAIIFACMIL-KKFPWKIPKLR---DNSFKLFCSGRDCDSLSSLVTRTPDP
Pim1       -----YHRYHGR-SADVWSLGILLYDMVC-GDIPF--------EHDEEIIRGQVFFVDGRQR-------
Ran1       E{24}SSSFATA-PNDVWALGIILINLCC-KRNPWK-------RACSQTDGTYRSYVHNPSTLLSILP-
Esk        DM{6}GKSKISP-KSDIWSLGVCILYYMTY-GKTPFQQII-----NQISKLHAIIDPNHEIEFPDIP----
Elm1       LG{4}DFVTDGF-KLDIWSLGVTLYCLLY-NELPFFGENEF---ETYHKIIEVSLSSKIN---------
Ykl516     -----GVPYDGH-LSDTWALGVILYSLFE-DRLPFDPPPNA{7}ATSHRIARFDWRWYRLSD-------
SpWee1     -----ANHLYDK-PADIFSLGITVFEAAA--NIVLPDN------GQSWQKLRSGDLSPRLSSTDNGSSLT
Wee1(Hs)   -----ENYTHLP-KADIFALALTVVCAAG--AEPLPRN------GDQWHEIRQGRLPRIPQV-------
PKR        -----SSQDYGK-EVDLYALGLILAELLH----VCDTA------FETSKFFTDLRDGIISDI-------
Gcn2       -----GTGHYNE-KIDMYSLGIIFFEMIY---PFSTG------MERVNILRKRSVSIEFPPDFDDN--
CK1α       -----GIEQSR--RDDMESLGYVLMYFNR-TSLPWQGLKAATLKQCKEKISEKKMSTPVEVLCKGFPAEF
Pkn1       -----LGRGVDG-RADLYALGVIAYQLLT-GRLPFNDE------GLAAQLVAHQLRPPPPPSSVYPA---
Mos        -----KGEGVTP-KADIYSFAITLWQMTT-KQAPYS-------GERQHILYAVVAYDLRPSLSAAVFED
ZmPK1      -----SSLPITA-KVDVYSYGVVLLELLT-GTRVSELV------GGTDEVHSMLRKLVRMLSAKLEGEEQ
Pelle      -----NFRQLST-GVDVYSGFIVLLEVFT-GRQVTDRVPEN---ETKKNLLDYVKQQWRQNRMELLEKHL
TGFβRII    SRMN-LENAESFKQTDVYSMALVLWEMT{13}PPFGS-------KVRDPVVESMKDNVLRDRGTRNSS-F
ActRII     GAIN-FQR-DAFLRIDMYAMGLVLWELA{14}LPFEEE------IGQHPSLEDMQEVVVHKKKRPVLRDY
Raf-1      --MQ-DNNPFSF-QSDVYSYGIVLYELMT-GELPYSHI------NNRDQIIFMVGRGYASPDLSKLYKN-
Sp1A       -----KGDSNSE-KSDVYSYGMVLFELLT-SDEPQQDM------KPMKMAHLAATESYRPPIPLT-----
Src        -----LYGRFTI-KSDVWSFGILLTELTTKGRVPYPG-------MVNREVLDQVERGYRMPCPPE-----
EGFR       -----LHRIYTH-QSDVWSYGVTVWELMTFGSKPYDG-------IPASEISSILEKGERLPQPPI-----
PDGFRβ     -----FNSLYTT-LSDVWSFGILLWEIFTLGGTPYPEL------PMNEQFYNAIKRGYRMAQPAH-----
```

conserving nonpolar residues; *, positions conserving polar residues; +, positions conserving small residues with near neutral polarity. Residues

```
                                                    XI
-----------------------------------o------oo--oo------R-+---------------o
                                    <-- aH -->           < aI >
                                    270        280       290
-------------------------------------------FSSDLKDLLRNLLQVDLTKRFGNLKNGVNDIKN--HKWF
-------------------------------------------IAKNAANLIKKLCRDNPSERLGNLKNGVKDIQK--HKWF
-------------------------------------------LSKEAVSICKGLMTKHPAKRLGCGPEGERDVRE--HAFF
-------------------------------------------FSPELRSLLEGLLQRDVNRRLGCLGRGAQEVKE--SPFF
-------------------------------------------LTQEARDLLKKLLLKRNAASRLGAGPGDAGEVQA--HPFF
-------------------------------------------LSTEAQSLLRALFKRNPANRLGSGPDGAEEIKR--HIFY
-------------------------------------------VPEEARDFIQRLLC-PPETRLGRGGAGDFRT----HPFF
-------------------------------------------VTPEAKDLINKMLTINPSKRITAAEALK-------HPWI
-------------------------------------------VSDNAKSFVKDLLQTDVVKRLNSKQGLK-------HIWI
-------------------------------------------YSDTVKDLVSRFLVVQPQKRYTAEEALA-------HPFF
-------------------------------------------LSIEVISLLSKMLVVDPKRRATLKQVVE-------HHWM
-------------------------------------------LSPGAAGLIKRMLIVNPLNRISIHEIMQ-------DDWF
-------------------------------------------LRKPAADMVIAMLQPNPESRPAIGQLLN-------FEFL
-------------------------------------------ISDEGKILIRDILSLDPIERPSLTEIMD-------YVWF
DYKPSFP--------KWARQDFSKVVPP-------LDEDGRSLLSQMLHYDPNKRISAKAALA-------HPFF
KARNYLLSLPHKNKVPWNRLFPN-----------ADSKALDLLDKMLTFNPHKRIEVEQALA-------HPYL
YTEFKFPQIKAH---PWTKVFKSR----------TPPEAIALCSSLLEYTPSSRLSPLEACA-------HSFF
DPRFNDILGRHSRK-RWERFVHSENQHL------VSPEALDFLDKLLRYDHQSRLTAREAME-------HPYF
YFHHDRLDWDEHSSAGRYVSRACKPLKEFMLSQDVEHERLFDLIQKMLEYDPAKRITLREALK-----HPFF
-------------------------------------------LIAEATDLISQMIDHDPLKRPTAMKVLR-------HPLF
LGFEASGLIWDKPNGYSNGLKEFVYDLLNKE{57}DHYWCFQVLEQCFEMDPQKRSSAEDLLK-------TPFF
D--------------------------------CSPGMRELIEASLERNPNHRPRAADLLK-------HEAL
IAG------------------------------VETAYTRFITDILGVSADSRPDSNEARL-------HEFL
-------------------------------------------FSLEFQDFVNKCLIKNPAERADLKQLMV-------HAFI
-------------------------------------------YSKEMTDFVNRCCIKNERERSSIHELLH-------HDLI
-------------------------------------------ATSEGKNFLRKAFELDYQYRPSALELLQ-------HPWL
-------------------------------------------YSYDLRSLLSQLFKRNPRDRPSVNSILE-------KGFI
-------------------------------------------YSSELKNVIASCLRVNPDHRPDTATLIN-------TPVI
-------------------------------------------LTCECRSFLQGLLEKDPGLRISWTQLLC-------HPFV
-------------------------------------------WSQQINDFISESLEKNAENRPMMVEMVE-------HPFL
-------------------------------------------LSSSLKKFLDWCLCVEPEDRASATELLH-------DEYI
-------------------------------------------FSEPLKDFLSKCFVKNMYKRPTADQLLK-------HVWI
PSYDESHSTEKKKPESSSNNVSDPNNVNIGPQ{5}LPEETQHIVGRMIDLAPACRGNIEEIME-------DPWI
-------------------------------------------VSSECQHLIRWCLALRPSDRPTFEEIQN-------HPWM
-------------------------------------------ISRELNSLLNRIFDRNPKTRITLPELST-------LVSN
-------------------------------------------EKDDLQDVLKCCLKRDPKQRISIPELLA-------HPYV
-------------------------------------------GNTLNDLVIKRLLYKDVTLRISIQDLVK-------VLSR
-------------------------------------------YKTNVGKQIVENTLTRKNQRWSINEIYE-------SPFV
SSSRETPANSII----------------------GQGGLDRVVEWMLSPEPRNRPTIDQILATD-----EVCW
-------------------------------------------LSQEFTELLKVMIHPDPERRPSAMALVK-------HSVL
-------------------------------------------FDKKEKTLLQKLLSKKPEDRPNTSEILR-------TLTV
-------------------------------------------KNKVEKKIIRLLIDHDPNKRPGARTLLN-------SGWL
AM----------------------------------------YLNYCRGLRFEEAPDYMYLRQLFRILFRTLNHQY-DYTF
-------------------------------------------VSAALEHVILRALAKKPEDRYASIAAFRNAL----QVAL
SL----------------------------------------PGQRLGDVIQRCWRPSAAQRPSARLLLV-------DLTS
SWIDGYLDSKLNRPVNYV---------------QARTLIKLAVSCLEEDRSKRPTMEHA--------VQTL
AAPMGKELD-------------------------MCMCAIEAGLHCTALDPQDRPSMNAVLKRF-----EPFV
WLNHQ----------------------------GIQMVCETLTECWDHDPEARLTAQCVAERFSE---LEHL
WQKHA----------------------------GMAMLCETIEECWDHDAEARLSAGCVGERITQ---MQRL
-------------------------------------------CPKAMKRLVADCVKKVKEERPLFPQILSSIELLQ-HSLP
-------------------------------------------TSSKWKEILTQCWDSNPDSRPTFKQI---------IVHL
-------------------------------------------CPESLHDLMCQCWRKEPEERPTFEYL---------QAFL
-------------------------------------------CTIDVYMIMVKCWMIDADSRPKFREL---------IIEF
-------------------------------------------ASDEIYEIMQKCWEEKFEIRPPFSQL---------VLLL
```

corresponding to the numbered β-strands (b) and ahelices (a) in PKA-Cα are indicated in the 2⁰ structure line.

The kinase domain of PKA-Cα folds into a two-lobed structure (Fig. 2). The smaller N-terminal lobe, which includes subdomains I–IV, is primarily involved in anchoring and orienting the nucleotide. This lobe has a predominantly antiparallel β-sheet structure which is unique among nucleotide binding proteins. The larger C-terminal lobe, which includes subdomains VIA–XI, is largely responsible for binding the peptide substrate and initiating phosphotransfer. It is predominantly α-helical in content. Subdomain V residues span the two lobes. The deep cleft between the two lobes is recognized as the site of catalysis. The crystal structures of two additional eukaryotic protein kinase superfamily members, cyclin-dependent protein kinase 2 (**Cdk2**)[19] and p42 MAP kinase (**Erk2**)[20] have more

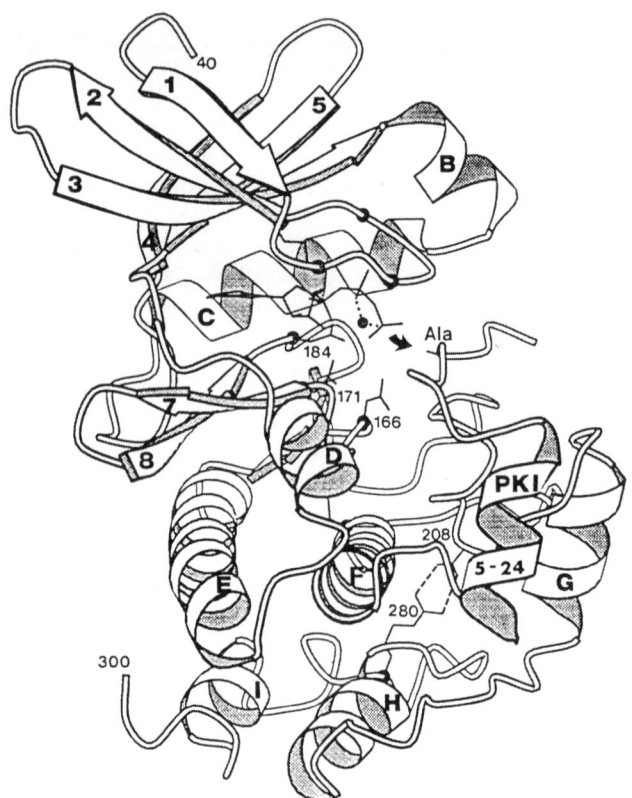

Figure 2. *Ribbon diagram of the catalytic core of **PKA**-Cα (residues 40–300) in a ternary complex with MgATP and pseudosubstrate peptide inhibitor (PKI524). Invariant or nearly invariant residues (G50, G52, G55, K72, E91, D166, N171, D184, E208, D220, and R280) are indicated by dots along the ribbon diagram. Side chains are shown for K72, D166, N171, D184, E208, and R280. β-strands and α-helices are indicated by flat arrows and helices, respectively, and are numbered according to Knighton et al.[15]. The small arrow indicates the site of phospho-transfer, with the alanine in PKI substituting for the phosphoacceptor serine in the true substrate. (Reproduced, with permission, from Taylor et al.[189]).*

recently been reported and, as expected, their kinase domains were found to fold into two-lobed structures topologically very similar to the catalytic core of PKA-Cα. Notable differences, however, were found in the regions corresponding to subdomain VIII, apparently reflecting the fact that the Cdk2 and Erk2 structures are of enzymes in an inactive state (see below).

The conserved kinase subdomains correspond quite well to precise units of higher order structure. These relationships will be briefly discussed below on a subdomain-by-subdomain basis, making reference to the crystal structure of PKA-Cα, and drawing attention to the proposed[15–18] roles of the nearly invariant amino acid residues and other residues of interest.

Subdomain I

Subdomain I, at the N-terminus of the kinase domain, is distinguished by the consensus motif GXGXXGXV (starting with G50 in PKA-Cα). The kinase domain N-terminal boundary occurs seven positions upstream of the first glycine in the consensus, where a hydrophobic residue is usually found. In the crystal structure of PKA-Cα, subdomain I residues fold into a β-strand, turn, β-strand structure encompassing β-strands 1 and 2. Backbone hydrogen bonds between G50 and V57 and between G52 and G55 stabilize the β-strands. This structure acts as a flexible flap or clamp that covers and anchors the non-transferable phosphates of ATP. The backbone amides of S53, F54 and G55 of PKA-Cα form hydrogen bonds with ATP β-phosphate oxygens. L49 and V57 contribute to a hydrophobic pocket that encloses the adenine ring of ATP. V51 and S53 also interact with the pseudo-substrate peptide.

Subdomain II

The outstanding feature of subdomain II is the invariant lysine residue, corresponding to PKA-Cα K72, which has long been recognized as being essential for maximal enzyme activity [reviewed in ref. 1]. This lysine lies within β-strand 3 of the small lobe. Its function is to help anchor and orient ATP by interacting with the α- and β-phosphates. In addition, K72 forms a salt bridge with the carboxyl group of the nearly invariant E91 in subdomain III. A70 contributes to the hydrophobic adenine ring pocket. In PKA-Cα, β-strand 3 is followed immediately by α-helix B which, judging from the sequence alignment, would appear to be a quite variable structure among the kinases. Indeed, this α-helix is absent in the Cdk2 and Erk2 crystal structures.

Subdomain III

Subdomain III represents the large α-helix C in the small lobe. The nearly invariant glutamic acid residue, equivalent to E91 in PKA-Cα, is centrally located in this helix. E91 forms an ion pair with the invariant K72 in subdomain II, thereby helping to stabilize the interactions between K72 and the α- and β-phosphates.

Subdomain IV

Subdomain IV corresponds to the hydrophobic β-strand 4 in the small lobe. This subdomain contains no invariant or nearly invariant residues and does not appear to be directly involved in catalysis or substrate recognition.

Subdomain V

Subdomain V links the small and large lobes of the catalytic subunit and consists of the very hydrophobic β-strand 5 in the small lobe, the small α-helix D in the large lobe, and an extended chain that connects them. Three residues in the connecting chain of PKA-Cα, E121, V123 and E127, help to anchor ATP by forming hydrogen bonds with either the adenine or the ribose ring. M120, Y122 and V123 contribute to the hydrophobic pocket surrounding the adenine ring. E127 also participates in peptide binding by forming an ion pair with an arginine in the pseudosubstrate site of the PKA inhibitor peptide. This represents the first arginine in the PKA substrate recognition consensus Arg–Arg–X–Ser*–hydrophobic (see **PKA**).

Subdomain VIA

Subdomain VIA folds into the large hydrophobic α-helix E that extends through the large lobe. None of the residues in helix E appear to interact directly with either MgATP or peptide substrate; hence this part of the molecule appears to act mainly as a support structure.

Subdomain VIB

Subdomain VIB folds into the small hydrophobic β-strands 6 and 7 with an intervening loop. Included here are two invariant residues, equivalent to D166 and N171 in PKA-Cα, which lie within the consensus motif HRDLKXXN. The loop has been termed the "catalytic loop" since D166 within the loop has emerged as the likely candidate for the catalytic base – accepting the proton from the attacking substrate hydroxyl group during an in-line phosphotransfer mechanism[21–23]. K168 in the loop (substituted by arginine in the conventional protein-tyrosine kinases) may help to facilitate phosphotransfer by neutralizing the negative charge of the γ-phosphates during transfer. The side-chain of N171 helps to stabilize the catalytic loop through hydrogen bonding to the backbone carbonyl of D166 and also acts to chelate the secondary Mg^{2+} ion that bridges the α and γ phosphates of the ATP. The carbonyl group of E170 forms a hydrogen bond with an ATP ribose hydroxyl group. E170 also participates in substrate binding by forming an ion pair with the second arginine of the peptide recognition consensus.

Subdomain VII

Subdomain VII also folds into a β-strand–loop–β-strand structure, encompassing β-strands 8 and 9. The highly conserved "DFG" triplet, corresponding to residues 184–186 in PKA-Cα, lies in the loop which is stabilized by a hydrogen bond between D184 and G186. A184 chelates the primary activating Mg^{2+} ions that bridges the β- and γ-phosphates of the ATP and thereby helps to orient the γ-phosphate for transfer. In Cdk2, β-strand 9 is replaced with a small α-helix designated aL12. However, it is unclear if this helical character is maintained when Cdk2 is in its active conformation.

Subdomain VIII

Subdomain VIII, which includes the highly conserved "APE" motif corresponding to PKA-Cα residues 206–208, folds into a tortuous chain that faces the cleft. Residues lying 7–10 positions immediately upstream of the APE motif are characteristically well-conserved among the members of different protein kinase families[1–3]. The nearly invariant glutamic acid residue corresponding to PKA-Cα E208 forms an ion pair with an invariant arginine (corresponding to R280) in subdomain XI, thereby helping to stabilize the large lobe.

Subdomain VIII

Subdomain VIII appears to play a major role in recognition of peptide substrates. Several PKA-Cα subdomain VIII residues participate in binding the pseudosubstrate inhibitor peptide. L198, C199, P202 and L205 of PKA-Cα provide a hydrophobic pocket in which rests the side chain of the hydrophobic residue (isoleucine in the case of the inhibitor peptide) at position P+1 of the substrate consensus. G200 forms a hydrogen bond with the same isoleucine residue. E203 forms two ion pairs with the arginine in the high-affinity binding region of the inhibitor peptide.

Many protein kinases are known to be activated by phosphorylation of residues in subdomain VIII. In PKA-Cα, maximal kinase activity requires phosphorylation of T197, probably occurring through an intermolecular autophosphorylation mechanism. In the crystal structure, phosphate oxygens of phospho-T197 form hydrogen bonds with the charged side-chains of R165, K189 and the hydroxyl group of T195, and thereby may act to stabilize the subdomain VIII loop in an active conformation, permitting proper orientation of the substrate peptide. For members of the **Erk1/2/MAP** kinase family, phosphorylation of both threonine and tyrosine residues in subdomain VIII (mediated by members of the **Mek** family) is required for activation. In the crystal structure determined for Erk2, these residues (T183 and Y185) were not phosphorylated and thus the enzyme was in an inactive state (unlike the PKA-Cα structure). The unphosphorylated Y185 is buried in a hydrophobic pocket and this would appear to inhibit binding of peptide substrate. During activation of Erk2, Y185 phosphorylation appears to precede T183 phosphorylation, and therefore it was proposed[20] that binding of Mek to Erk2 alters the conformation of the subdomain VIII loop, thereby exposing Y185 for phosphorylation by Mek which, in turn, would release the substrate-binding block. Subsequent phosphorylation of the exposed T183 may activate the enzyme by promoting correct alignment of the catalytic residues. From the crystal structure of Cdk2, likewise in an inactive unphosphorylated state, the subdomain VIII loop also appears to be in a conformation that would inhibit enzyme activity by blocking the presumed protein substrate binding site. Phosphorylation of T160 in the Cdk2 subdomain VIII, mediated by **Mo15** (CAK) would presumably act to remove this inhibition by stabilizing the loop in an active conformation similar to that found in **PKA-Cα**. Cyclin binding to the N-terminal lobe is also needed to activate Cdk2.

Subdomain IX

Subdomain IX corresponds to the large α-helix F of the large lobe. The nearly invariant aspartic acid residue, corresponding to PKA-Cα D220, lies in the N-terminal region of this helix and acts to stabilize the catalytic loop by hydrogen bonding to the backbone amides of R165 and Y164 that precede the loop. E230 of PKA-Cα forms an ion pair with the second arginine of the peptide recognition consensus. PKA-Cα residues 235–239 are all involved in hydrophobic interactions with the inhibitor peptide.

Subdomain X

Subdomain X is the most poorly conserved subdomain and its function is obscure. In the crystal structure of PKA-Cα, it corresponds to the small α-helix G that occupies the base of the large lobe. Members of the Cdk, Erk(Map), GSK3, and Clk families (the "C–M–G–C" group) all have rather large insertions between subdomains X and XI, the functional significance of which is presently unclear.

Subdomain XI

Subdomain XI extends to the C-terminal end of the kinase domain. The most notable feature here is the nearly invariant arginine corresponding to R280 in PKA-Cα, which lies between α-helices H and I. As mentioned above, its role is to help stabilize the large lobe by forming an ion pair with E208 in subdomain VIII. The C-terminal boundary of the kinase domain is still poorly defined. For many serine/threonine-specific kinases, the consensus motif His–X–aromatic–hydrophobic is found beginning 9–13 residues downstream of the invariant arginine. For protein-tyrosine kinases, a hydrophobic amino acid lying 10 positions downstream of the invariant arginine appears to define the C-terminal boundary.

The amphipathic α-helix A of PKA-Cα (residues 15–35; not shown in Fig. 2), while lying outside of the conserved catalytic core, appears to be an important feature found in many protein kinases[24]. This helix spans the surface of both lobes of the core structure and complements and stabilizes the hydrophobic cleft between the two lobes. The A-helix motif appears to be present in many other protein kinases including members of the **PKC** family and the Src family of protein-tyrosine kinases[24].

CLASSIFICATION OF EUKARYOTIC PROTEIN KINASES

The classification scheme shown in Table 1 represents an attempt to subdivide the known members of the eukaryotic protein kinase superfamily into distinct families that share basic structural and functional properties. Phylogenetic trees derived from an alignment of kinase domain amino acid sequences (essentially an expanded version of Fig. 1) served as the basis for this classification. Thus, the sole consideration was similarity in kinase domain amino acid sequence. However, this property, considered alone, has proven to be a good indicator of other characteristics held in common by the different members of the family.

Protein kinases whose entire kinase domain amino acid sequence had been published as of July 1993 were included in phylogenetic analysis (as well as a few others made available through sequence databases). If a given kinase domain sequence had been determined from more than one species among the vertebrates (i.e. orthologous gene products), only one representative (usually human) was included in the analysis. This policy was not employed for the other phyla, however, owing to greater divergences between the species and hence the sequences. The kinase domain phylogenies were inferred using the principle of maximum parsimony according to the PAUP software package developed by David Swofford[25]. Minimum length trees were found using PAUP's "heuristic" search method with branch swapping by the "tree bisection–reconnection" strategy. Equal weights were given for all amino acid substitutions. Since multiple minimum-length trees were found, a consensus tree was calculated according to the method of Adams[186] in order to show branching ambiguities.

Table 1. *Eukaryotic protein kinase superfamily classification. Protein kinases with individual entries in the FactsBook are indicated by bold type. More information on these (including references), can be obtained from the entries. Kinases marked with asterisks were not included as separate entries in the book, due either to space restrictions or to recent discovery, and references for these are given. Those also marked by the letter "T" were included in the phylogenetic analysis, however. The species used to represent the various vertebrate kinases in the phylogenetic analysis is indicated within parentheses. Protein kinase homologues from DNA viruses, and protein kinases not members of this superfamily (e.g. **Bcr, BKDK, PDK**) are not included in this classification*

AGC group
AGC-I. Cyclic nucleotide-regulated protein kinase family
   A. Cyclic AMP-dependent protein kinase (PKA) subfamily
Vertebrate:
      1. **PKA**-Cα: PKA catalytic subunit, α form (human)
      2. **PKA**-Cβ: PKA catalytic subunit, β form (human)
      3. **PKA**-Cγ: PKA catalytic subunit, γ form (human)
Drosophila melanogaster:
      1. **DmPKA**-C0: PKA catalytic subunit, C0 form
      2. **DmPKA**-C1: PKA catalytic subunit, C1 form
      3. **DmPKA**-C2: PKA catalytic subunit, C2 form
Caenorhabditis elegans:
      1. **CePKA**: PKA catalytic subunit homologue
Saccharomyces cerevisiae:
      1. **ScPKA**-Tpk1: PKA catalytic subunit homologue, type 1
      2. **ScPKA**-Tpk2: PKA catalytic subunit homologue, type 2
      3. **ScPKA**-Tpk3: PKA catalytic subunit homologue, type 3
Dictyostelium discoideum:
      1. **DdPKA**: PKA catalytic subunit
Aplysia californica:
  *T   1. AplC: PKA catalytic subunit homologue[25]
  *T   2. Sak: "Spermatozoon-associated kinase"[26]
   B. Cyclic GMP-dependent protein kinase (PKG) subfamily
Vertebrate:
      1. **PKG**-I: PKG, type I (human)
      2. **PKG**-II: PKG, type II (mouse)[191]
Drosophila melanogaster:
      1. **DmPKG**-G1: PKG homologue, type 1
      2. **DmPKG**-G2: PKG homologue, type 2
   C. Others
Dictyostelium discoideum:
  *T   1. DdPK1: PKA homologue[27]
AGC-II. Diacylglycerol-activated/phospholipid-dependent protein kinase C (PKC) family
   A. "Conventional" (Ca^{2+}-dependent) protein kinase C (cPKC) subfamily
Vertebrate:
      1. **cPKC**α: Protein kinase C, α form (rat)
      2. **cPKC**β: Protein kinase C, β form (rat)
      3. **cPKC**γ: Protein kinase C, γ form (rat)

Drosophila melanogaster:
    1. **DmPKC**-53Ebr: PKC homologue expressed in brain, locus 53E
    2. **DmPKC**-53Eey: PKC homologue expressed in eye, locus 53E
Aplysia californica:
*T   1. Apl-I: PKC homologue, type I[28]
B. "Novel" (Ca^{2+}-independent) protein kinase C (nPKC) subfamily
Vertebrate:
    1. **nPKCδ**: Protein kinase C, δ form (rat)
    2. **nPKCϵ**: Protein kinase C, ϵ form (rat)
    3. **nPKCη**: Protein kinase C, η form (human)
    4. **nPKCθ**: Protein kinase C, θ form (mouse)
Drosophila melanogaster:
    1. **DmPKC**-98F: PKC homologue, locus 98F
Aplysia californica:
*T   1. Apl-II: PKC homologue, type II[28]
Caenorhabditis elegans :
    1. **CePKC**: PKC homologue, product of *tpa-1* gene
*   2. **CePKC1B**: PKC homologue expressed in neurones and interneurones[29]
Saccharomyces cerevisiae:
    1. **ScPKC**: PKC homologue, product of *PKC1* gene
Schizosaccharomyces pombe:
*T   1. Pck1: "Pombe C-kinase", type 1[30]
*T   2. Pck2: "Pombe C-kinase", type 2[30, 31]
C. "Atypical" protein kinase C (aPKC) subfamily
Vertebrate:
    1. **aPKCζ**: Protein kinase C, ζ form (rat)
*   2. **aPKCι**: Protein kinase C, ι form (human)[32]
*   3. **aPKCμ**: Protein kinase C, μ form (human)[33]
D. Others
Vertebrate:
*   1. PKN: Protein kinase with PKC-related catalytic domain (human)[34]
AGC-III. Related to PKA and PKC (RAC) family
Vertebrate:
    1. **RAC**-α: RAC, α form; cellular homologue of v-Akt oncoprotein (human)
    2. **RAC**-β: RAC, β forms (human)
Drosophila melanogaster:
*   1. DmRAC: RAC homologue[35]
Caenorhabditis elegans :
*   1. CeRAC: RAC homologue[36]
AGCIV. Family of kinases that phosphorylate G protein-coupled receptors
Vertebrate:
    1. **βARK1**: β-Adrenergic receptor kinase, type 1 (bovine)
    2. **βARK2**: β-Adrenergic receptor kinase, type 2 (bovine)
    3. **RhK**: Rhodopsin kinase (bovine)
*   4. IT11: G protein-coupled receptor kinase homologue (human)[37]
*   5. GRK5: G protein-coupled receptor kinase, type 5 (human)[38] (bovine)[39]
*   6. GRK6: G protein-coupled receptor kinase, type 6 (human)[40]
Drosophila melanogaster :
*T   1. DmGPRK1: *Drosophila* G protein-coupled receptor kinase, type 1[41]

*T   2. **DmGPRK2**: *Drosophila* G protein-coupled receptor kinase, type 2[41]

AGC-V. Family of budding yeast AGC related kinases

   Saccharomyces cerevisiae:
1. **Sch9**: Suppressor of defects in cAMP effector pathway
2. **Ykr2**: AGC-related kinase
3. **Ypk1**: AGC-related kinase

AGC-VI. Family of kinases that phosphorylate ribosomal S6 protein

   Vertebrate:
1. **S6K**: 70 kDa S6 kinase with single catalytic domain (human)
2. **RSK**1(Nt): 90 kDa S6 kinase, type 1 (mouse)
3. **RSK**2(Nt): 90 kDa S6 kinase, type 2 (mouse)

(Note: The RSK enzymes have two distinct catalytic domains. The Nt-domain is closely related to S6K, but the Ct-domain is most closely related to phosphorylase kinase.)

AGC-VII. Budding yeast Dbf2/20 family

   Saccharomyces cerevisiae:
1. **Dbf**2: Product of gene periodically expressed in cell cycle
2. **Dbf**20: Close relative of Dbf2 not under cell cycle control

AGC-VIII. Flowering plant "PVPK1 Family" of protein kinase homologues

   Phylum Angiospermophyta (Kingdom Plantae):
1. **PvPK1**: Bean protein kinase homologue
2. **OsG11A**: Rice protein kinase homologue

*T   3. **ZmPPK**: Maize protein kinase homologue[42]
*T   4. **AtPK5**: *Arabidopsis* protein kinase homologue[43]
*T   5. **AtPK7**: *Arabidopsis* protein kinase homologue[44]
*T   6. **AtPK64**: *Arabidopsis* protein kinase homologue[45]
*T   7. **PsPK5**: Pea protein kinase homologue[46]

Other AGC-related kinases

   Vertebrate:
1. **DMPK**: "Myotonic dystrophy protein kinase" (human)

*T   2. **Sgk**: "Serum and glucocorticoid regulated kinase" (rat)[47]
*   3. **Mast205**: Spermatid "microtubule-associated serine/threonine kinase" (mouse)[48]

   Neurospora crassa:

*T   1. **NcCot-1**: Product of gene required for normal colonial growth[49]

   Dictyostelium discoideum:
1. **Ddk2**: Product of developmentally regulated gene

   Saccharomyces cerevisiae:
1. **ScSpk1**: Dual specificity kinase

CaMK group

CaMK-I. Family of kinases regulated by Ca^{2+}/calmodulin, and close relatives

   A. Subfamily including "multifunctional" Ca^{2+}/calmodulin kinases (CaMKs)

   Vertebrate:
1. **CaMKI**: CaMK, type I (rat)
2. **CaMKII**α: CaMK, type II, α subunit (rat)
3. **CaMKII**β: CaMK, type II, β subunit (rat)
4. **CaMKII**γ: CaMK, type II, γ subunit (rat)

     5. **CaMKIIδ**: CaMK, type II, δ subunit (rat)

     6. **EF2K**: Elongation factor-2 kinase or CaMK type III (not included in tree)

     7. **CaMKIV**: CaMK, type IV (rat)

Drosophila melanogaster:

     1. **DmCaMKII**: CaMK-II homologue

Saccharomyces cerevisiae:

     1. **ScCaMKII-1**: CaMK-II homologue, product of *CMK1* gene

     2. **ScCaMKII-2**: CaMK-II homologue, product of *CMK2* gene

Aspergillus nidulans:

*T   1. AnCaMKII: CamK-II homologue[50]

B. Subfamily including phosphorylase kinases

Vertebrate:

     1. **PhK-γM**: Skeletal muscle phosphorylase kinase catalytic subunit (rabbit)

     2. **PhK-γT**: Male germ cell phosphorylase kinase catalytic subunit (human)

     3. **RSK**1(Ct): 90 kDa S6 kinase, type 1; C-terminal catalytic domain (mouse)

     4. **RSK**2(Ct): 90 kDa S6 kinase, type 2; C-terminal catalytic domain (mouse)

C. Subfamily including myosin light chain kinases

Vertebrate:

     1. **SkMLCK**: Skeletal muscle MLCK (rabbit)

     2. **SmMLCK**: Smooth muscle MLCK (rabbit)

*T   3. Titin: Huge protein implicated in skeletal muscle development (human)[51]

Caenorhabditis elegans:

     1. **Twn**: "Twitchin" protein involved in muscle contraction or development

Dictyostelium discoideum:

     1. **DdMLCK**: Slime mould myosin light chain kinase

D. Subfamily of plant kinases with intrinsic calmodulin-like domain

Phylum Angiospermophyta (Kingdom Plantae):

     1. **CDPK**: Soybean Ca^{2+}-regulated kinase with intrinsic CaM-like domain

*T   2. AtAK1: *Arabidopsis* CDPK homologue[52]

*   3. OsSpk: Rice CDPK homologue[53]

*   4. DcPk431: Carrot CDPK homologue[54]

E. Subfamily of plant kinases with highly acidic domain

Phylum Angiospermophyta (Kingdom Plantae):

*   1. ASK1: *Arabidopsis* protein kinase homologue with highly acidic domain[55]

*   2. ASK2: *Arabidopsis* protein kinase homologue with highly acidic domain[55]

F. Other CaMK-related kinases

Vertebrate:

*T   1. PskH1: Putative protein-serine kinase (J.R. Woodgett, unpublished)[56]

*   2. MAPKAP2: "MAP kinase-activated protein kinase 2"[57, 58]

Saccharomyces cerevisiae:

*T   1. Mre4: Protein required for meiotic recombination[59]

*   2. Dun1: Protein required for DNA damage-inducible gene expression[60]

CaMK-II. Snf1/AMPK family

Vertebrate:

     1. **AMPK**: "AMP-activated protein kinase" (FactsBook entry not included in tree)

*T   2. **p78**: Protein lost in carcinomas of human pancreas (human)
       (Maheshwari *et al.* GenBank Acc. M80359)

Saccharomyces cerevisiae:
     1. **Snf1**: Kinase essential for release from glucose repression
     2. **Kin1**: Protein kinase with N-terminal catalytic domain
     3. **Kin2**: Close relative of Kin1
*T   4. **Ycl24**: Protein kinase homologue on chromosome III[61]
*   5. **Ycl453**: Protein kinase homologue on chromosome XI[62]

Schizosaccharomyces pombe:
     1. **SpKin1**: Product of gene important for growth polarity
     2. **Nim1**: Inducer of mitosis

Phylum Angiospermophyta (Kingdom Plantae):
     1. **PSnf1**-RKIN1: Rye putative protein kinase that complements yeast *snf1* mutants
     2. **PSnf1**-AKIN10: *Arabidopsis* putative protein kinase related to SNF1[63]
     3. **PSnf1**-BKIN12: Barley protein related to SNF1[64]
*   4. **PKABA1**: Wheat kinase induced by abscisic acid[65]
*   5. **WPK4**: Wheat kinase homologue regulated by light and nutrients[66]

Other CaMK group kinases

Plasmodium falciparum (malarial parasite):
*T   1. **PfCPK**: Ca^{2+}-regulated kinase with intrinsic CaM-like domain[67]
*T   2. **PfPK2**: Putative protein kinase[68]

CMGC group

CMGC-I. Family of cyclin-dependent kinases (Cdks)and other close relatives

Vertebrate:
     1. **Cdc2**: Inducer of mitosis; functional homologue of yeast $cdc2^+$/Cdc28 kinases (human)
     2. **Cdk2**: Type 2 cyclin-dependent kinase (human)
     3. **Cdk3**: Type 3 cyclin-dependent kinase (human)
     4. **Cdk4**: Type 4 cyclin-dependent kinase (human)
     5. **Cdk5**: Type 5 cyclin-dependent kinase (human)
     6. **Cdk6**: Type 6 cyclin-dependent kinase (human)
     7. **PCTAIRE1**: Cdc2-related protein (human)
     8. **PCTAIRE2**: Cdc2-related protein (human)
     9. **PCTAIRE3**: Cdc2-related protein (human)
   10. **Mo15**(Cak): "Cdk-activating kinase"; Negative regulator of meiosis (*Xenopus*)

Drosophila melanogaster:
     1. **DmCdc2**: Functional homologue of yeast $cdc2^+$/Cdc28 kinases
     2. **DmCdc2c**: Cdc2-cognate protein; Cdk2 homologue

Dictyostelium discoideum:
*T   1. **DdCdc2**: Functional homologue of yeast $cdc2^+$/Cdc28 kinases[69]
*T   2. **DdPRK**: "Cdc2-related PCTAIRE kinase"[70]

Plasmodium falciparum:
*T   1. **PfC2R**: Cdc2-related protein from human malarial parasite (Ross-Macdonald and Williamson, GenBank Acc X61921)

Entamoeba histolytica:
*T   1. **EhC2R**: Cdc2-related protein[71]

Crithidia fasciculata:
*T   1. **CfCdc2R**: Cdc2-related protein[72]
Leishmania mexicana :
*   1. **LmCRK1**: "Cdc2-related kinase"[73]
Saccharomyces cerevisiae:
    1. **Cdc28**: "Cell-division-cycle" gene product
    2. **Pho85**: Negative regulator of the PHO system and cell cycle regulator
    3. **Kin28**: CDC28-related protein
Schizosaccharomyces pombe:
    1. **SpCdc2**: "Cell-division-cycle" gene product
Phylum Angiospermophyta (Kingdom Plantae):
*T   1. **PCdc2**: Flowering plant Cdc2 homologues that complement yeast mutants (*Arabidopsis*)[74] (also cloned from corn[75], alfalfa[76] and soybean[77])
*   2. **MsCdc2B**: Alfalfa Cdc2 cognate gene products that complements G_1/S transition[78]
*T   3. **OsC2R**: More distantly related Cdc2 homologue (rice)[79]
CMGC-II. Erk(MAP kinase) family
Vertebrate:
    1. **Erk1**: "Extracellular signal-regulated kinase", type 1; p44 MAP kinase (rat)
    2. **Erk2**: "Extracellular signal-regulated kinase", type 2; p42 MAP kinase (rat)
    3. **Erk3**: Somewhat distant relative of the Erk/MAP kinases (rat)
*   4. **p63MAPK**: Another more distant relative of the Erk/MAP kinases (human)[80]
*   5. **Jnk1**: "Jun N-terminal kinase" (human)[81]
Drosophila melanogaster:
    1. **DmErkA**: Homologue of Erk/MAP kinases
Saccharomyces cerevisiae:
    1. **Kss1**: Suppressor of sst2 mutant, overcomes growth arrest
    2. **Fus3**: Product of gene required for growth and mating
*T   3. **Slt2**: Product of gene complementing *lyt2* mutants[82, 83]
*   4. **Hog1**: Product of gene required for osmoregulation[84]
Schizosaccharomyces pombe:
    1. **Spk1**: Product of gene that confers drug resistance to PKC inhibitor
Phylum Deuteromycota (Kingdom Fungi)
*T   1. **CaErk1**: Protein that interferes with mating factor-induced cell cycle arrest[85]
Phylum Angiospermophyta (Kingdom Plantae):
*T   1. **PErk**: Flowering plant Erk/MAP kinase homologue (alfalfa)[86] (Six more Erk/MAP homologues identified in *Arabidopsis*)[87, 88]
CMGC-III. Glycogen synthase kinase 3 (GSK3) family
Vertebrate:
    1. **GSK3α**: Glycogen synthase kinase 3, α form (rat)
    2. **GSK3β**: Glycogen synthase kinase 3, β form (rat)
Drosophila melanogaster:
    1. **Sgg**: Product of *shaggy/zeste-white 3* gene
Saccharomyces cerevisiae:
    1. **Mck1**: "Meiosis and centromere regulatory kinase"

* 2. Mds1: Dosage suppressor of mck1 mutant[89, 90]

Phylum Angiospermophyta (Kingdom Plantae):
* 1. ASK-α: "*Arabidopsis* shaggy-related protein kinase", type α[91]
* 2. ASK-γ: "*Arabidopsis* shaggy-related protein kinase", type γ[91]

CMGC-IV. Casein kinase II family

Vertebrate:
    1. **CK2α**: Casein kinase II, α subunit (human)
    2. **CK2α'**: Casein kinase II, α' subunit (human)

Drosophila melanogaster:
    1. **DmCK2**: Casein kinase II homologue

Caenorhabditis elegans:
    1. **CeCK2**: Casein kinase II homologue

Theileria parva (a protozoan parasite):
*T   1. **TpCK2**: Casein kinase II α subunit homologue[92]

Dictyostelium discoideum:
*T   1. **DdCK2**: Casein kinase II, α subunit[93]

Saccharomyces cerevisiae:
    1. **ScCK2α**: Casein kinase II, α subunit
    2. **ScCK2α'**: Casein kinase II, α' subunit

Schizosaccharomyces pombe:
* 1. **SpCka1**: Casein kinase II, α-subunit homologue[94]

Phylum Angiospermophyta (Kingdom Plantae):
    1. **ZmCK2**: Flowering plant casein kinase II, α-subunit homologue (maize)

CMGC-IV. Clk family

Vertebrate:
    1. **Clk**: "Cdc-like kinase" (human)
*T   2. **PskG1**: Putative protein kinase (human)[95]
*T   3. **PskH2**: Putative protein kinase (human)[95]

Drosophila melanogaster:
* 1. **Doa**: Kinase encoded by "Darkener of Apricot" locus[96]

Saccharomyces cerevisiae:
    1. **Yak1**: Suppressor of RAS mutant
*T   2. **Kns1**: Non-essential protein kinase homologue[97]

Schizosaccharomyces pombe:
*T   1. **Dsk1**: Dis1-suppressing protein kinase implicated in mitotic control[98]
* 2. **Prp4**: Pre-mRNA processing gene product; lacks subdomains X–XI[99]

Other CMGC group kinases

Vertebrate:
    1. **Mak**: "Male germ cell-associated kinase" (rat)
    2. **Ched**: "Cholinesterase-related cell division controller" (human)
    3. **PITSLRE**: Galactosyltransferase-associated kinase (human)
    4. **KKIALRE**: Cdc2-related protein (human)

Saccharomyces cerevisiae:
    1. **Sme1**: Product of gene essential for start of meiosis
    2. **Sgv1**: Kinase required for G protein-mediated adaptive response to pheromone
*T   3. **Ctk1**: Product of gene required for normal growth[100]

Phylum Angiospermophyta (Kingdom Plantae):
* 1. **Mhk**: *Arabidopsis thaliana* "Mak homologous kinase"[101]

PTK ("conventional" protein-tyrosine kinase) group
(I–IX: Non-membrane-spanning; X–XXII: Membrane-spanning)
PTK-I. Src family
    Vertebrate:
        1. **Src**: Cellular homologue of Rous sarcoma virus oncoprotein (human)
        2. **Yes**: Cellular homologue of Yamaguchi 73 sarcoma virus oncoprotein (human)
  *T   3. **Yrk**: Yes-related kinase (chicken)[102]
        4. **Fyn**: Protein related to Fgr and Yes (human)
        5. **Fgr**: Cellular homologue of Gardner–Rasheed sarcoma virus oncoprotein (human)
        6. **Lyn**: Protein related to Fgr and Yes (human)
        7. **Hck**: Haematopoietic cell protein-tyrosine kinase (human)
        8. **Lck**: Lymphoid T-cell protein-tyrosine kinase (human)
        9. **Blk**: Lymphoid B-cell protein-tyrosine kinase (mouse)
  *   10. **Fyk**: "Fyn and Yes-related kinase" (electric ray)[103]
Drosophila melanogaster:
        1. **DmSrc**: Src homologue, polytene locus 64B
Dugesia (Girardia) tigrina (Phylum Platyhelminthes):
  *   1. DtSpk1: "Src-like planarian kinase"[104]
Hydra vulgaris (Phylum Cnidaria):
  *T 1. Stk: Src-related protein
Spongilla lacustris (Phylum Porifera):
  *T 1. Srk1–4: Src-related kinases, types1–4[105]

PTK-II. Tec family
    Vertebrate:
        1. **Tec**: "Tyrosine kinase expressed in hepatocellular carcinoma" (mouse)
        2. **Emt**: "Expressed mainly in T cells" kinase' (mouse)
  *T 3. **Btk**: "Bruton's agammaglobulinaemia tyrosine kinase" (human)[106], (mouse)[107,108]
Drosophila melanogaster:
        1. **DmTec**: Tec homologue, polytene locus 28C

PTK-III. Csk family
    Vertebrate:
        1. **Csk**: "C terminal Src kinase"; negative regulator of Src (human)
  *   2. MatK: "Megakaryocyte-associated tyrosine kinase" (human)[109] (human[110];designated "HYL") (mouse[111]; designated "Ctk")

PTK-IV. Fes(Fps) family
    Vertebrate:
        1. **Fes/Fps**: Cellular homologue of feline and avian sarcoma viruses (human)
        2. **Fer**: "Fes/Fps-related" kinase (human)
Drosophila melanogaster:
        1. **DmFer**: Fer-related protein

PTK-V. Abl family
    Vertebrate:
        1. **Abl**: Cellular homologue of Abelson murine leukaemia virus (human)
        2. **Arg**: "Abl-related gene" product (human)

Drosophila melanogaster :
> 1. **DmAbl**: Abl-related protein

Caenorhabditis elegans:
> 1. **CeAbl**: Nematode Abl-related protein

PTK-VI. Syk/Zap70 family
> Vertebrate:
>> 1. **Syk**: "Spleen tyrosine kinase" (pig)
>> 2. **Zap70**: T-cell receptor "zeta chain-associated protein of 70 kDa" (human)
>
> *Hydra vulgaris* (Phylum Cnidaria):
> *   1. Htk16: Syk/Zap70-related protein[112]

PTK-VII. Tyk2/Jak1 family
> Vertebrate:
>> 1. **Tyk2**: Transducer of interferon α/β signals (human)
>> 2. **Jak1**: "Janus kinase", type 1 (human)
>> 3. **Jak2**: "Janus kinase", type 2 (mouse)
>
> *Drosophila melanogaster:*
> *   1. Hop: Product of *hopscotch* gene required for establishing segmental body plan[113]

PTK-VIII. Ack
> Vertebrate:
> *   1. Ack: "CDC42Hs-associated kinase" (human)[114]

PTK-IX. Fak
> Vertebrate:
>> 1. **Fak**: "Focal adhesion kinase" (mouse)

PTK-X. Epidermal growth factor receptor family
> Vertebrate:
>> 1. **EGFR**: Epidermal growth factor receptor (human)
>> 2. **ErbB2**: Normal homologue of oncogene activated in ENU-induced rat neuroblastoma (human)
>> 3. **ErbB3**: Protein closely related to EGFR (human)
> *T 4. **ErbB4**: Receptor tyrosine kinase related to EGFR (human)[115]
>
> *Drosophila melanogaster:*
>> 1. **DER**: Homologue of EGF receptor
>
> *Caenorhabditis elegans:*
>> 1. **LET-23**: Product of gene required for normal vulval development
>
> *Schistosoma mansoni* (Phylum Platyhelminthes):
> *T 1. SER: EGF receptor homologue (Shoemaker et al., unpublished, GenBank Acc M86396)

PTK-XI. Eph/Elk/Eck orphan receptor family
> Vertebrate:
>> 1. **Eph**: Kinase detected in "erythropoeitin-producing hepatoma" (human)
>> 2. **Eck**: "Epithelial cell kinase" (human)
>> 3. **Eek**: Eph and Elk related tyrosine kinase (human)
>> 4. **Hek**: Eph and Elk related tyrosine kinase (human)
> *T 5. **Sek**: "Segmentally-expressed kinase" (mouse)[116]
>> 6. **Elk**: "Eph-like kinase" detected in brain (rat)
> *   7. **Hek2**: "Human embryo kinase", type 2 (human)[117]
>> 8. **Cek5**/Nuk: "Chicken embryo kinase 5" (FactsBook entry; see also ref. 118)

   *   9. Ehk1: "Eph homologuey kinase-1" (rat)[119]
            (mouse[120]; designated "BSK")
            (chicken[121]; designated "Cek7")
   *   10. Ehk2: "Eph homologuey kinase2" (rat)[119]
   *   11. Cek9: "Chicken embryo kinase 9"[121]
   *   12. Cek10: "Chicken embryo kinase 10"[121]

PTK-XII. Axl family
Vertebrate:
       1. **Axl**: "Anexelekto" (*Gr.* "uncontrolled") tyrosine kinase (human)
 *T   2. Eyk: Cellular homologue of RPL30 avian oncoprotein (chicken)[122]
   *   3. Brt/Sky/Tif: "Brain tyrosine kinase" (mouse)[123] or "Sea-related protein tyrosine kinase" (human)[124] or "Tyrosine kinase with Ig-like and FN-III-like domains" (human)[125]

PTK-XIII. Tie/Tek family
Vertebrate:
       1. **Tie**: "Tyrosine kinase with Ig and EGF homology" (human)
       2. **Tek**: "Tunica interna endothelial cell kinase" (human)

PTK-XIV. Platelet-derived growth factor receptor family
A. Subfamily with 5 Ig-like extracellular domains
Vertebrate:
       1. **PDGFRα**: Platelet-derived growth factor receptor, type α (human)
       2. **PDGFRβ**: Platelet-derived growth factor receptor, type β (human)
       3. **CSF1R**: Colony-stimulating factor-1 receptor (human)
       4. **Kit**: Steel growth factor receptor (human)
       5. **Flk2**: "Fetal liver kinase 2" (mouse)
B. Subfamily with 7 Ig-like extracellular domains
Vertebrate:
       1. **Flt1**: "Fms-like tyrosine kinase", type 1 (human)
       2. **Flt4**: "Fms-like tyrosine kinase", type 4 (human)
       3. **Flk1**: "Fetal liver kinase-1" (mouse)

PTK-XV. Fibroblast growth factor receptor family
Vertebrate:
       1. **FGFR1**(Flg): Fibroblast growth factor receptor, type 1 (human)
       2. **FGFR2**(Bek): Fibroblast growth factor receptor, type 2 (human)
       3. **FGFR3**: Fibroblast growth factor receptor, type 3 (human)
       4. **FGFR4**: Fibroblast growth factor receptor, type 4 (human)
Drosophila melanogaster:
       1. **DmFGFR1**: Fibroblast growth factor receptor homologue, type 1
       2. **DmFGFR2**: Fibroblast growth factor receptor homologue, type 2

PTKX-VI. Insulin receptor family
Vertebrate:
       1. **InsR**: Insulin receptor (human)
       2. **IGF1R**: Insulin-like growth factor receptor (human)
       3. **IRR**: Insulin receptor-related protein (human)
Drosophila melanogaster:
       1. **DmInsR**: Homologue of insulin receptor

PTK-XVII. Ltk/Alk family
    Vertebrate:
        1. **Ltk**: "Leukocyte tyrosine kinase" (human)
    *   2. Alk: "Anaplastic lymphoma kinase" (human)[126]
PTK-XVIII. Ros/Sev family
    Vertebrate:
        1. **Ros**: Cellular homologue of UR2 avian sarcoma virus oncoprotein (human)
    Drosophila melanogaster:
        1. **Sev**: Product of *sevenless* gene required for R7 photoreceptor cell development
PTK-XIX. Trk/Ror family
    Vertebrate:
        1. **Trk**: High molecular weight nerve growth factor receptor (human)
        2. **TrkB**: Receptor for neurotrophic factor and neurotrophin-3 (mouse)
        3. **TrkC**: Trk-related protein; receptor for neurotrophin-3 (pig)
    *T   4. Ror1: "Ror" putative receptor, type 1 (human)[127]
    *T   5. Ror2: "Ror" putative receptor, type 2 (human)[127]
    *T   6. TcRTK: Trk-related receptor (electric ray)[128]
    Drosophila melanogaster:
    *   1. Dror: Putative neurotrophic receptor[129]
PTK-XX. Ddr/Tkt family
    *   1. Ddr: "Discoidin domain receptor" (human)[130] (human[131]; desig. "TrkE") (human[132]; desig. "CAK") (mouse[133]; desig. "NEP") (rat[134]; desig. "Ptk3")
    *   2. Tkt: "Tyrosine kinase related to Trk" (human)[135] (mouse[136]; desig. "Tyro10")
PTK-XXI. Hepatocyte growth factor receptor family
    Vertebrate:
        1. **HGFR**: Hepatocyte growth factor receptor (human)
        2. **Sea**: Cellular homologue of S13 avian erythroleukaemia virus oncoprotein (chicken)
    *T   3. Ron: "Recepteur d'Origine Nantaise" (human)[137]
PTK-XXII. Nematode Kin15/16 family
    Caenorhabditis elegans:
    *T   1. CeKin15: PTK expressed during hypodermal development[138]
    *T   2. CeKin16: PTK expressed during hypodermal development[138]
Other membrane-spanning protein-tyrosine kinases (each with no close relatives)
    Vertebrate:
        1. **Ret**: Normal homologue of oncoprotein activated by recombination (human)
        2. **Klg**: "Kinase-like gene" product (chicken)
    *T   3. Nyk/Ryk: "Novel tyrosine kinase-related protein" (mouse)[139] (mouse[140]) (mouse[141]; desig. "VIK") (mouse[142]; desig. "Mrk") (human)[143, 144]
    Drosophila melanogaster:
        1. **Torso**: Product of *torso* gene required for embryonic anterior/posterior determination
        2. **DmTrk**: Distant relative of the mammalian *trk* gene

Other protein kinase families (not falling into major groups)
O-I. Polo family
    Vertebrate:
    *T   1. **Plk**: "Polo-like kinase" (mouse)[145]
          (human[146]; desig. "STPK13")
         2. **Snk**: "Serum-inducible kinase" (mouse)
    Drosophila melanogaster:
         1. **Polo**: Protein kinase homologue required for mitosis
    Saccharomyces cerevisiae:
    *T   1. **Cdc5**: Product of gene required for cell cycle progression[147]
O-II. Mek/Ste7 family
    Vertebrate:
         1. **Mek**1: "MAP Erk kinase", type 1 (human)
         2. **Mek**2: "MAP Erk kinase", type 2 (human)
    Drosophila melanogaster:
         1. **Dsor1**: *Drosophila* Mek homologue
    Saccharomyces cerevisiae:
         1. **Ste7**: Kinase required for haploid-specific gene expression
         2. **Pbs2**: Kinase required for antibiotic drug resistance
         3. **Mkk**1: "MAP kinase kinase", type 1 (supresses lysis defect of *pkc1* mutant)
         4. **Mkk**2: "MAP kinase kinase", type 2 (supresses lysis defect of *pkc1* mutant)
    Schizosaccharomyces pombe:
         1. **Byr1**: Kinase that suppresses *ras1* mutant sporulation defect
         2. **Wis1**: Suppressor of cdc phenotype in triple mutant *cdc25/wee1/win1* strains
O-III. MekK/Ste11 family
    Vertebrate:
    *   1. **MekK**: "Mek kinase" (mouse)[148]
    Saccharomyces cerevisiae:
         1. **Ste11**: Protein required for cell-type-specific transcription
         2. **Bck1**: "Bypass of C kinase" kinase
    Schizosaccharomyces pombe:
         1. **Byr2**: Product of gene required for pheromone signal transduction
    Phylum Angiospermophyta (Kingdom Plantae):
    *   1. **NPK1**: Flowering plant (tobacco) homologue of Bck1[149]
O-IV. Pak/Ste20 family
    Vertebrate:
    *   1. **Pak**: "p21-(Cdc42/Rac) activated kinase" (rat)[150]
    Saccharomyces cerevisiae:
    *T   1. **Ste20**: Product of gene required for pheromone response[151]
O-V. NimA family
    Vertebrate:
         1. **Nek1**: "NimA-related kinase" (mouse)
    Aspergillus nidulans:
         1. **NimA**: Cell cycle control protein kinase
    Drosophila melanogaster:
         1. **Fused**: Product of gene required for segment polarity

Trypanosoma brucei (Phylum Zoomastigina, Kingdom Protoctista):
*T   1. **NrkA**: Trypanosome protein kinase related to Nek1 and NimA[152]
Saccharomyces cerevisiae:
       1. **Kin3**: Putative protein kinase
O-VI. wee1/mik1 family
    Vertebrate:
       1. **Wee1**: Gene product able to complement *S. pombe wee1* mutant (human)
Saccharomyces cerevisiae:
*   1. **Swe1**: Wee1 homologue from budding yeast[153]
Schizosaccharomyces pombe:
       1. **SpWee1**: "Wee" size at division kinase; Cdc2 negative regulator
       2. **Mik1**: "Mitosis inhibitory kinase"; negative regulator of Cdc2
O-VII. Family of kinases involved in translational control
    Vertebrate:
       1. **HRI**: "Haem-regulated eukaryotic initiation factor 2α kinase" (rabbit)
       2. **PKR(Tik)**: "Double-stranded RNA-dependent kinase" (human)
Saccharomyces cerevisiae:
       1. **Gcn2**: Protein required for translational derepression
O-VIII. Raf family
    Vertebrate:
       1. **Raf1**: Cellular homologue of retroviral oncogene product (human)
       2. **A-Raf**: Oncogenic protein closely related to c-Raf (human)
       3. **B-Raf**: Oncogenic protein closely related to c-Raf (human)
Drosophila melanogaster:
       1. **DmRaf**: Raf homologue
Caenorhabditis elegans:
       1. **CeRaf**: Raf homologue; product of *lin-45* gene required for vulval differ-
          entiation
    Phylum Angiospermophyta (Kingdom Plantae):
*T   1. **Ctr1**: Negative regulator of ethylene response pathway (*Arabidopsis*)[154]
O-IX. Activin/TGFβ receptor family
    Vertebrate:
       1. **ActRII**: Activin receptor II, type A (mouse)
       2. **ActRIIB**: Activin receptor II, type B (mouse)
       3. **TGFβRII**: TGFβ type II receptor (human)
*T   4. **Tsk7L/ActR-I**: Activin and/or TGFβ type I receptor (mouse)[155],
          (human)[156, 157], (rat)[158]
*   5. **TSR-1**: "TGFβ superfamily receptor", type 1 (human)[156, 159]
*   6. **ALK-3**: "Activin receptor-like kinase", type 3 (human)[156]
*   7. **ALK-4**: "Activin receptor-like kinase", type 4 (human)[156]
*   8. **ALK-5**: "Activin receptor-like kinase", type 5 (human)[160]
*   9. **ALK-6**: "Activin receptor-like kinase", type 4 (mouse)[161]
*   10. **C14**: Putative receptor kinase expressed in gonads (rat)[162]
Drosophila melanogaster:
*   1. **DmAtr-II**: Activin receptor homologue[163]
*   2. **DmSax**: Product of *saxophone* gene[164]
Caenorhabditis elegans:
       1. **DAF-1**: Product of gene required for vulval development
*   2. DAF-4: Larva development regulatory protein; BMP receptor[165]

O-X. Flowering plant putative receptor kinase family
   Phylum Angiospermophyta (Kingdom Plantae):
       1. **ZmPK1**: Putative receptor protein-serine kinase (maize)
       2. **Srk**: "S receptor kinase"; three distinct alleles: 2, 6, and 910 (*Brassica*)
   *T 3. **Tmk1**: Putative "Transmembrane receptor kinase" (*Arabidopsis*)[166]
   *T 4. **Apk1**: Kinase that phosphorylates Tyr, Ser, and Thr (*Arabidopsis*)[167]
   *   5. **Nak**: "Novel *Arabidopsis* kinase" (*Arabidopsis*)[101]
   *T 6. **Pro25**: Putative kinase selected for specificity to thylakoid membrane protein (*Arabidopsis*)[168]
   *   7. **Pto**: Product of gene conferring pathogen resistance (tomato)[169]
   *   8. **Tmk1**: Transmembrane protein with unusual kinase-like domain (*Arabidopsis*)[170]

O-XI. Family of "mixed-lineage" kinases with leucine zipper domain
   Vertebrate:
   *   5. **Mlk1**: "Mixed lineage kinase", type 1 (human)[171]
   *   6. **PTK1**: Putative protein kinase (human)[172]

O-XII. Casein kinase I family
   Vertebrate:
       1. **CK1α**: Casein kinase I, type α (bovine)
       2. **CK1β**: Casein kinase I, type β (bovine)
       3. **CK1γ**: Casein kinase I, type γ (bovine) (partial, not in tree)
   *T 4. **CK1δ**: Casein kinase I, type δ (rat)[173]
   Saccharomyces cerevisiae:
       1. **Yck1**: Yeast casein kinase I homologue, type 1
       2. **Yck2**: Yeast casein kinase I homologue, type 2
       3. **Hrr25**: Kinase required for DNA repair

O-XIII. PKN family of prokaryotic protein kinases
   Myxococcus xanthus (Phylum Myxobacteria; Kingdom Prokaryotae)
       1. **Pkn1**: Protein kinase homologous to eukaryotic kinases
   *T 2. Pkn2: Protein kinase required for maintenance of stationary phase cells and development
       (Munoz-Dorado et al., unpublished; GenBank M94857 M94858)

Other protein kinase family members (each with no known close relatives)
   Vertebrate:
       1. **Mos**: Cellular homologue of retroviral oncogene product (human)
       2. **Pim1**: Proto-oncogene activated by murine leukaemia virus (human)
       3. **Cot**: Product of oncogene expressed in human thyroid carcinoma (human)
       4. **Esk/Ttk**: "Embryonal carcinoma STY kinase"; dual specificity (human)
   Drosophila melanogaster:
       1. **NinaC**: Product of gene essential for photoreceptor function
   *T 2. **Pelle**: Product of gene required for dorsalventral polarity[174]
   Dictyostelium discoideum:
       1. **SplA**: Spore lysis A protein kinase
       2. **Dpyk2**: Developmentally regulated tyrosine kinase, type 2
   Ceratodon purpureus: (a moss)
   *T 1. **PhyCer**: Putative protein-tyrosine kinase encoded by a phytochrome gene[175]

Saccharomyces cerevisiae:
1. **Cdc7**: "Cell-division-cycle" control gene product
2. **Cdc15**: "Cell-division-cycle" control gene product
3. **Vps15**: Product of gene essential for sorting to lysosome-like vacuole

*T   4. **Npr1**: Product of gene required for activity of ammonia-sensitive amino acid permeases[176]

*T   5. **Elm1**: Product of gene required for yeast-like cell morphology[177]

*T   6. **Ire1**: Product of gene required for Myo-inositol synthesis and signalling from the endoplasmic reticulum to the nucleus[178, 179]

*T   7. **Ykl516**: Putative protein kinase gene on chromosome XI[180]

Schizosaccharomyces pombe:
1. **Ran1**: Product of gene required for normal meiotic function

*T   2. **Chk1**: "Checkpoint kinase" that links rad pathway to Cdc2[181]

*   3. **Csk1**: "Cyclin suppressing kinase"[182]

Entamoeba histolytica (Phylum Rhizopoda, Kingdom Protoctista):

*T   1. **Pstk1**: Distant relative of Mos (Lohia and Samuelson, unpublished; GenBank L05668)

Phylum Angiospermophyta (Kingdom Plantae):

*T   1. **GmPK6**: Protein kinase homologue (soybean)[183]

*   2. **Tsl**: Product of *Tousled* gene required for normal leaf and flower development (*Arabidopsis*)[184]

Yersinia pseudotuberculosis (Phylum Omnibacteria, Kingdom Prokaryotae):

*T   1. **YpkA**. Enterobacterial protein kinase essential for virulence[185]

In order to accommodate the large numbers of sequences, it was necessary to construct five separate trees. Initially, a "skeleton" tree of 99 kinases was obtained (Fig. 3A). The skeleton tree included only representative members from each of four large groups of kinases, each consisting of multiple related families known (from previous work) to cluster together in the tree. These four groups are designated: (1) the AGC group which includes the cyclic nucleotide-dependent family (**PKA** and **PKG**), the protein kinase C (**PKC**) family, the β-adrenergic receptor kinase (**βARK**) family, the ribosomal protein S6 kinase family, and other close relatives; (2) The CaMK group which includes the family of kinases regulated by Ca^{2+}/calmodulin, the **Snf1/AMPK** family, and other close relatives; (3) The CMGC group which includes the family of cyclin-dependent kinases, the **Erk1/2**/(MAP kinase) family, the glycogen synthase 3 (**GSK3**) family, the casein kinase II (**CK2**) family, the **Clk** (Cdk-like kinase) family, and other close relatives; and (4) the "conventional" protein-tyrosine kinase PTK group. Separate trees (Figs 3B–E) were later obtained for each of the four large kinase groups, and contain all members of the groups whose sequences were available at the time of analysis.

It can be reasonably surmised that the kinases having closely related catalytic domains, and thus defining a family, represent products of genes that have undergone relatively recent evolutionary separations. Given this, it should come as no surprise that members of a given family tend also to share related functions. This is manifest by similarities in overall structural topology, mode of regulation and substrate specificity. The details of the common properties exhibited by the members of the various kinase families can best be gained from study of the information outlined in the individual entries in this book. Some of the most salient relationships are discussed below.

AGC group

The AGC group kinases tend to be "basic amino acid-directed" enzymes, phosphorylating substrates at serine/threonine residues lying very near the basic residues arginine and lysine. For the cyclic nucleotide-dependent and ribosomal S6 kinase families the preferred substrates have basic residues lying in specific positions N-terminal to the phosphate acceptor. Preferred substrates for the **PKC** and **RAC** families have basic residues on both the N-terminal and C-terminal sides of the acceptor. The G protein-coupled receptor kinases (β**ARK** and **RhK**) appear to break this rule, however, as they are reported to prefer synthetic peptide substrate residues located within an acidic environment. Little substrate information is available for the other families falling within this group.

CaMK group

The CaMK group kinases also tend to be basic amino acid-directed, and in this regard it is notable that the AGC and CaMK groups fall near one another in the phylogenetic tree. **CaMKII**, **CaMKIV**, **SmMLCK**, **SkMLCK**, **CDPK** and **AMPK** are all reported to prefer substrates with basic residues at specific positions N-terminal to the acceptor site, while **EF2K** and **PhK** prefer sites with basic residues at both N- and C-terminal locations. Many, but not all, of the CaMK group kinases are known to be activated by Ca^{2+}/calmodulin binding to a small domain located just C-terminal to the catalytic domain; e.g. **CaMKII**, **CaMKIV**, **PhK**-γ, **SmMLCK** and **SkMLCK**. These enzymes and their close relatives are grouped together in a large family within the CaMK group. Also included in this family are a subfamily of plant enzymes (represented by **CDPK**) that contain an intrinsic calmodulin-like domain that confers Ca^{2+}-dependent activation. The other family within the CaMK group is the **Snf1/AMPK** family. Within this family substrate specificity determinant information has been obtained only for **AMPK**, which shows a requirement for an N-terminal basic residue, but has additional requirements for hydrophobic residues.

CMGC group

The other major category of protein-serine/threonine kinases is the CMGC group. For the most part, these are "proline-directed" enzymes, phosphorylating substrates at sites lying in proline-rich environments. Available data for **Cdc2** and **Cdk2** indicate that members of the cyclin-dependent kinase family require phosphate acceptors lying immediately N-terminal to a proline. A similar requirement is indicated for the **Erk1/2/(MAP kinase)** family. The situation for the **GSK3** family is more complicated, but most known acceptor sites lie within proline-rich regions. The **CK2** family enzymes fail to conform to the proline-directed specificity exhibited by the other major families of this group, showing, instead, a strong preference for serine residues located N-terminal to a cluster of acidic residues. The CMGC group kinases have larger-than-average kinase domains due to insertions between subdomains X and XI.

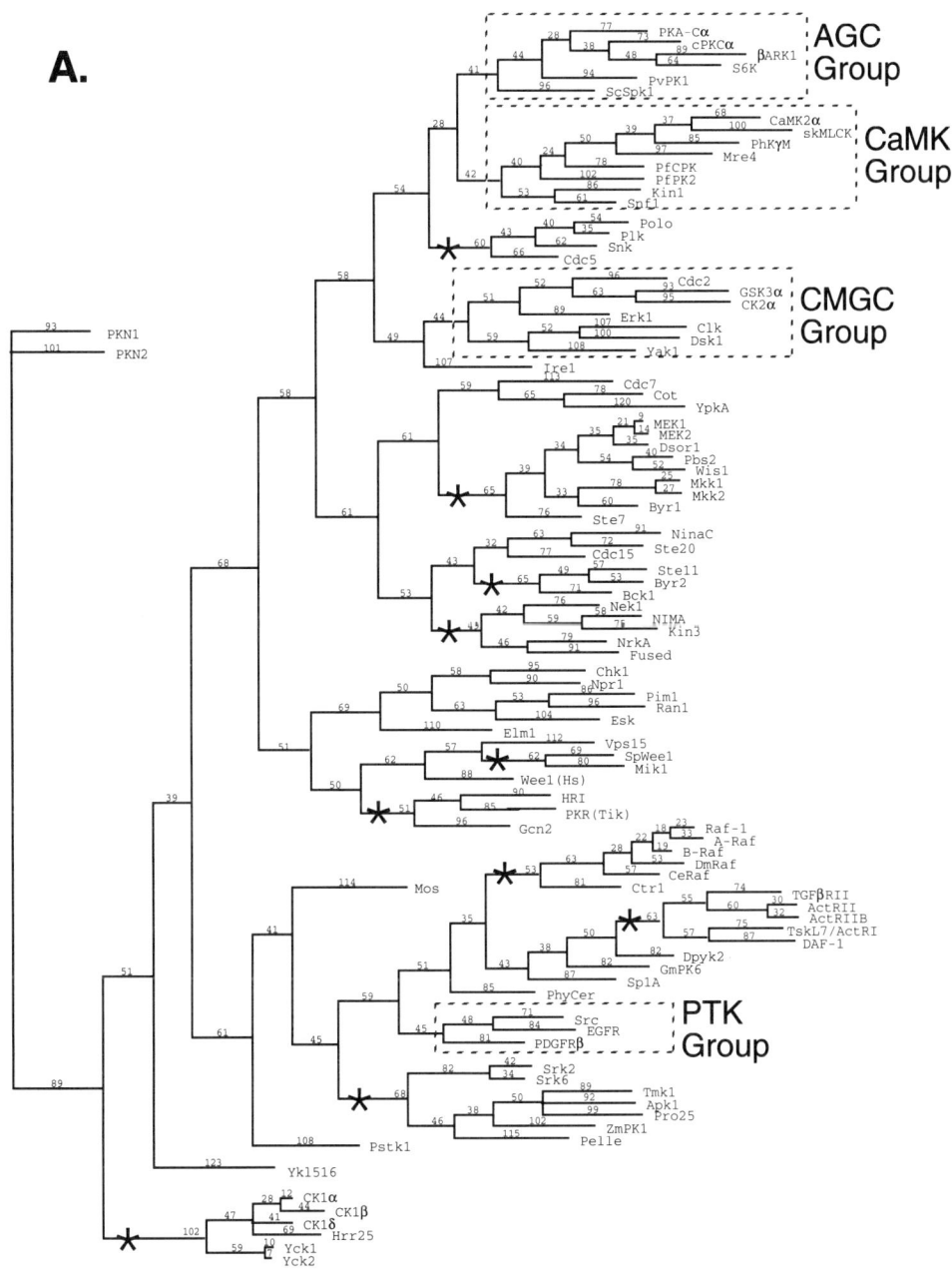

Figure 3. *Phylogenetic trees of the eukaryotic protein kinase superfamily inferred from kinase domain amino acid sequence alignments. The abbreviated nomenclature is the same as used in Table 1. (A) "Skeleton" tree*

B. AGC Group

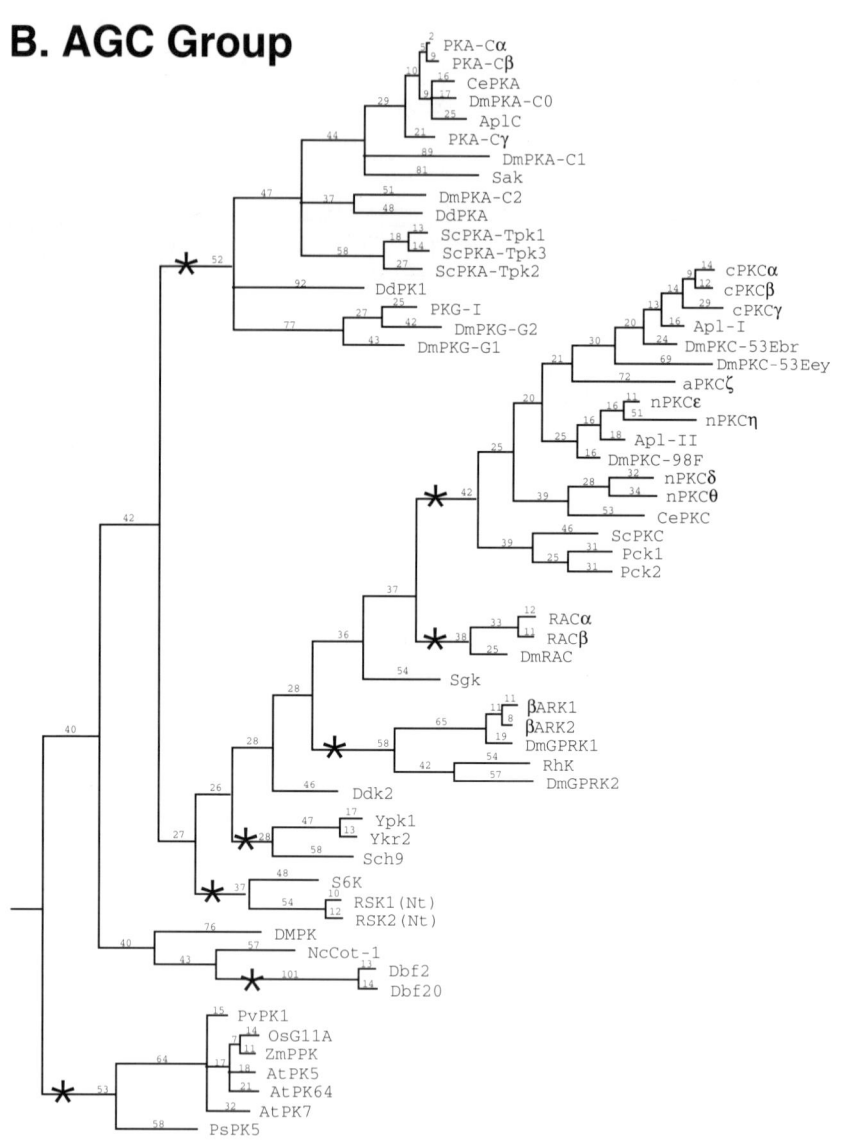

showing 99 kinases. Positions of 4 clusters (AGC, CaMK, CMGC, and PTK) containing kinases representative of larger groups are indicated in the skeleton tree. (B) AGC group tree of 59 kinases including **PKA**, **PKG**, *and* **PKC**

C. CaMK Group

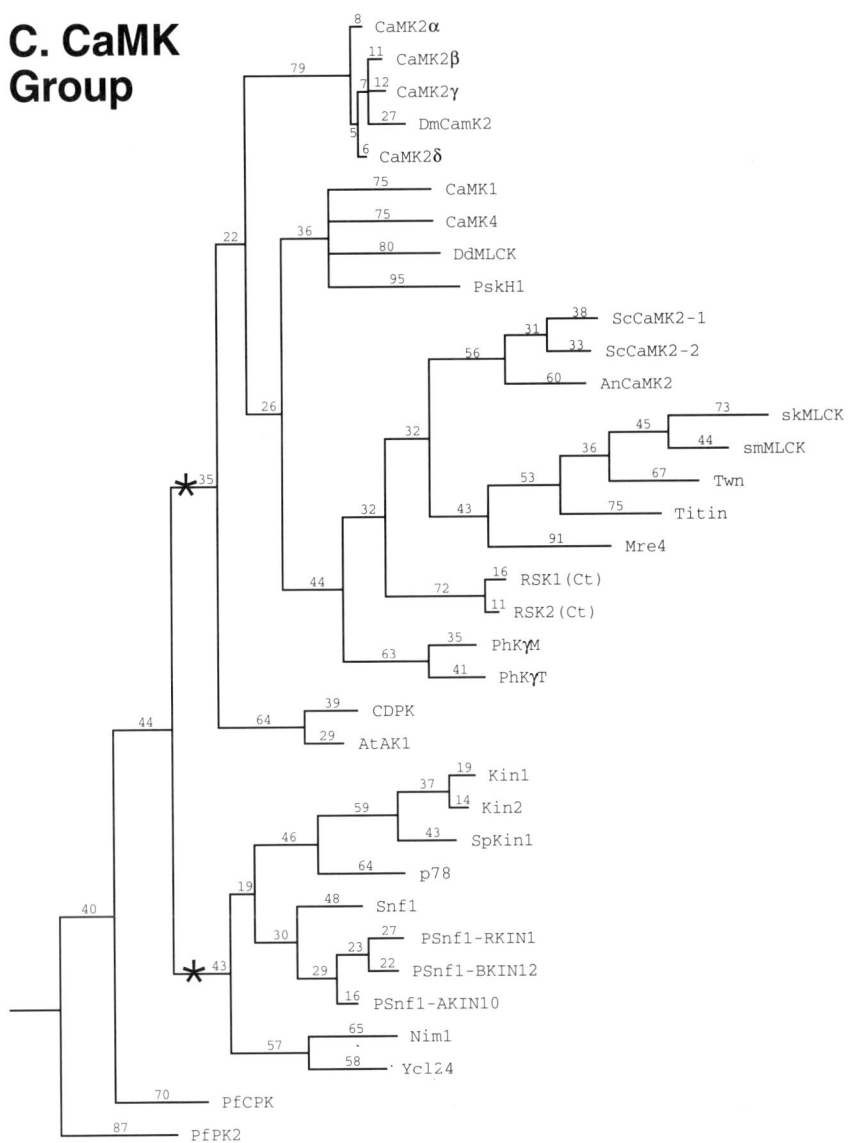

and other close relatives. (C) CaMK group tree of 35 kinases including the Ca²⁺/calmodulin-regulated enzymes. (D) CMGC group tree of 59 kinases including the cyclin-dependent protein kinases. (E) PTK group tree of 90

D. CMGC Group

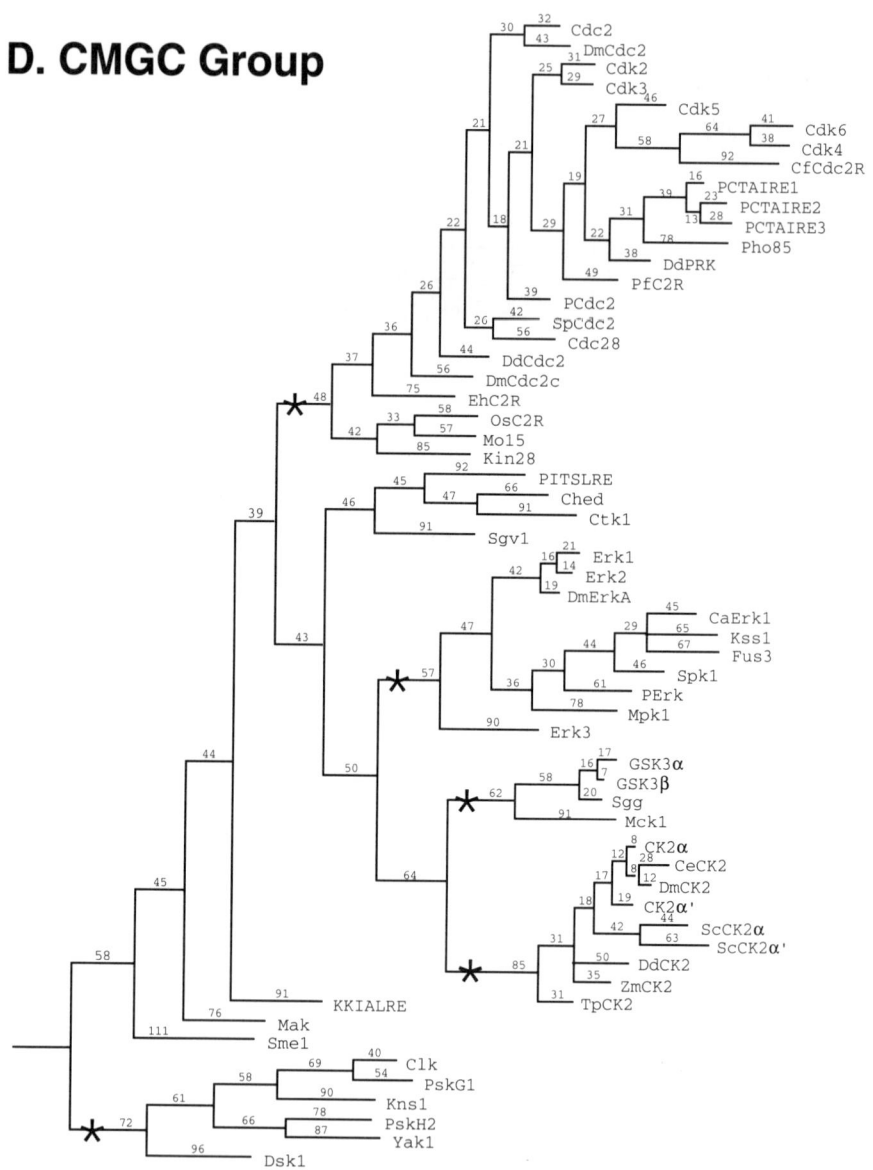

conventional protein-tyrosine kinases. Tree A is unrooted and drawn with **Pkn1** and **Pkn2** as outgroups. Outgroups of two or more distantly related kinases (not shown) were included in the analysis of Trees B-E to provide a rooting point.

E. PTK
Group

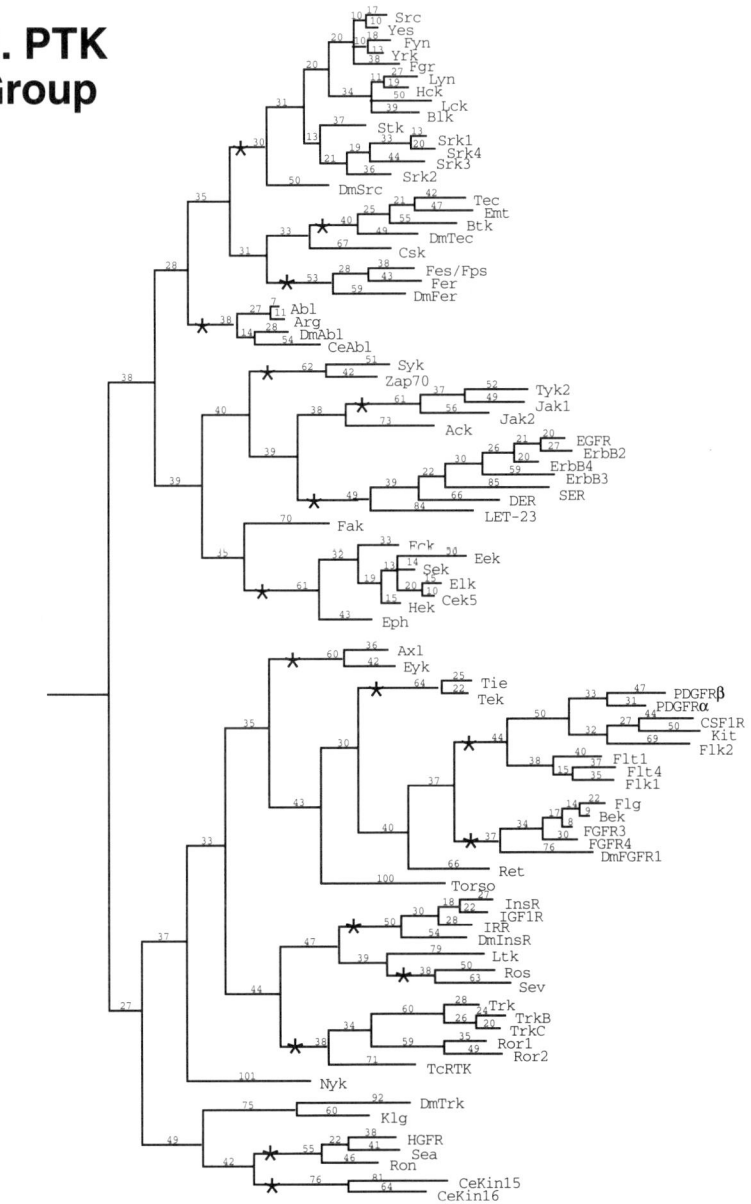

Asterisks (*) in all trees indicate branches leading to defined protein kinase families listed in Table 1. Branch lengths indicate number of amino acid substitutions required to reach hypothetical common ancestors at internal nodes.

PTK group

The "conventional protein-tyrosine" kinase PTK group includes a large number of enzymes with quite closely related kinase domains that specifically phosphorylate tyrosine residues (i.e. they cannot phosphorylate serine or threonine). These enzymes, first recognized among retroviral oncoproteins, have been found only in metazoan cells where they are widely recognized for their roles in transducing growth and differentiation signals. Included in this group are over a dozen distinct receptor families made up of membrane-spanning molecules that share similar overall structural topologies, and nine non-receptor families also composed of structurally similar molecules. The specificity determinants surrounding the tyrosine phosphoacceptor sites have yet to be firmly established for these enzymes, but glutamic acid residues of the N- or C-terminal side of the acceptor are often preferred. This group is labelled "conventional" to distinguish it from other kinases (including **ScSpk1**, **Clk**, the **Mek/Ste7** family, **SpWee1/Mik1**, **ActRII**, **Hrr25**, **Esk** and **Sp1A/DPyk2** reported to exhibit a "dual-specificity"; that is, being capable of phosphorylating both tyrosine and serine/threonine residues. Considered as a group, these dual-specificity kinases are not particularly closely related to the conventional PTKs. Indeed, they seem to map throughout the phylogenetic tree, suggesting that the ability to phosphorylate on tyrosine may have had numerous independent origins during the evolutionary history of the superfamily.

The kinases falling outside the four major groups are a rather mixed bag. Although the individual members within the defined families found in this "Other" category are clearly related to one another through both structure and function, it is difficult to make broader generalizations that could group any of these families together into a larger category. As far as substrate specificity determinants go, little is known about most of the "Other" category kinases, due primarily to their rather recent discovery and the paucity of known physiological substrates. The casein kinase I (**CK1**) family members, however, have been shown to prefer serine/ threonine residues located C-terminal to a phosphoserine or phosphothreonine, although a stretch of acidic residues may substitute. Also, the family of kinases involved in translational control (**HRI**, **PKR/Tik** and **Gcn2**) appear to be basic amino acid-directed enzymes preferring serine residues lying N-terminal to an arginine. Finally, as mentioned above, the **Mek/Ste7** family and **SpWee1/Mik1** family exhibit a dual-specificity.

FUTURE PROSPECTS

The rate of protein kinase discovery still shows no signs of abating. In addition to the continuing successes of homology-based approaches, genomic sequencing projects are beginning to make significant contributions. For instance, the sequences of an entire budding yeast chromosome[187] and a $\approx$2 Mb stretch of *C. elegans* chromosome III[188] have revealed a number of new putative protein kinase genes. As genome sequencing projects gather speed the number of new protein kinase genes discovered in this way will undoubtedly mushroom. This explosion of sequence data is making it increasingly difficult to manage protein kinase databases of the sort described here. Programs designed to align and derive relatedness trees are currently unable to handle the large number of available kinase domain sequences.

New data-handling programs will have to be developed to cope with large numbers of sequences like those of the eukaryotic protein kinase superfamily.

Protein kinase catalytic domain structures will continue to be solved. The first structure of a conventional protein-tyrosine kinase will shortly be available, and this should reveal how tyrosine versus serine/threonine is selected as the acceptor amino acid. Such structures will enable comparative analysis to be carried out at the three-dimensional level, and allow predictions of structures from primary sequences. Structural comparisons of catalytic domains with bound peptide substrates will also provide insights into substrate specificity. Most protein kinases show some degree of primary sequence specificity, and new methods are being developed to determine consensus sequence specificities for individual protein kinases. The structural basis for the binding of a preferred peptide sequence to the cognate substrate binding site can then be deduced. In the future it may be possible to model the three-dimensional structure of a novel protein kinase catalytic domain with sufficient accuracy to be able to predict the preferred primary sequence surrounding the hydroxyamino acid it phosphorylates, which in turn will allow one to predict what proteins might be its substrates from the increasingly complete database of protein sequences.

References

[1] Hanks, S.K. et al. (1988) Science 241, 42–52.

[2] Hanks, S.K., and Quinn, A.M. (1991) Meth. Enzymol. 200, 38–62.

[3] Hanks, S.K. (1991) Curr. Opin. Struct. Biol. 1, 369 383.

[4] Hunter, T. (1987) Cell 50, 823–829.

[5] Chang, C. et al. (1993) Science 262, 539–544.

[6] Ota, I.M., and Varshavsky, A. (1993) Science 262, 566–569.

[7] Beeler, J.F. et al. (1994) Mol. Cell. Biol. 14, 982–988.

[8] Huang et al. (1991) J. Biol. Chem. 266, 9023–9031.

[9] Cozzone, A.J. (1988) Annu. Rev. Microbiol. 42, 97–125.

[10] Bourret, R.B. et al. (1991) Annu. Rev. Biochem. 60, 401–441.

[11] Saier, M.H. Jr. (1993) J. Cell. Biochem. 51, 1–6.

[12] Galyov, E.E. et al. (1993) Nature 361, 730–732.

[13] Munoz-Dorado, J. et al. (1991) Cell 67, 995–1006.

[14] Munoz-Dorado, J. et al. (1993) J. Cell. Biochem. 51, 29–33.

[15] Knighton, D.R. et al. (1991) Science 253, 407–414.

[16] Knighton, D.R. et al. (1991) Science 253, 414–420.

[17] Bossemeyer, D. et al. (1993) EMBO J. 12, 849–859.

[18] Zheng, J. et al. (1993) Biochemistry 32, 2154–2161.

[19] De Bondt, H.L. et al. (1993) Nature 363, 595–602.

[20] Zhang, F. et al. (1994) Nature 367, 704–711.

[21] Bramson, H.N. et al. (1984) CRC Crit. Rev. Biochem. 15, 93–124.

[22] Yoon, M.-Y., and Cook, P.F. (1987) Biochemistry 26, 4118–4125.

[23] Ho, M.-F. et al. (1988) J. Am. Chem. Soc. 110, 2680–2681.

[24] Veron, M. et al. (1993) Proc. Natl Acad. Sci. USA 90, 10618–10622.

[25] Beushausen S. et al. (1988) Neuron 1, 853–864.

[26] Beushausen S., and Bayley, H. (1990) Mol. Cell. Biol. 10, 6775–6780.

[27] Burki, E. et al. (1991) Gene 102, 57–65.

[28] Kruger, K.E. et al. (1991) J. Neurosci. 11, 2303–2313.

[29] Land, M. et al. (1994) J. Biol. Chem. 269, 9234–9244.

[30] Toda, T. et al. (1993) EMBO J. 12, 1987–1995.
[31] Mazzei, G.J. et al. (1993) J. Biol. Chem. 10, 7401–7406.
[32] Selbie, L.A. et al. (1993) J. Biol. Chem. 268, 24296–24302.
[33] Johannes, F.J. et al. (1994) J. Biol. Chem. 269, 6140–6148.
[34] Mukai, H., and Ono, Y. (1994) Biochem. Biophys. Res. Commun. 199, 897–904.
[35] Franke, T.F., et al. (1994) Oncogene 9, 141–148.
[36] Waterson, R. et al. (1992) Nature Genet. 1, 114–123.
[37] Ambrose, C. et al. (1992) Human Mol. Genet. 1, 697–703.
[38] Kunapuli, P., and Benovic, J.L. (1993) Proc. Natl Acad. Sci. USA 90, 5588–5592.
[39] Premont, R.T. et al., 1994 J. Biol. Chem. 269, 6832–6841.
[40] Benovic, J.L., and Gomez, J. (1993) J. Biol. Chem. 268, 19521–19527.
[41] Cassill, J.A. et al. (1991) Proc. Natl Acad. Sci. USA 88, 11067–11070.
[42] Biermann, B.J. et al. (1990) Plant Physiol. 94, 1609–1615.
[43] Hayashida, N. et al. (1993) Gene 124, 251–255.
[44] Hayashida, N. et al. (1992) Gene 121, 325–330.
[45] Mizoguchi, T. et al. (1992) Plant Mol. Biol. 18, 809–812.
[46] Lin, X., and Watson, J.C. (1992) Plant Physiol. 100, 1072–1074.
[47] Webster, M.K. et al. (1993) Mol. Cell. Biol. 13, 2031–2040.
[48] Walden, P.D., and Cowan, N.J. (1993) Mol. Cell. Biol. 13, 7625–7635.
[49] Yarden, O. et al. (1992) EMBO J. 11, 2159–2166.
[50] Kornstein, L.B. et al. (1992) Gene 113, 75–82.
[51] Labeit, S. et al. (1992) EMBO J. 11, 1711–1716.
[52] Harper, J.F. et al. (1993) Biochemistry 32, 3282–3290.
[53] Kawasaki, T. et al. (1993) Gene 129, 183–189.
[54] Suen, K.L., and Choi, J.H. (1991) Plant. Mol. Biol. 17, 581–590.
[55] Park, Y.S. et al. (1993) Plant Mol. Biol. 22, 615–624.
[56] Hanks, S.K. (1987) Proc. Natl. Acad. Sci. USA 84, 388–392.
[57] Stokoe, D. et al. (1993) Biochem. J. 296, 843–849.
[58] Engel, K. et al. (1993) FEBS Lett. 336, 143–147.
[59] Leem, S.H., and Ogawa, H. (1992) Nucl. Acids Res. 20, 449–457.
[60] Zhou, Z., and Elledge, S.J. (1993) Cell 75, 1119–1127.
[61] Bolle, P.A. et al. (1992) Yeast 8, 205–213.
[62] Pallier, C. et al. (1993) Yeast 9, 1149–1155.
[63] Le Guen, L. et al. (1992) Gene 120, 249–254.
[64] Halford, N.G. et al. (1992) Plant J. 2, 791–797.
[65] Anderberg, R.J., and Walker-Simmons, M.K. (1992) Proc. Natl. Acad. Sci. USA 89, 10183–10187.
[66] Sano, H., and Youssefian, S. (1994) Proc. Natl Acad. Sci. USA 91, 2582–2586.
[67] Zhao, Y. et al. (1993) J. Biol. Chem. 268, 4347–4354.
[68] Zhao, Y. et al. (1992) Eur. J. Biochem. 207, 305–313.
[69] Michaelis, C., and Weeks, G. (1992) Biochim. Biophys. Acta 1132, 35–42.
[70] Michaelis, C., and Weeks, G. (1993) Biochim. Biophys. Acta 1179, 117–124.
[71] Lohia, A., and Samuelson, J. (1993) Gene 127, 203–207.
[72] Brown, L.M. et al. (1992) Nucl. Acids Res. 20, 5451–5456.
[73] Mottram, J.C. et al. (1993) J. Biol. Chem. 268, 21044–21052.
[74] Hirayama, T. et al. (1991) Gene 105, 159–165.
[75] Colasanti, J. et al. (1991) Proc. Natl Acad. Sci. USA 88, 3377–3381.
[76] Hirt, H. et al. (1991) Proc. Natl Acad. Sci. USA 88, 1636–1640.
[77] Miao, G.H. et al. (1993) Proc. Natl Acad. Sci. USA 90, 943–947.

[78] Hirt, H. et al. (1993) Plant J. 4, 61–69.
[79] Hatta, S. (1991) FEBS Lett. 279, 149–152.
[80] Gonzalez, F.A. et al. (1992) FEBS Lett. 304, 170–178.
[81] Derijard, B. et al. (1994) Cell 76, 1025–1037.
[82] Torres, L. et al. (1991) Mol. Microbiol. 5, 2845–2854.
[83] Lee, K.S. et al. (1993) Mol. Cell. Biol. 13, 3067–3075.
[84] Brewster, J.L. et al. (1993) Science 259, 1760–1763.
[85] Whiteway, M. et al. (1992) Proc. Natl Acad. Sci. USA 89, 9410–9414.
[86] Duerr, B. et al. (1993) Plant Cell 5, 87–96.
[87] Mizoguchi, T. et al. (1993) FEBS Lett. 336, 440–444.
[88] Mizoguchi, T. et al. (1994) Plant J. 5, 111–122.
[89] Bianchi, M.W. et al. (1993) Gene 134, 51–56.
[90] Puziss, J.W. et al. (1994) Mol. Cell. Biol. 14, 831–839.
[91] Bianchi, M.W. et al. (1994) Mol. Gen. Genet. 242, 337–345.
[92] ole-MoiYoi, O.K. et al. (1992) Biochemistry 31, 6193–6202.
[93] Kikkawa, U. et al. (1992) Mol. Cell. Biol. 12, 5711–5723.
[94] Roussou, I., and Draetta, G. (1994) Mol. Cell. Biol. 14, 576–586.
[95] Hanks, S.K., and Quinn, A.M. (1991) Method. Enzymol. 200, 38–62.
[96] Yun, B. et al. (1994) Genes Devel. 8, 1160–1173.
[97] Padmanabha, R. et al. (1991) Mol. Gen. Genet. 229, 1–9.
[98] Takeuchi, M., and Yanagida, M. (1993) Mol. Biol. Cell 4, 247–260.
[99] Alahari, S.K. et al. (1993) Nucl. Acids Res. 21, 4079–4083.
[100] Lee, J.M., and Greenleaf, A.L. (1991) Gene Expression 1, 149–167.
[101] Moran, T., and Walker, J.C. (1993) Biochim. Biophys. Acta 1216, 9–14.
[102] Sudol, M. et al. (1993) Oncogene 8, 823–831.
[103] Swope, S.L., and Huganir, R.L. (1993) J. Biol. Chem. 268, 25152–25161.
[104] Burgaya, F. et al. (1994) Oncogene 9, 1267–1272.
[105] Ottilie, S. et al. (1992) Oncogene 7, 1625–1630.
[106] Vetrie, D. et al. (1993) Nature 361, 226–233.
[107] Tsukada, S. et al. (1993) Cell 72, 279–290.
[108] Yamada, N. et al. (1993) Biochem. Biophys. Res. Commun. 192, 231–240.
[109] Bennett, B.D. et al. (1994) J. Biol. Chem. 269, 1068–1074.
[110] Sakano, S. et al. (1994) Oncogene 9, 1155–1161.
[111] Klages, S. et al. (1994) Proc. Natl Acad. Sci. USA 91, 2597–2601.
[112] Chan, T.A. et al. (1994) Oncogene 9, 1253–1259.
[113] Binari, R., and Perrimon, N. (1994) Genes Devel. 8, 300–312.
[114] Manser, E. et al. (1993) Nature 363, 364–367.
[115] Plowman, G.D. et al. (1993) Proc. Natl Acad. Sci. USA 90, 1746–1750.
[116] Gilardi-Hebenstreit, P. et al. (1992) Oncogene 7, 2499–2506.
[117] Bohme, B. et al. (1993) Oncogene 8, 2857–2862.
[118] Henkemeyer, M. et al. (1994) Oncogene 9, 1001–1014.
[119] Maisonpierre, P.C. et al. (1993) Oncogene 8, 3277–3288.
[120] Zhou, R. et al. (1994) J. Neurosci. Res. 37, 129–143.
[121] Sajjadi, F.G., and Pasquale, E.B. (1993) Oncogene 8, 1807–1813.
[122] Jia, R., and Hanafusa, H. (1994) J. Biol. Chem. 269, 1839–1844.
[123] Fujimoto, J., and Yamamoto, T. (1994) Oncogene 9, 693–698.
[124] Ohashi, K. et al. (1994) Oncogene 9, 699–705.
[125] Dai, W. et al. (1994) Oncogene 9, 975–979.
[126] Morris, S.W. et al. (1994) Science 263, 1281–1284.

[127] Masiakowski, P., and Carroll, R.D. (1992) J. Biol. Chem. 267, 26181–26190.
[128] Jennings, C.G.B. et al. (1993) Proc. Natl Acad. Sci. USA 90, 2895–2899.
[129] Wilson, C. et al. (1993) Proc. Natl Acad. Sci. USA 90, 7109–7113.
[130] Johnson, J.D. et al. (1993) Proc. Natl Acad. Sci. USA 90, 5677–5681.
[131] Di Marco, E. et al. (1993) J. Biol. Chem. 268, 24290–24295.
[132] Perez, J. et al. (1994) Oncogene 9, 211–219.
[133] Zerlin, M. et al. (1993) Oncogene 8, 2731–2739.
[134] Sanchez, M.P. et al. (1994) Proc. Natl Acad. Sci. USA 91, 1819–1823.
[135] Karn, T. et al. (1993) Oncogene 8, 3433–3440.
[136] Lai, C., and Lemke, G. (1994) Oncogene 9, 877–883.
[137] Ronsin, C. et al. (1993) Oncogene 8, 1195–1202.
[138] Morgan, W.R., and Greenwald, I. (1993) Mol. Cell. Biol. 13, 7133–7143.
[139] Paul, S.R. et al. (1992) Int. J. Cell Cloning 10, 309–314.
[140] Hovens, C.M. et al. (1992) Proc. Natl. Acad. Sci. USA 89, 11818–11822.
[141] Kelman, Z. et al. (1993) Oncogene 8, 37–44.
[142] Yee, K. et al. (1993) Blood 82, 1335–1343.
[143] Stacker, S.A. et al. (1993) Oncogene 8, 1347–1356.
[144] Tamagnone, L. et al. (1993) Oncogene 8, 2009–2014.
[145] Clay, F.J. et al. (1993) Proc. Natl Acad. Sci. USA 90, 4882–4886.
[146] Lake, R.J., and Jelinek, W.R. (1993) Mol. Cell. Biol. 13, 7793–7801.
[147] Kitada, K. et al. (1993) Mol. Cell. Biol. 13, 4445–4457.
[148] Lange–Carter, C.A. et al. (1993) Science 260, 315–319.
[149] Banno, H. et al. (1993) Mol. Cell. Biol. 13, 4745–4752.
[150] Manser, E. et al. (1994) Nature 367, 40–46.
[151] Leberer, E. et al. (1992) EMBO J. 11, 4815–4824.
[152] Gale, M. Jr., and Parsons, M. (1993) Mol. Biochem. Parasitol. 59, 111–122.
[153] Booher, R.N. et al. (1993) EMBO J. 12, 3417–3426.
[154] Kieber, J.J. et al. (1993) Cell 72, 427–441.
[155] Ebner, R. et al. (1993) Science 260, 1344–1348.
[156] ten Dijke, P. et al. (1993) Oncogene 8, 2879–2887.
[157] Matsuzaki, K. et al. (1993) J. Biol. Chem. 268, 12719–12723.
[158] Tsuchida, K. et al. (1993) Proc. Natl Acad. Sci. USA 90, 11242–11246.
[159] Attisano, L. et al. (1993) Cell 75, 671–680.
[160] Franzen, P. et al. (1993) Cell 75, 681–692.
[161] ten Dijke, P. et al. (1994) Science 264, 101–104.
[162] Baarends, W.M. et al. (1994) Development 120, 189–197.
[163] Childs, S.R. et al. (1993) Proc. Natl Acad. Sci. USA. 90, 9475–9479.
[164] Xie, T. et al. (1994) Science 263, 1756–1759.
[165] Estevez, M. et al. (1993) Nature 365, 644–649.
[166] Chang, C. et al. (1992) Plant Cell 4, 1263–1271.
[167] Hirayama, T., and Oka, A. (1992) Plant Mol. Biol. 20, 653–662.
[168] Kohorn, B.D. et al. (1992) Proc. Natl Acad. Sci. USA 89, 10989–10992.
[169] Martin, G.B. et al. (1993) Science 262, 1432–1436.
[170] Valon, C. et al. (1993) Plant Mol. Biol. 23, 415–421.
[171] Dorow, D.S. et al. (1993) Eur. J. Biochem. 213, 701–710.
[172] Ezoe, K., et al. (1994) Oncogene 9, 935–938.
[173] Graves, P.R. et al. (1993) J. Biol. Chem. 268, 6394–6401.
[174] Shelton, C.A., and Wasserman, S.A. (1993) Cell 72, 515–525.
[175] Thümmler, F. et al. (1992) Plant Mol. Biol. 20, 1003–1017.

[176] Vandenbol, M. et al. (1990) Mol. Gen. Genet. 222, 393–399.

[177] Blacketer, M.J. et al. (1993) Mol. Cell. Biol. 13, 5567–5581.

[178] Nikawa, J., and Yamashita, S. (1992) Mol. Microbiol. 6, 1441–1446.

[179] Mori, K. et al. (1993) Cell 74, 743–756.

[180] Jacquier, A. et al. (1992) Yeast 8, 121–132.

[181] Walworth, N. et al. (1993) Nature 363, 368–371.

[182] Molz, L., and Beach, D. (1993) EMBO J. 12, 1723–1732.

[183] Feng, X.H. et al. (1993) Biochim. Biophys. Acta 1172, 200–204.

[184] Roe, J.L. et al. (1993) Cell 75, 939–950.

[185] Galyov, E.E. et al. (1993) Nature 361, 730–732.

[186] Adams, N.E., III (1986) J. Classification 3, 299–317.

[187] Koonin, E.V. et al. (1994) EMBO J. 13, 493–503.

[188] Wilson, R. et al. (1994) Nature 368, 32–38.

[189] Taylor, S.S. et al. (1993) Phil. Trans. R. Soc. Lond. B340, 315–324.

[190] Swofford, D.L. (1991) PAUP: Phylogenetic Analysis Using Parsimony, Version 3.1. Computer program distributed by the Illinois Natural History Survey, Champaign, Illinois.

[191] Uhler, M.D. (1993) J. Biol. Chem. 268, 13586–13591.

3 Cellular Functions of Protein Kinases

D. Grahame Hardie (Dundee University, UK)

In the previous chapter, Steve Hanks and Tony Hunter discussed the structure of protein kinase catalytic domains, and the classification of protein kinases according to the sequences of those domains. In this chapter I will provide a brief overview of the physiological functions of protein kinases, and classify them according to the mechanism by which they are regulated. This division into functional groups often, but not always, correlates with the subfamilies defined by sequence similarity. In this chapter, I have given a limited set of references to original papers or reviews, but references regarding individual protein kinases can be found by consulting the appropriate entry later in the book.

Along with the cycles catalysed by G proteins and their associated factors (Fig. 1B), protein phosphorylation cycles (Fig. 1A) appear to represent the primary means for rapidly switching the activities of cellular proteins from one state to another. Unlike a modulation of protein function induced by non-covalent binding of a ligand such as a substrate or allosteric effector (Fig. 1C), a protein phosphorylation cycle involves *distinct* reactions in the forward and backward directions, catalysed by protein kinases and protein phosphatases respectively. The equilibrium constants for protein kinase and protein phosphatase reactions, and the prevailing intracellular concentrations of ATP, ADP and P_i, ensure that a protein kinase will effectively only catalyse protein phosphorylation *in vivo*, while a protein

(A) Protein phosphorylation cycle

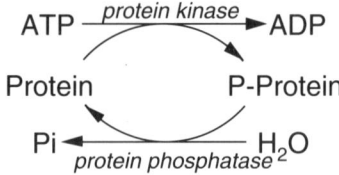

(B) G protein cycle

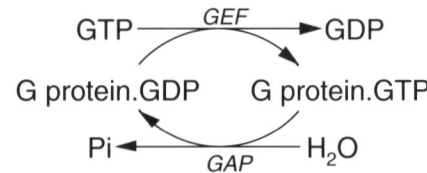

(C) Allosteric (non-covalent) regulation

Protein + ligand ⇌ Protein.ligand

Figure 1 *Comparison of protein phosphorylation, G protein and protein-ligand cycles. KEY:* GEF: *guanine nucleotide exchange factor;* GAP: *GTPase activator protein.*

phosphatase will only catalyse dephosphorylation. Therefore the proportion of phospho- to dephospho-protein can be varied from zero to one by changing the *rate* of the protein kinase and/or protein phosphatase reaction, i.e. by modulating the catalytic efficiency of the interconverting enzyme(s). By contrast, in a control system involving only non-covalent binding of an effector, the system is constrained by the equilibrium constant for binding, and very large changes in concentration of the effector would be necessary to effect a switch from one form of the protein to the other. Non-covalent systems are therefore more suited to achieving constancy of conditions, i.e. homeostasis, while protein phosphorylation cycles are additionally capable of *switching* a system from one state to another. An apposite example of the necessity for a switch mechanism is in the control of the timing of DNA synthesis and mitosis during the cell cycle. In this case an all-or-nothing response is clearly required: a cell which only replicated half of its DNA, or did not completely segregate its sister chromatids at mitosis, would not produce successful progeny.

The hydrolysis of ATP which occurs in a protein phosphorylation cycle is the small price which is paid to allow the control system to escape from the constraints of equilibrium, and can be thought of as equivalent to the small amount of energy required to activate a control switch on a piece of electrical apparatus. G proteins, with their associated exchange factors (GEFs) and GTPase activating proteins (GAPs), can also act as molecular switches, and once again a small expenditure of energy (in this case, GTP hydrolysis) is required. Both G protein and protein phosphorylation cycles are capable of *signal amplification*, because in the former case the guanine nucleotide exchange factors, and in the latter case both the protein kinase and protein phosphatase, act catalytically.

As well as signal amplification, protein phosphorylation cycles can also exhibit *high sensitivity*. These two phenomena are distinct. *Amplification* refers to the case where the absolute concentration of the final target protein is much higher than that of the input signal: a downstream target of a protein kinase cascade, such as glycogen phosphorylase, can be present at concentrations several orders of magnitude higher than the hormone which triggers the effect. *Sensitivity* refers to the case where a small *percentage* change in the initial signal produces a much larger *percentage* change in the final response. One way in which this can be produced is a mechanism whereby the protein kinase and protein phosphatase acting on a target protein are modulated simultaneously in opposite directions. An example of this is the effect of cyclic AMP-dependent protein kinase **(PKA)** on glycogen phosphorylase. PKA simultaneously phosphorylates and activates the protein kinase **(PhK)** which activates phosphorylase, and inactivates the protein phosphatase which inactivates it (protein phosphatase-1G), in the latter case by phosphorylation of the regulatory G subunit[1]. An additional, more hypothetical mechanism by which sensitivity can be achieved in a protein phosphorylation cycle is known as *zero order ultrasensitivity*[2]. This refers to the situation where the protein kinase (and/or the protein phosphatase) is saturated with its substrate, i.e. the reaction is zero order with respect to substrate. If the interconverting enzyme was not saturated, then the reaction it catalysed would run progressively more slowly as it proceeded, because the concentration of its substrate (the dephosphoprotein in the case of a protein kinase) would be dropping. If, however, the protein kinase was initially saturated, then the conversion would proceed at a constant rate until the concentration of the dephosphoprotein dropped to levels approaching K_m.

Under zero order conditions a very small change in protein kinase activity could produce a large change in the ratio of dephospho- to phospho-protein. Sensitivity is enhanced if the protein phosphatase is also saturated with its substrate, and can be enhanced even further in a system involving more than one protein phosphorylation cycle (a *multicyclic* system, or protein kinase cascade). It should be stressed that zero order ultrasensitivity remains a theoretical concept and it is not yet clear that it is exhibited by any protein phosphorylation system *in vivo*.

An additional feature exhibited by protein phosphorylation cycles, but not yet demonstrated for G protein cycles, is their ability to act in chains known as *protein kinase cascades*. The original term "cascade" is in fact more apt than the alternative "chain", because it is now clear, particularly from the example of the MAP kinase (**Erk1/2**) system, that such cascades can branch in both the convergent and divergent directions (Fig. 2). Protein kinase cascades should therefore be thought of as branching networks rather than linear chains. The existence of such networks allows the summation of many different inputs into the cell to be processed into numerous different outputs. Another feature of such cascades is that they can contain feedback loops, which can act in both a negative and positive direction. Thus, for example, **Erk1/2** (MAP kinase) has been shown to phosphorylate the protein kinase which lies upstream of it, i.e. **Mek**[3]. This may be involved in down-regulation of the pathway. On the other hand, **Cdc2** phosphorylates and activates Cdc25[4], the protein tyrosine phosphatase which reverses the inhibitory phosphorylation of Cdc2 by **Wee1**. This positive feedback loop may ensure that any minor activation of Cdc2 by dephosphorylation is amplified into a full-blown response.

In what systems do protein phosphorylation cycles act as molecular switches? This question can be addressed in two ways, by looking at inputs and outputs. The outputs are the cellular proteins whose activities are modified as a result of phosphorylation. These targets, where known, are covered in detail in individual entries in the book, and it is sufficient to state here that proteins of almost every functional class have now been shown to be modified in this way. Thus the known targets include metabolic enzymes, regulatory proteins (including other protein kinases and protein phosphatases), receptors, cytoskeletal proteins, ion channels and pumps, and transcription factors.

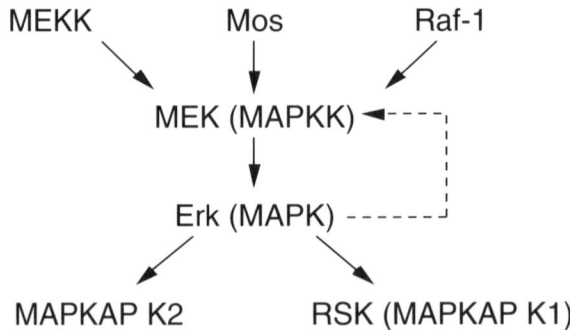

Figure 2 *A converging and diverging protein kinase cascade: the MAP kinase (Erk) cascade. A putative feedback loop is shown using a dashed line. Two recently described protein kinases which do not have separate entries in the FactsBook are MAPKAP K2 (MAP kinase-activated protein kinase-2)[25] and MEKK (MEK kinase)[29].*

In the remainder of this chapter, I will discuss the various types of input which trigger protein phosphorylation cycles or cascades, and consider how they bring this about by reviewing the mechanisms by which individual protein kinases are regulated.

INPUTS WHICH MODULATE PROTEIN PHOSPHORYLATION CYCLES

The three major classes of input which are currently known to modulate protein phosphorylation cycles are extracellular signals, cell cycle checkpoints and environmental or nutritional stresses. Each of these are discussed briefly below, but it should be stated at this point that they are inter-linked rather than independent. Many mammalian growth factors can trigger quiescent cells to enter the cell cycle, while yeast mating factors[5], and some mammalian growth factors (e.g. TGFβ)[6], can cause cell cycle arrest in their target cells. Similarly, it is well known that nutritional stress in the yeast *Saccharomyces cerevisiae* causes an arrest at the *START* point in the G1 phase in the cell cycle[7]. Clearly the three systems must interact at various levels, but the molecular details of how this operates are only just emerging.

Extracellular signals

The first known protein kinase (phosphorylase kinase, **PhK**) was discovered by Krebs and Fischer[8] during studies of the mechanism by which nervous stimulation and adrenaline activated glycogen breakdown in skeletal muscle. The response to extracellular signals can now be regarded as probably the major function of protein phosphorylation cycles, and the majority of the protein kinases in this book play a part in this response. Classes of extracellular signal molecules which have been shown to modulate protein phosphorylation cycles include hormones, neurotransmitters, growth and differentiation factors, local mediators (responsible for *paracrine* regulation), extracellular matrix molecules, and membrane proteins involved in adhesion and communication of neighbouring cells (which has been termed *juxtacrine* regulation). This aspect of cellular regulation is particularly associated with multicellular organisms, but unicellular organisms also display it, as exemplified by the response of yeasts to mating factors.

Cell cycle checkpoints

Many events required for cell division, particularly DNA replication and mitosis, must occur at discrete points in the cell cycle. For each daughter cell to receive the full complement of DNA at mitosis (M phase), duplication of the sister chromatids must have been completed in the preceeding S phase. A cell cycle checkpoint[9] is envisaged as a stage in the cycle at which a pause will occur if the previous process has not been completed. How process completion is monitored remains unclear, but there is now general agreement that once the process has been completed, the output (entry into the next phase of the cell cycle) is triggered by protein phosphorylation cycles, particularly those utilizing members of the cyclin-dependent protein kinase family[9]. As well as checkpoints at which completion of DNA replication and mitosis are monitored, these protein phosphorylation cycles are clearly also modulated by extracellular signals such as growth and differentiation factors (see above), and nutritional status (see below).

Nutritional and environmental stress

This represents perhaps the least widely known of the roles for protein phosphorylation, although it may be one of the most ancient. Monitoring of the nutritional status of the medium is particularly important for unicellular eukaryotes such as the yeast, *Saccharomyces cerevisiae*, and several protein kinase systems appear to be required for the response to starvation for particular nutrients in this organism. These include **Snf1** which is required for the response to starvation for glucose, **Pho85** which is involved in the response to starvation for phosphate, and **Gcn2**, which appears to be activated by amino acid starvation. Cyclic AMP-dependent protein kinase **ScPKA** also appears to be primarily regulated by availability of glucose in this organism. In multicellular organisms, most cells are bathed constantly in nutrient-rich medium, and at first sight this role for protein phosphorylation may seem less important. However it is becoming clear that related protein kinases function in the response to environmental stress. In vertebrates the AMP-activated protein kinase (**AMPK**), which is related to **Snf1**, appears to be involved in the response to heat shock and other stresses which deplete ATP, while relatives of **Gcn2**, i.e. **HRI** and **PKR**, inhibit protein synthesis in response to haem deficiency in reticulocytes, and viral infection, respectively. HRI/Gcn2-like protein kinases may also inhibit protein synthesis in response to heat shock[10].

MECHANISMS FOR REGULATION OF INDIVIDUAL PROTEIN KINASES

In the subsections below I have discussed the various mechanisms by which protein kinases are regulated. As with almost any system of classification, difficulties arise because the criteria are not mutually exclusive, and an individual protein kinase may be regulated by more than one of these mechamisms. A case in point is the cyclin-dependent kinase family, which while being activated by cyclins, can often also be regulated both positively and negatively by upstream protein kinases, and negatively by interaction with proteins other than cyclins[5, 6].

Second-messenger dependent protein kinases

Many extracellular signals *(first messengers)* bind to cell surface receptors which cause changes in the concentration of intracellular signals or *second messengers*. The known or probable second messengers include cyclic AMP, cyclic GMP, inositol trisphosphate (IP_3), phosphatidylinositol 3,4,5-trisphosphate (PIP_3), cyclic ADP ribose[11] (cADPR), arachidonic acid and diacylglycerol. Ca^{2+} may also be regarded as a second messenger: it becomes elevated in the cytoplasm due to opening of voltage-gated channels, or due to binding of messengers to ligand-gated ion channels, which can either span the plasma membrane (e.g. the nicotinic acetylcholine receptor) or membranes of internal organelles (e.g. the IP_3 and cADPR (ryanodine) receptors[12]). Many, but not all, second messengers exert their effects through allosteric binding to second messenger-dependent protein kinases. These include the cyclic AMP- and cyclic GMP-dependent protein kinases (**PKA and PKG**) which, with the exception of certain cyclic nucleotide-gated ion channels, are thought to mediate all of the effects of cyclic AMP and cyclic GMP. The various isoforms of protein kinase C (**PKC**) are activated by lipid second messengers, including diacylglycerol, fatty acids and phosphatidylinositol 3,4,5-trisphosphate[13] (PIP_3), with some also requiring Ca^{2+}. Most effects of Ca^{2+} are mediated by Ca^{2+}-binding proteins of the calmodulin family, which bind to calmodulin-activated proteins. An important subgroup within the latter are the calmodulin-dependent protein kinases (e.g. **CaMKII**, **SmMLCK** and **Twn**). Variations on this theme are seen both in vertebrates, where phosphorylase kinase (**PhK**) contains endogenous calmodulin as a tightly bound subunit, and in higher plants, where Ca^{2+}-dependent protein kinases (**CDPK**) have a calmodulin-like domain on the same polypeptide as the kinase domain.

A common, if not universal feature, of this class of protein kinases is that they are thought to be maintained in an inactive state in the absence of second messenger by association with autoinhibitory regions which resemble the phosphorylation sites on exogenous proteins[14]. These may be actual autophosphorylation sites or, if they do not contain a phosphorylatable residue, are referred to as *pseudosubstrate sequences*. The autoinhibitory regions may be on the same subunit as the kinase domain (e.g. **PKC**, **CaMKII**, **SmMLCK**, or they may occur on separate regulatory subunits (e.g. **PKA**).

Other ligand-activated protein kinases

A limited number of protein kinases are regulated by small molecules which are normal metabolites rather than second messengers. These belong to the group of kinases involved in the response to nutritional or environmental stress. The AMP-activated protein kinase (**AMPK**) is activated allosterically by 5'-AMP, which is elevated in response to heat shock and other stresses. AMP binding to AMPK also promotes phosphorylation and activation by an upstream AMPK kinase[15], a rather novel mechanism. **Gcn2** is thought to be activated by binding of uncharged tRNA (a signal for amino acid starvation) to a domain related to those found in aminoacyl tRNA synthetases. **PKR/Tik** is activated by double stranded RNA, an intermediate in the life cycle of certain viruses against whom it is thought to provide protection against infection.

Cyclin-dependent protein kinases

This important group of protein kinases generally contain catalytical subunits which are little more than a canonical kinase domain. However the catalytic subunits are completely inactive unless complexed with regulatory subunits known as *cyclins*[16]. The occupancy of various phosphorylation sites which regulate in both positive and negative directions are also crucial. The classical examples of this group are Cdc2 and its close relatives (e.g. **Cdc2**, **Cdk2**, **Cdc28**, **SpCdc2**) which are involved in triggering various phases of the cell cycle. The catalytic subunits are generally synthesized throughout the cell cycle but, as suggested by their name, cyclins are often found to be synthesized, and then degraded, in specific phases of the cell cyle. Although most of the cyclin-dependent protein kinases are thought to be involved in regulation of the cell cycle, some appear to have other roles, e.g. **Cdk5**, which is only expressed in non-dividing, terminally differentiated cells, and **Pho85**, which is involved in the response to phosphate starvation in yeast.

Receptor protein-tyrosine kinases

This book contains entries for around 40 protein kinases which have the characteristics of receptor-linked PTKs. All contain an extracellular ligand-binding domain, a single transmembrane helix, and an intracellular domain containing a single protein-tyrosine kinase domain. In the case of so-called *orphan* receptors, the ligand has not yet been identified. The available evidence now favours the idea that receptor PTKs are activated by ligand-induced aggregation, usually dimerization, leading to juxtaposition and intermolecular autophosphorylation of the intracellular regions[17]. This phosphorylation may activate the PTK activity against exogenous substrates, and/or create docking sites for other components of signalling complexes, such as non-receptor PTKs, phospholipase Cγ, phosphatidyl-inositol 3-kinase, GTPase activator proteins, and adapter proteins such as Grb2 and Shc[18]. These accessory proteins bind to phosphorylation sites on the receptor PTKs via SH2 domains (see below).

Non-receptor protein-tyrosine kinases

Many PTKs do not contain transmembrane regions and therefore cannot represent cell surface receptors. However evidence is accumulating suggesting that in most cases these protein kinases form complexes with the intracellular regions of cell surface receptors, which may or may not themselves be PTKs. The best understood are members of the **Src** family[19]. These have myristoylated N-termini which cause association with the membrane and which are essential for function. Members of this subfamily contain, in addition to their kinase (or Src homology 1, SH1) domains, additional regions known as SH2 and SH3 domains. Different SH2 domains are now known to bind to specific sequence motifs containing phospho-tyrosine[18], while SH3 domains bind to motifs rich in proline residues[20]. The SH2 domains can cause inhibition (e.g. in **Src** itself) due to self-association with negative regulatory phosphotyrosine sites on the same subunit[21]. When the latter sites are dephosphorylated, interactions of the SH2 or SH3 domains with other proteins are thought to cause assembly of the non-receptor PTKs into signalling complexes with cell surface receptors or associated polypeptides. **Zap70**, a member of a different non-receptor PTK family with two SH2 domains, binds via the latter to the ζ chain of the T cell receptor when the latter is phosphorylated by other receptor-associated PTKs, probably **Fyn** or **Lck**.

Another group of non-receptor PTKs, represented by **Tyk2** and **Jak1/2**, do not contain SH2 or SH3 domains, but via unknown mechanisms associate with cell surface cytokine receptors which are not themselves PTKs. Genetic evidence suggests that the Jak family are essential components of the signal transduction pathway from interferon-α/β and -γ receptors[22].

Receptor protein-serine/threonine kinases

Representatives of this group, originally reported in *C. elegans* **DAF-1**, have also now been described in higher plants (**ZmPK1** and **Srk**), and vertebrates (**ActRII** and **TGFβRII**). In vertebrates, they act as receptors for extracellular messengers of the TGF-β superfamily. Binding of extracellular ligand causes the assembly of hetero-multimeric receptors, more than one polypeptide of which may be a protein serine/threonine kinase[23]. The protein kinase activity is essential for signal transduction, although whether this involves autophosphorylation only, or whether there are downstream substrates for the kinase domains is not yet known. The ultimate downstream targets for members of the TGF-β superfamily include cell cycle-controlling kinases such as the cyclin E–**Cdk2** complex, probably via effects on expression of inhibitor proteins[6].

Protein kinases regulated by phosphorylation

Many protein kinases are activated by phosphorylation, either by autophos-phorylation or by phosphorylation by upstream protein kinases in a cascade. The activating phosphorylation sites commonly occur in the loop between the conserved DFG and APE motifs (subdomains VII and VIII; see Chapter 2). In **PKA**, T197 appears to be modified by autophosphorylation, and the phosphate group does not turn over once formed. Several PTKs (e.g. **Src**, **InsR**) also autophosphorylate in the DFG..APE loop, and this leads to activation of kinase activity against exogenous substrates. In other cases, such as the **Cdc2** subfamily, **RSK**, **GSK3**, and members of the MAP kinase cascade such as **Erk1/2** and **Mek**, phosphorylation by upstream kinase kinases in this loop activates the enzymes. Protein kinases can also be activated by phosphorylation outside the catalytic domain, as occurs for phos-phorylase kinase (**PhK**)[24] where the phosphorylation occurs on different subunits, and MAPKAP kinase-2, where it occurs just beyond the C-terminal end of the kinase domain[25]. Protein kinases can also be inhibited by phosphorylation, as illustrated by inactivation of **SpCdc2** by **SpWee1**, the β isoform of **GSK3** by **RSK** and **S6K**[26], and the **Src** subfamily by **Csk**. The inhibitory sites can be in the N-terminal tail (**GSK3**), in the ATP-binding loop (subdomain I) within the kinase domain (**Cdc2** subfamily), or in the C-terminal tail (**Src** subfamily).

Constitutively active protein kinases

Some protein kinases have no known system of physiological regulation. While this may merely indicate that the regulation has yet to be discovered, there are cases where sites phosphorylated by the kinases *in vitro* appear to be constitutively phos-phorylated *in vivo*. This appears to be the case, for example, for the casein kinase II (**CK2**) sites on glycogen synthase[27]. In other cases where there is no known mechanism for regulation of the protein kinase itself, recognition sites for the kinase may be created by phosphorylation at neighbouring residues by other protein kinases, leading to so-called *hierarchical* phosphorylation where phos-phorylation by one protein kinase may be secondary to regulation of another protein kinase acting on a neighbouring site[28]. This appears to be the case for casein kinase I (**CK1**), which can phosphorylate the second serine in the sequence SXX<u>S</u> only when the first is phosphorylated.

Finally, although this book concentrates on protein kinases, the protein phosphatases clearly play roles of equal importance in any protein phosphorylation cycle. As they now appear to be represented in the genome almost as frequently as the protein kinases, and may have regulatory mechanisms every bit as complicated, they have been omitted from this book mainly for reasons of space. Hopefully they will be covered in other volumes in the FactsBook series.

References

1   Hubbard, M.J. and Cohen, P. (1993) Trends Biochem. Sci. 18, 172–177.
2   Koshland, D.E. et al. (1982) Science 217, 220–225.
3   Wu, J. et al. (1993) Mol. Cell. Biol. 13, 4539–4548.
4   Izumi, T. and Maller, J.L. (1993) Mol. Biol. Cell 4, 1337–1350.
5   Chang, F. and Herskowitz, I. (1990) Cell 63, 999–1011.
6   Polyak, K. et al. (1994) Gene Devel. 8, 9–22.
7   Thompson-Jaeger, S. et al. (1991) Genetics 129, 697–706.
8   Krebs, E.G. and Fischer, E.H. (1956) Biochim. Biophys. Acta 20, 150–157.
9   Murray, A.W. (1992) Nature 359, 599–604.
10 Murtha-Riel, P. et al. (1993) J. Biol. Chem. 268, 12946–12951.
11 Lee, H.C. et al. (1991) J. Biol. Chem. 264, 1608–1615.
12 Berridge, M.J. and Irvine, R.F. (1984) Nature 312, 315–321.
13 Nakanishi, H. et al. (1993) J. Biol. Chem. 268, 13–16.
14 Kemp, B.E. and Pearson, R.B. (1991) Biochim. Biophys. Acta 1094, 67–76.
15 Weekes, J. et al. (1994) Eur. J. Biochem. 219, 751–757.
16 Morgan, D.O. and de Bondt, H.L. (1994) Curr. Opin. Cell Biol. 6, 239–246.
17 Wells, J.A. (1994) Curr. Opin. Cell Biol. 6, 163–173.
18 Johnson, G.L. and Vaillancourt, R.R. (1994) Curr. Opin. Cell Biol. 6, 230–238.
19 Rudd, C.E. et al. (1993) Biochim. Biophys. Acta 1155, 239–266.
20 Zhou, S.Y. et al. (1993) Cell 72, 767–778.
21 Ren, R. et al. (1993) Science 259, 1157–1161.
22 Muller, M. et al. (1993) Nature 366, 129–135.
23 Attisano, L. et al. (1993) Cell 75, 671–680.
24 Heilmeyer, L., Jr. (1991) Biochim. Biophys. Acta 1094, 168–174.
25 Stokoe, D. et al. (1993) Biochem. J. 296, 843–849.
26 Sutherland, C. et al. (1993) Biochem. J. 296, 15–19.
27 Poulter, L. et al. (1988) J. Biol. Chem. 175, 497–510.
28 Roach, P.J. (1991) J. Biol. Chem. 266, 14139–14142.
29 Lange-Carter, C.A. et al. (1993) Science 260, 315–319.

THE
PROTEIN
KINASES

PKA cAMP-dependent protein kinase (vertebrates)

Kjetil Taskén, Rigmor Solberg, Kari Bente Foss, Bjørn S. Skålhegg, Vidar Hansson and Tore Jahnsen (Institute of Medical Biochemistry, University of Oslo, Oslo, Norway)

The second messenger cAMP is produced in response to a wide range of hormones and neurotransmitters, and is known to be implicated in the regulation of numerous cellular processes including metabolism, gene regulation, and cell division. With the exception of certain ion channels which are directly regulated by cAMP, all known effects of cAMP in the mammalian cell are mediated by cAMP-dependent protein kinases. PKA phosphorylates serine or threonine residues on a number of specific substrate proteins (enzymes, structural proteins, transcription factors, ion channels, etc.) in response to elevated levels of cAMP.

Subunit structure and isoforms

SUBUNIT	AMINO ACIDS	MOL. WT	SDS-PAGE
Cα	351	40 589	40 kDa
Cβ	351	40 622	41 kDa
Cγ	351	40 412	39 kDa
RIα	381	42 981	49 kDa
RIβ	380	43 027	53·5 kDa
RIIα	404	45 518	51 kDa (54 kDa*)
RIIβ	418	46 346	53 kDa

*phosphorylated form

PKA consists of distinct catalytic (C) and regulatory (R) subunits[1]. The holoenzyme contains two C subunits bound to homo- or heterodimers of R subunits[1,2]. Activation proceeds by the cooperative binding of two molecules of cAMP to each R subunit, which causes dissociation of two free active C subunits from the regulatory subunit dimer. The two major PKA isozymes, type I (PKAI) and type II (PKAII), consist of RI and RII, respectively, complexed with C. At present four different R subunits (RIα, RIβ, RIIα, RIIβ) and three different C subunits (Cα, Cβ, Cγ) have been identified as separate gene products[1]. Splice variants (Cα2, Cβ2[3,4]) further add to this complexity.

Genetics

Chromosomal locations of human genes:

SUBUNIT	LOCUS	LOCATION
Cα	PRKACA	–
Cβ	PRKACB	1p36.1
Cγ	PRKACG	9q13
RIα	PRKAR1A	17q23–24
RIα pseudo	PRKAR1AP	1p21–31
RIβ	PRKAR1B	7p22–pter
RIIα	PRKAR2A	–
RIIβ	PRKAR2B	7q22

Domain structures

The crystal structure of Cα^5, and crystallization of RIα^6, have recently been reported.

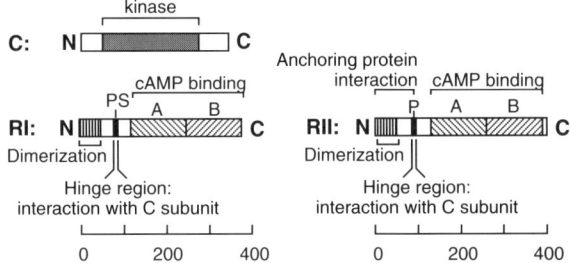

Database accession numbers

		PIR	SWISSPROT	EMBL/GENBANK	REF
Cα	Human	S01404	P17612	X07767	7
	Ox	A00620	P00517	X67154	
	Pig	S00085		X07617	
	Rat	S16240	P27791	X57986	
	Hamster	A40384	P25321	M63311	
	Mouse	A25125	P05132	M12303 (M18240)	
	Mouse	A28619			
Cα2	Human			M80335	
Cβ	Human	A34724	P22694	M34181	8
	Ox	A25334	P05131	J02647	
	Pig	S00086	P05383	X05998	
	Rat	JQ1139		D10770, D01144	
	Hamster	B40384	P05206	M63312	
	Mouse	A24596	P05206	J02626 (M21096)	
Cβ2	Ox	A23716	P24256	M60482	
Cγ	Human	B34724	P22612	M34182	8
RIα	Human	A34627	P10644	M18648, M33336	9
	Ox	A00617	P00514	K00833	
	Pig	S00083	P07802	X05942 (X05943)	
	Rat	A26910	P09456	M17086	
RIβ	Human	JH0392		M65066	10
	Mouse	A30205	P12849	M20473	
RIIα	Human	S03885	P13861	X14968	11
	Ox	A00618, S17058		P00515	
	Pig	A25652	P05207	X04709	
	Rat	A28325	P12368	J02934	
	Mouse	B28325	P12367	J02935	
RIIβ	Human	A40915		M31158	12
	Ox	A36528		J05692	
	Rat	A25201	P12369	M12492, M21194 (M75151)	
	Rat	A28893			
	Mouse	PQ0161		M68861 (M68861)	

(): indicates upstream genomic sequence where reported.

Amino acid sequences of human PKA

Cα

```
MGNAAAAKKG SEQESVKEFL AKAKEDFLKK WESPAQNTAH LDQFERIKTL  50
GTGSFGRVML VKHKETGNHY AMKILDKQKV VKLKQIEHTL NEKRILQAVN 100
FPFLVKLEFS FKDNSNLYMV MEYVPGGEMF SHLRRIGRFS EPHARFYAAQ 150
IVLTFEYLHS LDLIYRDLKP ENLLIDQQGY IQVTDFGFAK RVKGRTWTLC 200
GTPEYLAPEI ILSKGYNKAV DWWALGVLIY EMAAGYPPFF ADQPIQIYEK 250
IVSGKVRFPS HFSSDLKDLL RNLLQVDLTK RFGNLKNGVN DIKNHKWFAT 300
TDWIAIYQRK VEAPFIPKFK GPGDTSNFDD YEEEEIRVSI NEKCGKEFSE 350
F
```

Sites of interest: G2, N-terminal myristoylation; T198, autophosphorylation; S339, phosphorylation

Cβ

```
MGNAATAKKG SEVESVKEFL AKAKEDFLKK WENPTQNNAG LEDFERKKTL  50
GTGSFGRVML VKHKATEQYY AMKILDKQKV VKLKQIEHTL NEKRILQAVN 100
FPFLVRLEYA FKDNSNLYMV MEYVPGGEMF SHLRRIGRFS EPHARFYAAQ 150
IVLTFEYLHS LDLIYRDLKP ENLLIDHQGY IQVTDFGFAK RVKGRTWTLC 200
GTPEYLAPEI ILSKGYNKAV DWWALGVLIY EMAAGYPPFF ADQPIQIYEK 250
IVSGKVRFPS HFSSDLKDLL RNLLQVDLTK RFGNLKNGVS DIKTHKWFAT 300
TDWIAIYQRK VEAPFIPKFR GSGDTSNFDD YEEEDIRVSI TEKCAKEFGE 350
F
```

Sites of interest: G2, N-terminal myristoylation; T198, autophosphorylation; S339, phosphorylation

Cγ

```
Upstream, in-frame translation start: (MAAPAAATA-)
MGNAPAKKDT EQEESVNEFL AKARGDFLYR WGNPAQNTAS SDQFERLRTL  50
GMGSFGRVML VRHQETGGHY AMKILNKQKV VKMKQVEHIL NEKRILQAID 100
FPFLVKLQFS FKDNSYLYLV MEYVPGGEMF SRLQRVGRFS EPHACFYAAQ 150
VVLAVQYLHS LDLIHRDLKP ENLLIDQQGY LQVTDFGFAK RVKGRTWTLC 200
GTPEYLAPEI ILSKGYNKAV DWWALGVLIY EMAVGFPPFY ADQPIQIYEK 250
IVSGRVRFPS KLSSDLKDLL RSLLQVDLTK RFGNLRNGVG DIKNHKWFAT 300
TSWIAIYEKK VEAPFIPKYT GPGDASNFDD YEEEELRISI NEKCAKEFSE 350
F
```

Sites of interest: G2, N-terminal myristoylation; T198, autophosphorylation; S339, phosphorylation

RIα

```
MESGSTAASE EARSLRECEL YVQKHNIQAL LKDSIVQLCT ARPERPMAFL  50
REYFERLEKE EAKQIQNLQK AGTRTDSRED EISPPPPNPV VKGRRRRGAI 100
SAEVYTEEDA ASYVRKVIPK DYKTMAALAK AIEKNVLFSH LDDNERSDIF 150
DAMFSVSFIA GETVIQQGDE GDNFYVIDQG ETDVYVNNEW ATSVGEGGSF 200
GELALIYGTP RAATVKAKTN VKLWGIDRDS YRRILMGSTL RKRKMYEEFL 250
SKVSILESLD KWERLTVADA LEPVQFEDGQ KIVVQGEPGD EFFIILEGSA 300
AVLQRRSENE EFVEVGRLGP SDYFGEIALL MNRPRAATVV ARGPLKCVKL 350
DRPRFERVLG PCSDILKRNI QQYNSFVSLS V
```

Sites of interest: E2, N-terminal acetylation; C18, C39, interchain disulphide bridge in dimer; A99, pseudosubstrate site

RIβ

```
ASPPACPSEE  DESLKGCELY  VQLHGIQQVL  KDCIVHLCIS  KPERPMKFLR   50
EHFEKLEKEE  NRQILARQKS  NSQSDSHDEE  VSPTPPNPVV  KARRRRGGVS  100
AEVYTEEDAV  SYVRKVIPKD  YKTMTALAKA  ISKNVLFAHL  DDNERSDIFD  150
AMFPVTHIAG  ETVIQQGNEG  DNFYVVDQGE  VDVYVNGEWV  TNISEGGSFG  200
ELALIYGTPR  AATVKAKTDL  KLWGIDRDSY  RRILMGSTLR  KRKMYEEFLS  250
KVSILESLEK  WERLTVADRL  EPVQFEDGEK  IVVQGEPGDD  FYIITEGTAS  300
VLQRRSPNEE  YVEVGRLGPS  DYFGEIALLL  NRPRAATVVA  RGPLKCVKLD  350
RPRFERVLGP  CSEILKRNIQ  RYNSFISLTV
```
Sites of interest: A1, N-terminal acetylation; C17, C38, interchain
disulphide bridge in dimer; G98, pseudosubstrate site

RIIα

```
MSHIQIPPGL  TELLQGYTVE  VLRQQPPDLV  EFAVEYFTRL  REARAPASVL   50
PAATPRQSLG  HPPPEPGPDR  VADAKGDSES  EEDEDLEVPV  PSRFNRRVSV  100
CAETYNPDEE  EEDTDPRVIH  PKTDEQRCRL  QEACKDILLF  KNLDQEQLSQ  150
VLDAMFERIV  KADEHVIDQG  DDGDNFYVIE  RGTYDILVTK  DNQTRSVGQY  200
DNRGSFGELA  LMYNTPRAAT  IVATSEGSLW  GLDRVTFRRI  IVKNNAKKRK  250
MFESFIESVP  LLKSLEVSER  MKIVDVIGEK  IYKDGERIIT  QGEKADSFYI  300
IESGEVSILI  RSRTKSNKDG  GNQEVEIARC  HKGQYFGELA  LVTNKPRAAS  350
AYAVGDVKCL  VMDVQAFERL  LGPCMDIMKR  NISHYEEQLV  KMFGSSVDLG  400
NLGQ
```
Sites of interest: S2, N-terminal acetylation; S99, autophosphorylation

RIIβ

```
MSIEIPAGLT  ELLQGFTVEV  LRHQPADLLE  FALQHFTRLQ  QENERKGTAR   50
FGHEGRTWGD  LGAAAGGGTP  SKGVNFAEEP  MQSDSEDGEE  EEAAPADAGA  100
FNAPVINRFT  RRASVCAEAY  NPDEEEDDAE  SRIIHPKTDD  QRNRLQEACK  150
DILLFKNLDP  EQMSQVLDAM  FEKLVKDGEH  VIDQGDDGDN  FYVIDRGTFD  200
IYVKCDGVGR  CVGNYDNRGS  FGELALMYNT  PRAATITATS  PGALWGLDRV  250
TFRRIIVKNN  AKKRKMYESF  IESLPFLKSL  EFSERLKVVD  VIGTKVYNDG  300
EQIIAQGDSA  DSFFIVESGE  VKITMKRKGK  SEVEENGAVE  MPRCSRGQYF  350
GELALVTNKP  RAASAHAIGT  VKCLAMDVQA  FERLLGPCME  IMKRNIATYE  400
EQLVALFGTN  MDIVEPTA
```
Sites of interest: S2, N-terminal acetylation; S114, autophosphorylation

Homologues in other species

C and R subunits have been cloned and sequenced from ox, pig, rat and
mouse. The Cα and Cβ sequences are also available from hamster.
Homologues have also been cloned from *D. melanogaster* (**DmPKA**), *S.
cerevisiae* (**ScPKA**), *S. pombe* (**SpPKA**), *D. discoideum* (**DdPKA**) and *C.
elegans* (CePKA).

Physiological substrates and specificity determinants

The consensus sequence for phosphorylation by PKA is RRXS/TY, where Y
tends to be a hydrophobic residue. A phenylalanine in the nearby sequence
tends to be a negative determinant for phosphorylation by PKA. Some
variations with regard to spacing and basic residues are permissible as
seen from the list below, in which sequences from some substrates

phosphorylated *in vivo* by PKA are given (basic residues: underlined, phosphorylated residue: **bold**):

Phosphorylase kinase α subunit:	VEF<u>RR</u>LSI
Phosphorylase kinase β subunit:	RT<u>KR</u>SGSV
Pyruvate kinase (L type):	GYL<u>RR</u>ASV
6-Phosphofructo-2-kinase:	LQ<u>RRR</u>GSS
Hormone-sensitive lipase:	PM<u>RR</u>SV
Tyrosine hydroxylase:	FIG<u>RR</u>QSL
Cardiac phospholamban:	AI<u>RR</u>AST
Protein phosphatase inhibitor-1:	I<u>RRR</u>PTP
Δ-CREB:	ILS<u>RR</u>PSY

Assay

Phosphotransferase activity is assayed as a cAMP-dependent transfer of phosphate from $[\gamma\text{-}^{32}P]ATP$ to the synthetic heptapeptide LRRASLG (kemptide)[13]. Total amount of R subunits is usually determined as specific $[^3H]$-cAMP binding by a single dose saturation assay[14]. PKAI and PKAII holoenzymes can be separated by DEAE-cellulose chromatography employing a linear salt gradient (PKAI elutes at 25–50 and PKAII at 125--250 mM NaCl)[15]. However, recent data from our laboratory indicate that a PKAI isozyme (RIαRIβC$_2$) elutes as PKAII[2].

Enzyme activators and inhibitors

Both RI and RII contain two tandem cAMP binding sites designated sites A and B. Cyclic AMP analogues with selectivity for the A and B sites can complement each other in the synergistic activation of either PKAI or PKAII in both cell-free assays and intact cells[16, 17]. The stereoisomeric phosphorothioate analogues Sp-cAMPS and Rp-cAMPS are more resistant to degradation than their natural homologues. Sp-cAMPS is an agonist, and Rp-cAMPS a competitive inhibitor and antagonist that inhibits activation of PKA[18, 20].

	SPECIFICITY	K_a (nM)	K_i (nM)	CELL PERMEABILITY
cAMP agonists[19]:				
cAMP	Excellent	80		Poor
8-Br-cAMP	Good	50		Rel. good
8-CPT-cAMP	Poor vs. PKG	50		Good
DCL-cBiMPS	Excellent	30		Good
Sp-cAMPS	Excellent	1800		Poor
cAMP antagonists[18, 20]:				
Rp-cAMPS	Good	No activation		Poor
8-BrRpcAMPS	Good	No activation		Good
Inhibitors[21]:				
PKI	Excellent		0.09	None
PKI(5–22)	Excellent		0.83	None

KEY: 8-Br-cAMP, 8-bromo-cAMP; 8-CPT-cAMP, 8-(4-chlorophenylthio)-cAMP; DCL-cBiMPS, Sp-5,6-dichloro-1-b-D-furanosylbenzimidazole-3′,5′-

monophosphorothioate; Sp-cAMP, Rp-cAMP, phosphorothioate stereoisomers; PKI, skeletal muscle protein kinase inhibitor; PKI(5–22), peptide corresponding to PKI residues 5–22.

Pattern of expression

The PKAI holoenzymes are primarily soluble and cytoplasmic, whereas the particulate PKAII holoenzymes have been reported to localize to the Golgi–centrosomal area, probably due to interaction with RIIα- and RIIβ-specific binding proteins[1]. Upon activation, released C is translocated to the nucleus.

The Cα, RIα, and RIIα subunits are present in most cells and tissues investigated. The Cβ, RIβ and RIIβ subunits are present predominantly in brain and gonadal tissues, although unpublished data from our laboratory indicate a low-level expression of β subunits in most human tissues. Cβ is probably the predominant catalytic subunit in quiescent human T lymphocytes. Expression of the Cγ subunit has so far only been demonstrated in human testicular tissue.

References

1 Scott, J.D. (1991) Pharmacol. Ther. 50, 123–145.
2 Taskén, K. et al. (1993) J. Biol. Chem. 268, 21276–21283.
3 Thomis, D.C. et al. (1992) J. Biol. Chem. 267, 10723–10728.
4 Wiemann, S. et al. (1991) J. Biol. Chem. 266, 5140–5146.
5 Knighton, D.R. et al. (1991) Science 253, 407–414.
6 Su, Y. et al. (1993) J. Mol. Biol. 230, 1091–1093.
7 Maldonado, F. and Hanks, S.K. (1988) Nucl. Acids Res. 16, 8189–8190.
8 Beebe, S.J. et al. (1990) Mol. Endocrinol. 4, 465–475.
9 Sandberg, M. et al. (1987) Biochem. Biophys. Res. Commun. 149, 939–945.
10 Solberg, R. et al. (1991) Biochem. Biophys. Res. Commun. 176, 166–172.
11 Øyen, O. et al. (1989) FEBS Lett. 246, 57–64.
12 Levy, F.O. et al. (1988) Mol. Endocrinol. 2, 1364–1373.
13 Roskoski, R. (1983) Methods Enzymol. 99, 3–6.
14 Cobb, C.E. and Corbin, J.D. (1988) Methods Enzymol. 159, 202–208.
15 Corbin, J.D. et al. (1975) J. Biol. Chem. 250, 218–225.
16 Øgreid, D. et al. (1989) Eur. J. Biochem. 181, 19–31.
17 Skålhegg, B.S. et al. (1992) J. Biol. Chem. 267, 15707–15714.
18 Dostmann, W.R. et al. (1990) J. Biol. Chem. 265, 10484–10491.
19 Sandberg, M. et al. (1991) Biochem. J. 279, 521–527.
20 Dostmann, W.C. and Taylor, S.S. (1991) Biochemistry 30, 8710–8716.
21 Van Patten, S.M. et al. (1991) Proc. Natl Acad. Sci. USA 88, 5383–5387.

 **DmPKA** cAMP-dependent PK (*D. melanogaster*)

Daniel Kalderon (Columbia University, USA)

Although there have been some assays of the biochemical activity of PKA in *D. melanogaster*[1], most studies have concentrated on gene structure[2,3] and biological function[4]. The biochemical properties of DmPKA have largely been assumed to mirror those of the mammalian enzyme, because of the high degree of sequence similarity between the homologous PKA subunits.

Subunit structure and isoforms

SUBUNIT	AMINO ACIDS	MOL. WT	SDS-PAGE
DC0	353	40 707	40 kDa
DC1a	354	41 468	
DC1b	376	44 109	
DC2	502	56 690	65 kDa
DRI class I	377	42 367	48 kDa
DRI class II	320	35 479	
DRI class III/IV	297	32 896	

An R_2C_2 holoenzyme structure is assumed by analogy with vertebrate **PKA**. DC0 is 82% identical to mouse Cα and encodes most or all of the catalytic subunit activity as judged by biochemical assays of extracts from flies with loss of function mutations. Two other *D. melanogaster* gene products, DC1 and DC2, are more closely related to **PKA** (44% and 49% identical to mouse Cα) than to any other PK. DC1 mRNAs exist as two alternately spliced forms encoding distinct C-termini. DC2 protein, expressed from baculovirus vectors, phosphorylates kemptide, is inhibited by PKI(5–24), and is inhibited by excess RI subunit, an effect reversed by cAMP. It therefore acts like a PKA catalytic subunit. Transcripts of the DRI gene initiate at four distinct sites and could encode three distinct protein products (classes I, II and III/IV)[2]. The class I protein is similar in size to mammalian RI. The smaller protein products would lack the N-terminal dimerization domain, but have not yet been detected as proteins. An RII protein is expressed in *D. melanogaster*, but the corresponding gene has not yet been cloned.

Genetics
Chromosomal locations:
DC0:	30C1–6
DC1:	100A
DC2:	72A
DRI:	77F

Database accession numbers

	PIR	SWISSPROT	EMBL/GENBANK	REF
DC0	A28269	P12370	X16969	2, 3
DC1a	D31751	P16910	X16960	2
DC1b	E31751	P16911	X16960	2
DC2		P16912	X16960	2
DRI	A31751	P16905	X16970, X16971	2
			X16963, X16964	2
			X16966, X16968	2

Domain structure

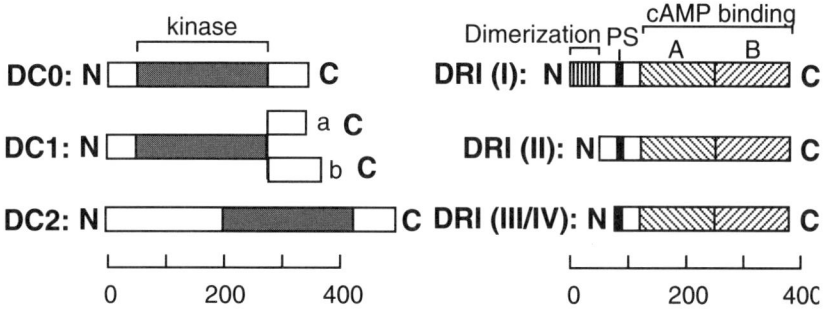

All of the sites of covalent modification of the catalytic subunit of vertebrate **PKA** are conserved in DC0, but have not yet been shown to be modified. Sequences related to the pseudosubstrate site and cAMP-binding domains of the vertebrate RI subunit are found in all forms of DRI, but the dimerization domain is only present in the class I form.

Amino acid sequences of *D. melanogaster* PKA

DC0

```
MGNNATTSNK KVDAAETVKE FLEQAKEEFE DKWRRNPTNT AALDDFERIK  50
TLGTGSFGRV MIVQHKPTKD YYAMKILDKQ KVVKLKQVEH TLNEKRILQA 100
IQFPFLVSLR YHFKDNSNLY MVLEYVPGGE MFSHLRKVGR FSEPHSRFYA 150
AQIVLAFEYL HYLDLIYRDL KPENLLIDSQ GYLKVTDFGF AKRVKGRTWT 200
LCGTPEYLAP EIILSKGYNK AVDWWALGVL VYEMAAGYPP FFADQPIQIY 250
EKIVSGKVRF PSHFGSDLKD LLRNLLQVDL TKRYGNLKAG VNDIKNQKWF 300
ASTDWIAIFQ KKIEAPFIPR CKGPGDTSNF DDYEEEALRI SSTEKCAKEF 350
AEF
```

DC1a

```
MSQHTSQYVF NSKEDYNVIL DNMSREFEER WNHQTQSPYT NLENYITRAV  50
LGNGSFGTVM LVREKSGKNY YAAKMMSKED LVRLKQVAHV HNEKHVLNAA 100
RFPFLIYLVD STKCFDYLYL ILPLVNGGEL FSYHRRVRKF NEKHARFYAA 150
QVALALEYMH KMHLMYRDLK PENILLDQRG YIKITDFGFT KRVDGRTSTL 200
CGTPEYLAPE IVQLRPYNKS VDWWAFGILV YEFVAGRSPF AIHNRDVILM 250
YSKICICDYK MPSYFTSQLR SLVESLMQVD TSKRLGNSND GSSDVKSHPW 300
FQGVDWFGIL NQEVTAPYQP TISGAEDLSN FENFEFKDRY KSRINRHPEL 350
FANF
```

Sites of interest: L285–F354, replaced in DC1b by residues below

DC1b (alternate C-terminus)

```
                              KLERRL QRREESSVVP 300
GRRLVWHSQP GSHRPLPAHH FRRRRSVELR ELRVQGSVQV PNKPPSRIVC 350
EFLNVNVSFE IVVFFSVCLI SRGALS
```

DC2

```
MSTATCARFC TPLSSGTAGS TSKLTTGNGS GNTMTSAYKK KIPSNNSTTA  50
NDSSNTETTF TFKLGRSNGR SSSNVASSES SDPLESDYSE EDPEQEQQRP 100
DPATKSRSSS TATTTTTSSA DHDNDVDEED EEDDEDEGEG NGRDADDATH 150
DSSESIEEDD GNETDDEEDD DESEESSSVQ TAKGVRKYHL DDYQIIKTVG 200
TGTFGRVCLC RDRISEKYCA MKILAMTEVI RLKQIEHVKN ERNILREIRH 250
PFVISLEWST KDDSNLYMIF DYVCGGELFT YLRNAGKFTS QTSNFYAAEI 300
VSALEYLHSL QIVYRDLKPE NLLINRDGHL KITDFGFAKK LRDRTWTLCG 350
TPEYIAPEII QSKGHNKAVD WWALGVLIYE MLVGYPPFYD EQPFGIYEKI 400
LSGKIEWERH MDPIAKDLIK KLLVNDRTKR LGNMKNGADD VKRHRWFKHL 450
NWNDVYSKKL KPPILPDVHH DGDTKNFDDY PEKDWKPAKA VDQRDLQYFN 500
DF
```

DRI (class I)

```
MSYMMAKTLE EQSLRECEHY IQTHGIQRVL KDCIVQLCVC RPENPVQFLR  50
QYFQKLEREQ VKLDASRQVI SPDDCEDLSP MPQTAAPPVR RRGGISAEPV 100
TEEDATNYVK KVVPKDYKTM NALSKAIAKN VLFAHLDESE RSDIFDAMFP 150
VNHIAGENII QQGDEGDNFY VIDVGEVDVF VNSELVTTIS EGGSFGELAL 200
IYGTPRAATV RAKTDVKLWG IDRDSYRRIL MGSTIRKRKM YEEFLSRVSI 250
LESLDKWERL TVADSLETCS FDDGETIVKQ GAAGDDFYII LEGCAVVLQQ 300
RSEQGEDPAE VGRLGSSDYF GEIALLLDRP RAATVVARGP LKCVKLDRAR 350
FERVLGPCAD ILKRNITQYN SFVSLSV
```

Sites of interest: R58, start of class II polypeptide (preceded by a single M);
M81, start of class III/IV polypeptide

Homologues in other species

The predicted products of DC0 and DRI are 82% and 71% identical to the C and RI subunits of mammalian **PKA**. DRI is only 32% identical to the RII subunit of mammalian **PKA**.

Pattern of expression

Four DC0 mRNAs were detected, differing in their 3′-untranslated regions. The shortest species was most prominent in the embryo and the longest in the adult. DC1 transcripts are present at all stages of development but are more abundant in adult body than adult head. Two DC2 transcripts were detected, differing in their 5′-untranslated regions. The shorter mRNA was only detected post-pupation. Class I DRI mRNAs were detected at all stages of development. Class II and IV DRI mRNAs were only reproducibly detected in adults, whereas class III was undetectable at any stage and was identified only as a cDNA clone. In adults, all forms of DRI transcript are enriched 5-fold in head as opposed to body.

References

[1] Foster, J.L. et al. (1984) J. Biol. Chem. 259, 13049–13055.
[2] Kalderon, D. and Rubin, G.M. (1988) Genes Devel. 2, 1539–1556.
[3] Foster, J.L. et al. (1988) J. Biol. Chem. 263, 1676–1681.
[4] Lane, M.E. and Kalderon, D. (1993) Genes Devel. 7, 1229–1243.

Stevan Marcus (University of Texas M.D. Anderson Cancer Center, USA)

In *S. cerevisiae*, Ras proteins activate adenylyl cyclase[1,2]. Three genes *(TPK1, TPK2* and *TPK3)* encode catalytic subunits[3], while *BCY1* encodes the regulatory subunit of ScPKA[4]. Yeast strains carrying deletions of all three *TPK* genes grow very slowly, but are not completely inviable, a phenotype indistinguishable from that resulting from deletion of the gene encoding adenylyl cyclase, *CYR1*. The yeast PKAs mediate cellular responses to various extracellular stimuli, including nutrients and heat shock. Both Bcy1 and the Tpk proteins exhibit substantial homology to their mammalian counterparts.

Subunit structure and isoforms

SUBUNIT	AMINO ACIDS	MOL. WT	SDS-PAGE
Tpk1	397	46 076	
Tpk2	380	44 210	
Tpk3	398	45 977	
Bcy1	416	47 088	

Genetic studies suggest that Tpk1–3 are functionally redundant. Bcy1 appears to be the sole regulatory subunit and, like its mammalian counterparts, is a dimer.

Genetics

Chromosomal locations:
  Tpk1:   10, left arm
  Bcy1:   9, left arm

Domain structure

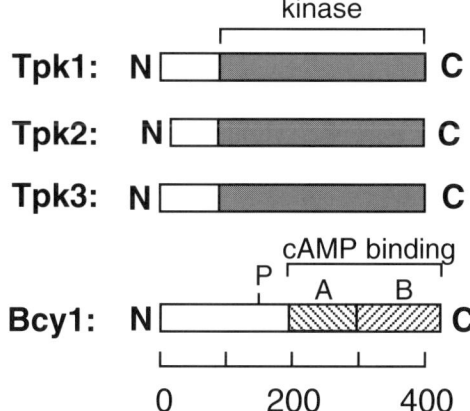

Residues 194–295 and 302–412 of Bcy1 are about 35% identical. These domains presumably comprise cAMP binding sites, based on homology to the mammalian PKA regulatory subunits.

Database accession numbers

	PIR	*SWISSPORT*	*EMBL/GENBANK*	*REF*
Tpk1	A27070	P06244	M17072	3
Tpk2	B27070	P06245	M17073	3
Tpk3	C27070	P05986	M17074	3
Bcy1	A25868	P07278	M15756	4

Amino acid sequences of *S. cerevisiae* PKA

Tpk1
```
MSTEEQNGGG QKSLDDRQGE ESQKGETSER ETTATESGNE SKSVEKEGGE  50
TQEKPKQPHV TYYNEEQYKQ FIAQARVTSG KYSLQDFQIL RTLGTGSFGR 100
VHLIRSRHNG RYYAMKVLKK EIVVRLKQVE HTNDERLMLS IVTHPFIIRM 150
WGTFQDAQQI FMIMDYIEGG ELFSLLRKSQ RFPNPVAKFY AAEVCLALEY 200
LHSKDIIYRD LKPENILLDK NGHIKITDFG FAKYVPDVTY TLCGTPDYIA 250
PEVVSTKPYN KSIDWWSFGI LIYEMLAGYT PFYDSNTMKT YEKILNAELR 300
FPPFFNEDVK DLLSRLITRD LSQRLGNLQN GTEDVKNHPW FKEVVWEKLL 350
SRNIETPYEP PIQQGQGDTS QFDKYPEEDI NYGVQGEDPY ADLFRDF
```

Tpk2
```
MEFVAERAQP VGQTIQQQNV NTYGQGVLQP HHDLQQRQQQ QQQRQHQQLL  50
TSQLPQKSLV SKGKYTLHDF QIMRTLGTGS FGRVHLVRSV HNGRYYAIKV 100
LKKQQVVKMK QVEHTNDERR MLKLVEHPFL IRMWGTFQDA RNIFMVMDYI 150
EGGELFSLLR KSQRFPNPVA KFYAAEVILA LEYLHAHNII YRDLKPENIL 200
LDRNGHIKIT DFGFAKEVQT VTWTLCGTPD YIAPEVITTK PYNKSVDWWS 250
LGVLIYEMLA GYTPFYDTTP MKTYEKILQG KVVYPPYFQP DVVDLLSKLI 300
TADLTRRIGN LQSGSRDIKA HPWFSEVVWE RLLAKDIETP YEPPITSGIG 350
DTSLFDQYPE EQLDYGIQGD DPYAEYFQDF
```

Tpk3
```
MYVEPMNNNE IRKLSITAKT ETTPDNVGQD IPVNAHSVHE ECSSNTPVEI  50
NGRNSGKLKE EASAGICLVK KPMLQYRDTS GKYSLSDFQI LRTLGTGSFG 100
RVHLIRSNHN GRFYALKTLK KHTIVKLKQV EHTNDERRML SIVSHPFIIR 150
MWGTFQDSQQ VFMVMDYIEG GELFSLLRKS QRFPNPVAKF YAAEVCLALE 200
YLHSKDITYR DLKPENILLD KNGHIKITDF GFAKYVPDVT YTLCGTPDYI 250
APEVVSTKPY NKSVDWWSFG VLIYEMLAGY TPFYNSNTMK TYENILNAEL 300
KFPPFFHPDA QDLLKKLITR DLSERLGNLQ NGSEDVKNHP WFNEVIWEKL 350
LARYIETPYE PPIQQGQGDT SQFDRYPEEE FNYGIQGEDP YMDLMKEF
```

Bcy1
```
MVSSLPKESQ AELQLFQNEI NAANPSDFLQ FSANYFNKRL EQQRAFLKAR  50
EPEFKAKNIV LFPEPEESFS RPQSAQSQSR SRSSVMFKSP FVNEDPHSNV 100
FKSGFNLDPH EQDTHQQAQE EQQHTREKTS TPPLPMHFNA QRRTSVSGET 150
LQPNNFDDWT PDHYKEKSEQ QLQRLEKSIR NNFLFNKLDS DSKRLVINCL 200
EEKSVPKGAT IIKQGDQGDY FYVVEKGTVD FYVNDNKVNS SGPGSSFGEL 250
ALMYNSPRAA TVVATSDCLL WALDRLTFRK ILLGSSFKKR LMYDDLLKSM 300
PVLKSLTTYD RAKLADALDT KIYQPGETII REGDQGENFY LIEYGAVDVS 350
KKGQGVINKL KDHDYFGEVA LLNDLPRQAT VTATKRTKVA TLGKSGFQRL 400
LGPAVDVLKL NDPTRH
```

Site of interest: S145, autophosphorylation

Homologues in other species

Homologues of ScPKA have been identified in mammals **(PKA)** and in the fission yeast, *S. pombe*.

References
1   Toda, T. et al. (1985) Cell 40, 27–36.
2   Field, J. et al. (1988) Mol. Cell. Biol. 8, 2159–2165.
3   Toda, T. et al. (1987) Cell 50, 277–287.
4   Toda, T. et al. (1987) Mol. Cell. Biol. 7, 1371–1377.

**Christophe Reymond (Université de Lausanne, Switzerland)
and Michel Véron (Institut Pasteur, France)**

DdPKA has been shown by molecular genetics to play a key role during *D. discoideum* cell type differentiation and fruiting body morphogenesis. Lack of PKA activity blocks aggregate formation, whereas overactivation results in rapid development and facilitated spore formation.

Subunit structure and isoforms

SUBUNIT	AMINO ACIDS	MOL. WT	SDS-PAGE
C	648	74 458	73 000
R	327	36 794	42 000

The holoenzyme (RC) consist of a single Type I R subunit and a single C subunit. cAMP binds to the R subunit and liberates the active catalytic subunit.

Genetics

Upregulation of PKA activity either by mutation in the R subunit gene (*rdeC*) or overexpression of the C subunit (K cells) results in rapid development and facilitated spore formation (sporogenous phenotype). Overexpression of PkaC in prespore cells (PsA-C) results in sporogeny, whereas overexpression in prestalk cells (ecmA-C) blocks development at the aggregation stage.

   Downregulation of PKA activity, achieved by gene disruption, results in failure to aggregate after starvation. Likewise, expression of a mutated PkaR (Rm) unable to bind cAMP (dominant inhibitor of PkaC) under a constitutive promoter prevents aggregation, whereas its expression under a prestalk promoter favours prolonged migration of the slug. When expressed under a prespore promoter, Rm prevents spore differentiation, but a normally shaped and proportioned fruit is formed.

Domain structure

The N-terminal region of the C subunit (function unknown) contains stretches of poly(Asn), poly(Gln) and poly(Thr), shown with black boxes below:

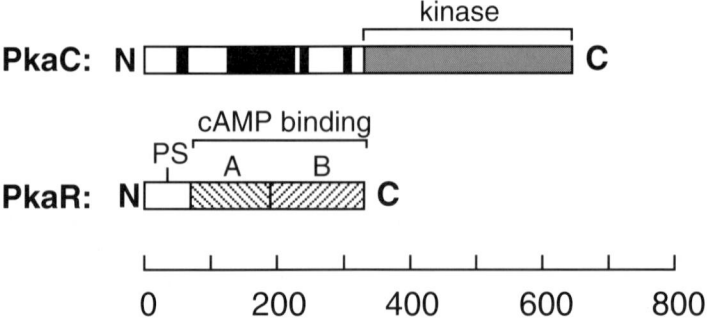

Database accession numbers

	PIR	SWISSPROT	EMBL/GENBANK	REF
PkaC	Q1150	P34099	M38703	3
PkaR	A29076	P05987	M15081	6

Amino acid sequences of *D. discoideum* PKA

PkaC

```
MSNSNNNSSS GNHNSTTINN PKVNVYSNIP NSTTYTYGSG GGGTLSGNNT  50
NNNNTNNNNN NNNNSSGDNK NHSPVTSATD RLTKMDIEEK WDNKNYEKDE 100
REKSPLFHIL ASNLNSFGNF KVPSTFSLTP PEPNKQQQPQ QQPQQQQPQQ 150
QQPQQQQPQQ QQQQQPQQQQ QPQQQLQQNN QQQQQQLQQQ QLQQQLQQQQ 200
QQQQQQQQQQ QQKQQKQQQQ QQQHLHQDGI VNTPSTTQTS TTTTTTTTTT 250
NPHTSGLSLQ HAHSSYTPSN VLHSPTHFQS SLPTRLDTNP ITTPIRQQQQ 300
SQQQLQQQQL QQIPPPTVNS FFLPPPVNAR ERLKEFKQIR VLGTGTFGKV 350
YLIQNTKDGC YYAMKCLNKA YVVQLKQVEH LNSEKSILSS IHHPFIVNLY 400
QAFQDEKKLY LLFEYVAGGE VFTHLRKSMK FSNSTAKFYA AEIVLALEFL 450
HKQNIVYRDL KPENLLIDNQ GHIKITDFGF AKRVEDRTFT LCGTPEYLAP 500
EIIQSKGHGK AVDWWALGIL IFEMLAGYPP FYDDDTFAIY NKILAGRITF 550
PLGFDVDAKD LIKRLLTADR TRRLGALKDG ALDVKNHRWF SDINWERLYQ 600
RRDNGPFIPK IQHQGDSSNF EMYDEEEMVE EPPSSNYVDP YAHLFKDF
```

PkaR

```
MTNNISHNQK ATEKVEAQNN NNITRKRRGA ISSEPLGDKP ATPLPNIPKT  50
VETQQRLEQA LSNNIMFSHL EEEERNVVFL AMVEVLYKAG DIIIKQGDEG 100
DLFYVIDSGI CDIYVCQNGG SPTLVMEVFE GGSFGELALI YGSPRAATVI 150
ARTDVRLWAL NGATYRRILM DQTIKKRKLY EEFLEKVSIL RHIDKYERVS 200
LADALEPVNF QDGEVIVRQG DPGDRFYIIV EGKVVVTQET VPGDHSTSHV 250
VSELHPSDYF GEIALLTDRP RAATVTSIGY TKCVELDRQR FNRLCGPIDQ 300
MLRRNMETYN QFLNRPPSSP NLTSQKS
```

Sites of interest: A30, pseudosubstrate site

Homologues in other species

The C-terminal half of the catalytic subunit shows 54% sequence identity with mammalian **PKA** C subunits. PkaR shows a high similarity with mammalian **PKA** RI subunits. The N-terminal domain responsible for R dimerization in mammals is missing in PkaR.

Physiological substrates and specificity determinants

None known as yet. It has a K_m for kemptide of about $12\,\mu\text{M}$.

Assay

Enzyme activity is measured as transfer of radiolabelled phosphate from $[\gamma\text{-}^{32}\text{P}]\text{ATP}$ into kemptide (LRRASVL), which can be inhibited by either mammalian protein kinase inhibitor protein (PKI) or PKI peptides 6–22 or 5–24.

Enzyme activator and inhibitors
PKI(5–24) K_i 0.4 μM

Pattern of expression
PkaC is expressed in vegetatively growing cells as well as during *Dictyostelium* development. The R subunit level is low in growing cells and increases, both at the RNA and protein level, after starvation at the aggregation stage.

References
[1] Anjard, C. et al. (1992) Development 115, 785–790.
[2] Anjard, C. et al. (1993) Biochemistry 32, 9532–9538.
[3] Burki, E. et al. (1991) Gene 102, 57–65.
[4] Harwood, A.J. et al. (1992) Cell 69, 615–624
[5] Mann, S.K. et al. (1992) Proc. Natl Acad. Sci. USA 89, 10701–10705.
[6] Mutzel, R. et al. (1987) Proc. Natl Acad. Sci. USA 84, 6–10.

 **cGMP-dependent protein kinase (vertebrates)**

Franz Hofmann (Technische Universität München, Germany)

To date three specific high-affinity cellular receptors for cyclic cGMP have been found: (a) cGMP-regulated phosphodiesterases, which hydrolyse cAMP; (b) cGMP and cAMP-regulated cation channels which are present in the outer segment of retinal rods and cones and in the olfactory epithelia; and (c) cGMP-dependent protein kinases which are highly enriched in cerebellar Purkinje cell, platelets and all types of smooth muscle cells[1]. Active cGMP-dependent protein kinase lowers elevated intracellular Ca^{2+} levels and hence decreases smooth muscle tone and platelet aggregation. A few physiological substrate proteins with mostly unknown functions have been identified, some of which are associated with cytoskeletal proteins or structures.

Subunit structure and isoforms

SUBUNIT	AMINO ACIDS	MOL. WT	SDS-PAGE
Iα	670	76787	75–78 kDa
Iβ	686	77803	78–80 kDa
II	762	87084	84–86 kDa

All isoforms are the dimer of identical chains arranged in parallel manner. Each subunit contains a regulatory and a catalytic part which are homologous to the regulatory and catalytic subunit of cAMP-dependent kinases[2]. Activation of the enzyme by cGMP does not result in dissociation of the enzyme into catalytic and regulatory subunits.

Genetics
Chromosomal location:
   Type I (human): 10p11.2–q11.2[3].

Domain structure

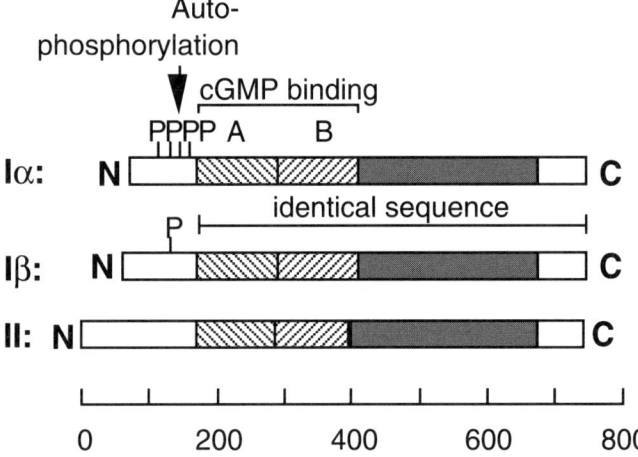

The deduced amino acid sequence of the Iα isozyme cloned from bovine trachea[4] is identical to that reported for PKG purified from bovine lung[2]. The cDNA clone for the Iβ isozyme differs from isozyme Iα only at the N-terminus (amino acids 1–104). The Iα and Iβ isozymes contain identical cGMP-binding sites and catalytic centres. A second form of PKG, termed cGMP kinase II, has been cloned from mouse brain[5]. The cGMP-dependent protein kinase from intestinal epithelia (termed by some authors type II) is identical with the brain type II enzyme[12].

Database accession numbers

		PIR	SWISSPROT	EMBL/GENBANK	REF
Iα	Bovine	A00619	P00516	X16086	2
Iβ	Bovine	S05035	P21136	X54289	2
	Human	S05702	P14619	Y07512	4
II	Mouse	A46590		L12460	5
II	Rat			Z36276	12

Amino acid sequences of bovine PKG

Isoform Iα

```
SELEEDFAKI LMLKEERIKE LEKRLSEKEE EIQELKRKLH KCQSVLPVPS  50
THIGPRTTRA QGISAEPQTY RSFHDLRQAF RKFTKSERSK DLIKEAILDN 100
DFMKNLELSQ IQEIVDCMYP VEYGKDSCII KEGDVGSLVY VMDGKVEVT 150
KEGVKLCTMG PGKVFGELAI LYNCTRTATV KTLVNVKLWA IDRQCFQTIM 200
MRTGLIKHTE YMEFLKSVPT FQSLPEEILS KLADVLEETH YENGEYIIRQ 250
GARGDTFFII SKGKVNVTRE DSPNEDPVFL RTLGKGDWFG EKALQGEDVR 300
TANVIAAEAV TCLVIDRDSF KHLIGGLDDV SNKAYEDAEA KAKYEAEAAF 350
FANLKLSDFN IIDTLGVGGF GRVELVQLKS EESKTFAMKI LKKRHIVDTR 400
QQEHIRSEKQ IMQGAHSDFI VRLYRTFKDS KYLYMLMEAC LGGELWTILR 450
DRGSFEDSTT RFYTACVVEA FAYLHSKGII YRDLKPENLI LDHRGYAKLV 500
DFGFAKKIGF GKKTWTFCGT PEYVAPEIIL NKGHDISADY WSLGILMYEL 550
LTGSPPFSGP DPMKTYNIIL RGIDMIEFPK KIAKNAANLI KKLCRDNPSE 600
RLGNLKNGVK DIQKHKWFEG FNWEGLRKGT LTPPIIPSVA SPTDTSNFDS 650
FPEDNDEPPP DDNSGWDIDF
```

Sites of interest: S1, N-terminal acetylation; S50, T58, S72, T84, phosphorylation; S1–K104, replaced in Iβ isoform by the sequence below

Isoform Iβ (unique N-terminal residues only)

```
MGTLRDLQYA LQEKIEELRQ RDALIDELEL ELDQKDELIQ KLQNELDKYR  50
SVIRPATQQA QKQSASTLQG EPRTKRQAIS AEPTAFDIQD LSHVTLPFYP 100
KSPQSKDLIK EAILDNDFMK
```

Sites of interest: S64, phosphorylation

Isoform II

```
MGNGSVKPKH AKHPDGHSGN LSNEALRSKV LELERELRRK DAELQEREYH  50
LKELREQLAK QTVAIAELTE ELQSKCIQLN KLQDVIHVQG GSPLQASPDK 100
VPLDVHRKTS GLVSLHSRRG AKAGVSAEPT TRTYDLNKPP EFSFEKARVR 150
KDSSEKKLIT DALNKNQFLK RLDPQQIKDM VECMYGEKLS TGSYVIKQGE 200
PGNHIFVLAE GRLEVFQGEK LLSSIPMWTT FGELAILYNC TRTASVKAIT 250
NVKTWALDRE VFQNIMRRTA QARDEEYRNF LRSVSLLKNL PEDKLTKIID 300
CLEVEYYDKG DYIIREGEEG STFFILAKGK VKVTQSTEGH DQPQLIKTLQ 350
KGEYFGEKAL ISDDVRSANI IAEENDVACL VIDRETFNQT VGTFDELQKY 400
LEGYVATLNR DDEKRHAKRS MSSWKLSKAL SLEMIQLKEK VARFSSTSPF 450
QNLEIIATLG VGGFGRVELV KVKNENVAFA MKCIRKKHIV DTKQQEHVYS 500
EKRILEELCS PFIVKLYRTF KDNKYVYMLL EACLGGELWS ILRDRGSFDE 550
PTSKFCVACV TEAFDYLHLL GIIYRDLKPE NLILDADGYL KLVDFGFAKK 600
IGSGQKTWTF CGTPEYVAPE VILNKGHDFS VDFWSLGILV YELLTGNPPF 650
SGIDQMMTYN LILKGIEKMD FPRKITRRPE DLIRRLCRQN PTERLGNLKN 700
GINDIKKHRW LNGFNWEGLK ARSLPSPLRR ELSGPIDHSY FDKYPPEKGV 750
PPDEMSGWDK DF
```

Homologues in other species

The Iβ isozyme has been cloned from human placenta[6]. This sequence contains only two changes when compared to the bovine lung and tracheal Iα and Iβ enzymes (K279T and N289S). The sequence of the rat intestinal PKGII differs from the mouse brain PKGII between amino acid 187–191, but is otherwise homologeous. Two PKG genes have been cloned from *D. melanogaster*, which give rise to several mRNAs differing mainly in their N-terminal regions (**DmPKG**).

Physiological substrates and specificity determinants

The consensus sequence for phosphorylation by PKG is RR**X**S/T**X**, where in contrast to PKA a basic residue (R/K) after S/T is tolerated quite well. However, some variation of the spacing and number of basic residues is permissible, as shown below, where a selection of sequences around sites (bold type) believed to be phosphorylated *in vivo* by the kinase are compared. Basic residues are underlined:

cGMP-dependent protein kinase Iα	GP**R**T**R**AQ
	F**R**K**F**T**K**SE
cGMP-binding cGMP-specific phosphodiesterase	**RK**I**S**A**S**E
G substrate (cerebellum)	**RR**K**D**T**P**AL
Histone H2B	**RKR**S**RK**E
Chromosomal high mobility group protein 14	**KRK**V**S**S**A**E
Vasodilator-stimulated phosphoprotein (VASP)	L**RK**V**SK**QE

Assay

The enzyme is usually assayed as a cGMP-dependent transfer of radioactivity from [γ-^{32}P]ATP to histone 2B. However, since histone can itself cause inhibition of the enzyme, it is preferable to use as substrate the synthetic peptide GRTGRRNSI-amide, derived from the protein kinase inhibitor, PKI[7]. In cell-free assays, the Iα isozyme is activated at 10-fold lower cGMP concentrations than the Iβ isozyme[7].

Enzyme activators and inhibitors

A specific inhibitor for PKG is not known[8]. KT 5822[9] is the only protein kinase inhibitor reported to have a higher affinity for PKG than for PKA. This inhibitor is commercially available under the name KT 5823.

	SPECIFICITY	K_i (μM)
KT 5822	PKG≫PKA	0.0024
H-9	PKA>PKG	0.9
H-8	PKA>PKG	0.5
H–88	PKA>PKG	0.8

Pattern of expression

Both type I isozymes generally have a cytosolic distribution. On stimulation of certain cells apparently a relocalization of the enzyme occurs[10]. The intestinal enzyme appears to be located in or close to the plasma membrane of the intestinal crypts. The type Iα cGMP-dependent protein kinase is highly expressed in the Purkinje cell of the cerebellum, in platelets and lung, whereas the type Iβ isozyme is expressed in all smooth muscle cells together with the Iα isozyme[11]. The mRNA of the mouse brain enzyme is expressed in brain, kidney[5] and intestines[12].

References

1   Hofmann, F. et al. (1992) Biochim. Biophys. Acta 1135, 51–60.
2   Takio, K. et al. (1984) Biochemistry 23, 4207–4218.
3   Orstavik, S. et al. (1992) Cytogenet. Cell Genet. 59, 270–273.
4   Wernet, W. et al. (1989) FEBS Lett. 251, 191–196.
5   Uhler, M.D. (1993) J. Biol. Chem. 263, 13586–13591.
6   Sandberg, M. et al. (1989) FEBS Lett. 255, 321–329.
7   Ruth, P. et al. (1991) Eur. J. Biochem. 202, 1339–1344.
8   Hidaka, H. and Kobayashi, R. (1992) Annu. Rev. Pharmacol. Toxicol. 32, 377–379.
9   Kase, H. et al. (1987) Biochem. Biophys. Res. Commun. 142, 436–440.
10 Wyatt, T.A. et al. (1991) J. Biol. Chem. 266, 21274–21280.
11 Keilbach, A. et al. (1992) Eur. J. Biochem. 208, 467–473.
12 Jorchan, et al. (1994) Proc. Natl Acad. Sci. 91, 9426–9430.

DmPKG — cGMP-dependent PK (*D. melanogaster*)

Daniel Kalderon (Columbia University, USA)

Two DmPKG genes were cloned from a genomic library by screening at low stringency with a probe derived from **DmPKA** DNA[1]. The predicted gene products are clearly related to **PKG**, but as yet there have been no biochemical or genetic studies regarding their activity or function.

Subunit structure and isoforms

SUBUNIT	AMINO ACIDS	MOL. WT	SDS-PAGE
DG1	768	86 758	
DG2 (T1)	1088	121 270	
DG2 (T2)	894	100 998	
DG2 (T3b)	742	83 335	

DG1 and DG2 are distinct gene products. Expression of DG1 from a baculovirus expression vector gives high levels of cGMP-dependent protein kinase activity (J.L. Foster, personal communication). Several mRNAs are produced from DG2 by alternate splicing, encoding up to six putative polypeptides[1]: however these have not yet been identified at the protein level. Only the major mRNA forms (T1, T2 and T3b) are discussed in detail in this entry; other forms were observed only as cDNA clones.

Genetics
Chromosomal locations:

    DG1:   21D
    DG2:   24A

Domain structure

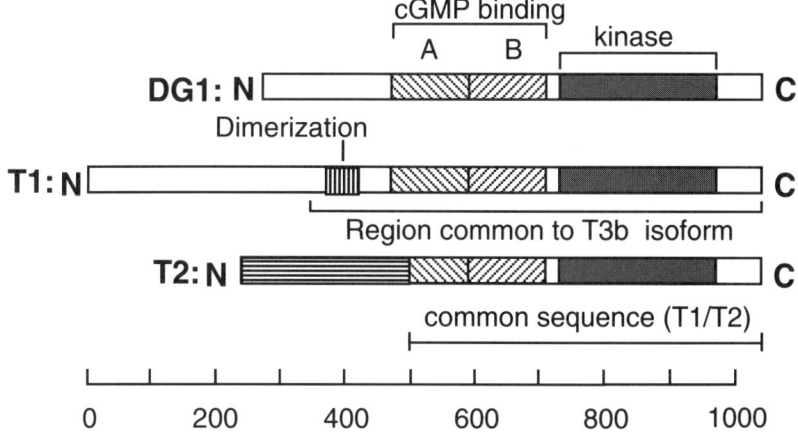

DG1 has regions homologous with all of the domains identified in vertebrate **PKG** except the dimerization domain. The T1 and T3b forms of DG2 have domain structure similar to vertebrate **PKG** but with extended N-terminal regions. The T2 form has an unrelated N-terminal region and lacks homology with the dimerization region, the hinge region, and the N-terminal part of the first cGMP-binding domain of vertebrate **PKG**.

Database accession numbers

		PIR	SWISSPROT	EMBL/GENBANK	REF
DG1				J04816, M27114	1
DG2	T1			M27115–20	1
	T2			M30147, M30148	1
	T3b			M30413	1

Amino acid sequences of *D. melanogaster* PKG

DG1

```
MAAGMLTDRE REAIVSNLTK DVQALREMVR SRESELVKLH REIHKLKSVL  50
QQTTNNLNVT RNEKAKKKLY SLPEQCGEQE SRNQNPHLCS SCGMVLPTSP 100
EFALEALSLG PLSPLASTSS ASPSGRTSAD EVRPKAMPAA IKKQGVSAES 150
CVQSMQQSYS IPIPKYEKDF SDKQQIKDAI MDNDFLKNID ASQVRELVDS 200
MYSKSIAAGE FVIREGEVGA HLYVSAAGEF AVMQQGKVLD KMGAGKAFGE 250
LAILYNCTRT ASIRVLSEAA RVWWVLDRRVF QQIMMCTGLQ RIENSVNFLR 300
SVPLLMNLSE ELLAKIADVL ELEFYAAGTY IIRQGTAGDS FFLISQGNVR 350
VTQKLTPTSP EETELRTLSR GDYFGEQALI NEDKRTANII ALSPGVECLT 400
LDRDSFKRLI GDLCELKEKD YGDESRKLAM KQAQESCRDE PKEQLQQEFP 450
DLKLTDLEVV STLGIGGFGR VELVKAHHQD RVDIFALKCL KKRHIVDTKQ 500
EEHIFSERHI MLSSRSPFIC RLYRTFRDEK YVYMLLEACM GGEIWTMLRD 550
RGSFEDNAAQ FIIGCVLQAF EYLHARGIIY RDLKPENLML DERGYVKIVD 600
FGFAKQIGTS SKTWTFCGTP EYVAPEIILN KGHDRAVDYW ALGILIHELL 650
NGTPPFSAPD PMQTYNLILK GIDMIAFPKH ISRWAVQLIK RLCRDVPSER 700
LGYQTGGIQD IKKHKWFLGF DWDGLASQLL IPPFVRPIAH PTDVRYFDRF 750
PCDLNEPPDE LSGWDADF
```

DG2 (T1 isoform)

```
MRFCFDRLCF ATKRPAQNSN SNAPHSSTTV DAPPRPADVD VATVPVATTA   50
PPPQQPVSNL FYADYQKLQP AIIDRDWERD RDTDTDTRSE AKPPDIVEHI  100
EPVEEQRQIH TQIQSPAEIQ IQIPPTPPAP SIQIQIQQRY RRHSSAEDRN  150
LNTRRNDSNI TEALRKAASM QQEPNANYQF PTDLGLVSIV NNNNNTNTHP  200
SGSNSGTNNN SNINNNLVGG IVTLPAAGGL IGLEHTASGL RLIPAPPTHS  250
DVLTHTLIYG TPPSGAQQLN QDPRSLLHQQ ELQLQQRYQQ LQQLQAQTQG  300
LYTSQGSPVL YHQPSPGSSQ PVAIPGATCH SPTQLQPPNT LNLQQQMQSL  350
RISGCTPSGT GGSATPSPVG LVDPNFIVSN YVAASPQEER FIQIIQAKEL  400
KIQEMQRALQ FKDNEIAELK SHLDKFQSVF PFSRGSAAGC AGTGGASGSG  450
AGGSGGSGPG TATGATRKSG QNFQRQRALG ISAEPQSESS LLLEHVSFPK  500
YDKDERSREL IKAAILDNDF MKNLDLTQIR EIVDCMYPVK YPAKNLIIKE  550
GDVGSIVYVM EDGRVEVSRE GKYLSTLSGA KVLGELAILY NCQRTATITA  600
ITECNLWAIE RQCFQTIMMR TGLIRQAEYS DFLKSVPIFK DLAEDTLIKI  650
SDVLEETHYQ RGDHIVRQGA RGDTFFIISK GKVRVTIKQQ DRQEEKFIRM  700
LGKGDFFGEK ALQGDDLRTA NIICESADGV SCLVIDRETF NQLISNLDEI  750
KHRYDDEGAM ERRKINEEFR DINLTDLRVI ATLGVGGFGR VELVQTNGDS  800
SRSFALKQMK KSQIVETRQQ QHIMSEKEIM GEANCQFIVK LFKTFKDKKY  850
LYMLMESCLG GELWTILRDK GNFDDSTTRF YTACVVEAFD YLHSRNIIYR  900
DLKPENLLLN ERGYGKLVDF GFAKKLQTGR KTWTFCGTPE YVAPEVILNR  950
GHDISADYWS LGVLMFELLT GTPPFTGSDP MRTYNIILKG IDAIEFPRNI 1000
TRNASNLIKK LCRDNPAERL GYQRGGISEI QKHKWFDGFY WWGLQNCTLE 1050
PPIKPAVKSV VDTTNFDDYP PDPEGPPPDD VTGWDKDF
```

Sites of interest: M347, first amino acid of T3b isoform; 563–1088, region common to T2a isoform

DG2 (T2a , unique N-terminal region)

```
MKIKHYPGKA VDASLSLEGS SAMGALYEAN WLRAANQPAA PATTGTKLSR  50
QSSSAGSSFL IEGISALSKY QMTLENIRQL ELQSRDKRIA STIKELSGYR 100
PSALQHHQQQ QMHNVWVAED QDQEHEELED ASEGKEKLAS IQEPPAVNHY 150
VLDPTERPRV PRPRQQFSVK PPSLRRSQTM SQPPSYATLR SPPKIKENLS 200
KSSSAYSTFS SAAEDSQDHV VICQQPQRLM APPPREPPPE PPKRVSKPLS 250
RSQTSVQRYA TVRMPNQTTS FSRSVVRSRD STASQRRLSL EQAIEGLKLE 300
GEKAVRQKSP QISPAASSNG SSKDLNGEGF CIPRPRLIVP VHTYARRRRT 350
GNLKEQSSGG QEEEAEKD
```

Homologues in other species

DG1 is 51% and 70% identical to bovine lung **PKG** within the cGMP-binding and kinase domains respectively. The corresponding figures for DG2 are 64% and 75%. Identity with the equivalent regions of vertebrate **PKA** are lower (30–40%).

Pattern of expression

DG1 mRNA is expressed throughout development, and is found predominantly in adult head as opposed to body. All of the major DG2 transcripts (T1, T2 and T3) are detectable throughout development, but the T3 transcript is relatively low in embryos.

Reference

[1] Kalderon, D. and Rubin, G.M. (1989) J. Biol. Chem. 264, 10738–10748.

PKC
Protein kinase C (vertebrates)
(Ca^{2+}-and phospholipid-dependent PK)

Shigeo Ohno (Yokohama City University, Japan)
and Koichi Suzuki (Tokyo University, Japan)

Protein kinase C (PKC) was originally identified as a Ca^{2+}-and phospholipid-dependent protein kinase, and subsequently shown to be activated by diacylglycerols and phorbol esters. This category is now termed conventional PKC (cPKC) to distinguish it from the other subfamilies. PKC is the only known receptor for tumour-promoting phorbol esters such as TPA (12-O-tetradecanoylphorbol 13-acetate). This property is shared not only by cPKC but also another PKC subfamily, novel PKC (nPKC). The major difference between the cPKC and nPKC subfamilies is that the latter are not Ca^{2+}-dependent. The third sub-family of PKC, atypical PKC (aPKC), shows phospholipid-dependent kinase activity. The physiological importance of cPKC and nPKC is clear because of their properties as phorbol ester/diacylglycerol receptors, but that of aPKC remains to be clarified.

Subunit structure and isoforms

SUBUNIT	AMINO ACIDS	MOL. WT	SDS-PAGE
cPKCα	672	76 704	
cPKCβI	671	76 663	
cPKCβII	673	76 806	
cPKCγ	697	78 268	
nPKCδ	674	77 458	
nPKCε	737	83 492	
nPKCη	683	77 884	
nPKCθ	707	81 479	
aPKCζ	592	67 605	
aPKCλi	586	67 200	

All forms are monomeric.

cPKCβI and βII differ in their C-terminal 50 (51) amino acid residues and are encoded by mRNAs generated from a single gene by alternative splicing of the 3′-exons.

Genetics

Chromosomal locations:

Human α:	17 q22–q24
Human βI/βII:	16 p12–q11.1
Human γ:	19 q13.2–q13.4
Mouse α:	11

Site-directed mutants:

Rabbit cPKCα	K368R (kinase negative)
Bovine cPKCα:	R22A, K23A, R27A (kinase constitutive)
Rabbit cPKCβII:	K371R (kinase negative)
Mouse nPKCδ:	K376A (kinase negative)
Mouse nPKCδ:	R144A, R145A (kinase constitutive)
Rabbit nPKCε:	K436R (kinase negative)

Domain structure

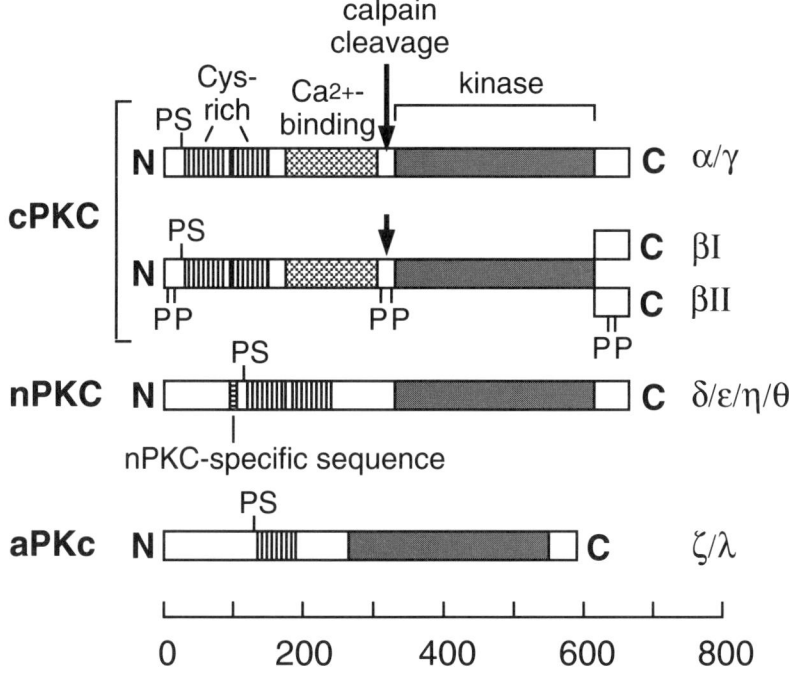

PKC isoforms consist of conserved domains, shown above, separated by variable sequences. The Cys-rich sequences binds two molecules of zinc, and are responsible for diacylglycerol/phorbol ester binding. Cleavage by calpain gives rise to a constitutively active kinase fragment. The central variable sequence in nPKCs varies in length from 50 to 100 amino acids.

Database accession numbers

		PIR	*SWISSPROT*	*EMBL/GENBANK*	*REF*
α	Human	S09496	P17252	X52479	2
	Rat	S02620	P05696	X07286	3
	Mouse	P20444		M25811	4
	Bovine	A00621	P04409	M13973	5
	Rabbit	C2603	P10102	X04796	6
βI	Human	S00159	P05771	X06318/M27545	7
	Rat	A00622	P04410	X04439/M19007	8, 9
	Rabbit	B26037	P05772	X04795	6
βII	Human	B24664	P05127	X07109/M13975	7, 10
	Rat	A00623	P04411	X04440/M13706	8, 11
	Rat	S11213			
	Mouse			X53532	12
	Bovine	A24664	P05126	M13974	10
	Rabbit	A26037	P05773	X04793	6

γ	Human	D24664	P05129	M13977/X62533	10
	Rat	A05105	P05697	X07287/M13707/M55417	3, 11
	Mouse			X67129	13
	Bovine	C24664	P05128	M13976	10
	Rabbit	A28708	P10829	M19338	14
δ	Rat	A28163	P09215	M18330/J03204	15
	Mouse			X60304/M69042/J05335	16
ε	Human			X65293	17
	Rat	B28163	P09216	M18331	15
	Mouse	S02270	P16054		18
	Rabbit	A29880	P10830	M20014	19
η	Human		P24723	M55284	20
	Mouse	A23690	P23298	D90242/J05703/M62980	21
θ	Human			L01087	22, 23
	Mouse			D11091	24
ζ	Human			Z15108	25
	Rat	P09217	A30314/C28163	J04532/M18332	26
		S00217			
	Mouse			M94632	27
λ	Human			L18964	28
	Mouse			D28577	29

Amino acid sequences of rat PKC

cPKCα

```
MADVYPANDS TASQDVANRF ARKGALRQKN VHEVKDHKFI ARFFKQPTFC  50
SHCTDFIWGF GKQGFQCQVC CFVVHKRCHE FVTFSCPGAD KGPDTDDPRS 100
KHKFKIHTYG SPTFCDHCGS LLYGLIHQGM KCDTCDMNVH KQCVINVPSL 150
CGMDHTEKRG RIYLKAEVTD EKLHVTVRDA KNLIPMDPNG LSDPYVKLKL 200
IPDPKNESKQ KTKTIRSTLN PQWNESFTFK LKPSDKDRRL SVEIWDWDRT 250
TRNDFMGSLS FGVSELMKMP ASGWYKLLNQ EEGEYYNVPI PEGDEEGNVE 300
LRQKFEKAKL GPAGNKVISP SEDRKQPSNN LDRVKLTDFN FLMVLGKGSF 350
GKVMLADRKG TEELYAIKIL KKDVVIQDDD VECTMVEKRV LALLDKPPFL 400
TQLHSCFQTV DRLYFVMEYV NGGDLMYHIQ QVGKFKEPQA VFYAAEISIG 450
LFFLHKRGII YRDLKLDNVM LDSEGHIKIA DFGMCKEHMM DGVTTRTFCG 500
TPDYIAPEII AYQPYGKSVD WWAYGVLLYE MLAGQPPFDG EDEDELFQSI 550
MEHNVSYPKS LSKEAVSICK GLMTKHPAKR LGCGPEGERD VREHAFFRRI 600
DWEKLENREI QPPFKPKVCG KGAENFDKFF TRGQPVLTPP DQLVIANIDQ 650
SDFEGFSYVN PQFVHPILQS AV
```

Sites of interest: A25, pseudosubstrate; H37, C50, C53, C67, C70, H75, C78, C86, H102, C115, C118, C132, C135, H140, C143, C151, conserved sites in Cys-rich sequences

cPKCβI

```
MADPAAGPPP SEGEESTVRF ARKGALRQKN VHEVKNHKFT ARFFKQPTFC  50
SHCTDFIWGF GKQGFQCQVC CFVVHKRCHE FVTFSCPGAD KGPASDDPRS 100
KHKFKIHTYS SPTFCDHCGS LLYGLIHQGM KCDTCMMNVH KRCVMNVPSL 150
CGTDHTERRG RIYIQAHIDR EVLIVVVRDA KNLVPMDPNG LSDPYVKLKL 200
IPDPKSESKQ KTKTIKCSLN PEWNETFRFQ LKESDKDRRL SVEIWDWDLT 250
```

cPKCβI *continued*

```
SRNDFMGSLS FGISELQKAG VDGWFKLLSQ EEGEYFNVPV PPEGSEGNEE 300
LRQKFERAKI GQGTKAPEEK TANTISKFDN NGNRDRMKLT DFNFLMVLGK 350
GSFGKVMLSE RKGTDELYAV KILKKDVVIQ DDDVECTMVE KRVLALPGKP 400
PFLTQLHSCF QTMDRLYFVM EYVNGGDLMY HIQQVGRFKE PHAVFYAAEI 450
AIGLFFLQSK GIIYRDLKLD NVMLDSEGHI KIADFGMCKE NIWDGVTTKT 500
FCGTPDYIAP EIIAYQPYGK SVDWWAFGVL LYEMLAGQAP FEGEDEDELF 550
QSIMEHNVAY PKSMSKEAVA ICKGLMTKHP GKRLGCGPEG ERDIKEHAFF 600
RYIDWEKLER KEIQPPYKPK ARDKRDTSNF DKEFTRQPVE LTPTDKLFIM 650
NLDQNEFAGF SYTNPEFVIN V
```

Sites of interest: A25, pseudosubstrate; H37, C50, C53, C67, C70, H75, C78, C86, H102, C115, C118, C132, C135, H140, C143, C151, conserved sites in Cys-rich sequences; R622–V671, different residues between βI and βII

cPKCβII

```
MADPAAGPPP SEGEESTVRF ARKGALRQKN VHEVKNHKFT ARFFKQPTFC  50
SHCTDFIWGF GKQGFQCQVC CFVVHKRCHE FVTFSCPGAD KGPASDDPRS 100
KHKFKIHTYS SPTFCDHCGS LLYGLIHQGM KCDTCMMNVH KRCVMNVPSL 150
CGTDHTERRG RIYIQAHIDR EVLIVVVRDA KNLVPMDPNG LSDPYVKLKL 200
IPDPKSESKQ KTKTIKCSLN PEWNETFRFQ LKESDKDRRL SVEIWDWDLT 250
SRNDFMGSLS FGISELQKAG VDGWFKLLSQ EEGEYFNVPV PPEGSEGNEE 300
LRQKFERAKI GQGTKAPEEK TANTISKFDN NCNRDRMKLT DFNFLMVLCK 350
GSFGKVMLSE RKGTDELYAV KILKKDVVIQ DDDVECTMVE KRVLALPGKP 400
PFLTQLHSCF QTMDRLYFVM EYVNGGDLMY HIQQVGRFKE PHAVFYAAEI 450
AIGLFFLQSK GIIYRDLKLD NVMLDSEGHI KIADFGMCKE NIWDGVTTKT 500
FCGTPDYIAP EIIAYQPYGK SVDWWAFGVL LYEMLAGQAP FEGEDEDELF 550
QSIMEHNVAY PKSMSKEAVA ICKGLMTKHP GKRLGCGPEG ERDIKEHAFF 600
RYIDWEKLER KEIQPPYKPK ACGRNAENFD RFFTRHPPVL TPPDQEVIRN 650
IDQSEFEGFS FVNSEFLKPE VKS
```

Sites of interest: A25, pseudosubstrate; S16, T17, T314, T324, T634, T641, autophosphorylation; H37, C50, C53, C67, C70, H75, C78, C86, H102, C115, C118, C132, C135, H140, C143, C151, conserved sites in Cys-rich sequences, different residues between βI and βII

cPKCγ

```
MAGLGPGGGD SEGGPRPLFC RKGALRQKVV HEVKSHKFTA RFFKQPTFCS  50
HCTDFIWGIG KQGLQCQVCS FVVHRRCHEF VTFECPGAGK GPQTDDPRNK 100
HKFRLHSYSS PTFCDHCGSL LYGLVHQGMK CSCCEMNVHR RCVRSVPSLC 150
GVDHTERRGR LQLEIRAPTS DEIHITVGEA RNLIPMDPNG LSDPYVKLKL 200
IPDPRNLTKQ KTKTVKATLN PVWNETFVFN LKPGDVERRL SVEVWDWDRT 250
SRNDFMGAMS FGVSELLKAP VDGWYKLLNQ EEGEYYNVPV ADADNCSLLQ 300
KFEACNYPLE LYERVRMGPS SSPIPSPSPS PTDSKRCFFG ASPGRLHISD 350
FSFLMVLGKG SFGKVMLAER RGSDELYAIK ILKKDVIVQD DDVDCTLVEK 400
RVLALGGRGP GGRPHFLTQL HSTFQTPDRL YFVMEYVTGG DLMYHIQQLG 450
KFKEPHAAFY AAEIAIGLFF LHNQGIIYRD LKLDNVMLDA EGHIKITDFG 500
MCKENVFPGS TTRTFCGTPD YIAPEIIAYQ PYGKSVDWWS FGVLLYEMLA 550
GQPPFDGEDE EELFQAIMEQ TVTYPKSLSR EAVAICKGFL TKHPGKRLGS 600
```

83

cPKCγ *continued*

```
GPDGEPTIRA HGFFRWIDWE RLERLEIAPP FRPRPCGRSG ENFDKFFTRA 650
APALTPPDRL VLASIDQADF QGFTYVNPDF VHPDARSPTS PVPVPVM
```

Sites of interest: A24, pseudosubstrate; H36, C49, C52, C66, C69, H74, C77, C85, H101, C114, C117, C131, C134, H139, C142, C150, conserved sites in Cys-rich sequences

Amino acid sequences of mouse PKC

nPKCδ

```
MAPFLRISFN SYELGSLQVE DEASQPFCAV KMKEALSTER GKTLVQKKPT  50
MYPEWKTTFD AHIYEGRVIQ IVLMRAAEDP VSEVTVGVSV LAERCKKNNG 100
KAEFWLDLQP QAKVLMCVQY FLEDGDCKQS MRSEEEAKFP TMNRRGAIKQ 150
AKIHYIKNHE FIATFFGQPT FCSVCKEFVW GLNKQGYKCR QCNAAIHKKC 200
IDKIIGRCTG TATNSRDTIF QKERFNIDMP HRFKVYNYMS PTFCDHCGSL 250
LWGLVKQGLK CEDCGMNVHH KCREKVANLC GINQKLLAEA LNQVTQRSSR 300
KLDTTESVGI YQGFEKKPEV SGSDILDNNG TYGKIWEGST RCTLENFTFQ 350
KVLGKGSFGK VLLAELKGKD KYFAIKCLKK DVVLIDDDVE CTMVEKRVLA 400
LAWESPFLTH LICTFQTKDH LFFVMEFLNG GDLMFHIQDK GRFELYRATF 450
YAAEIICGLQ FLHSKGIIYR DLKLDNVMLD RDGHIKIADF GMCKENIFGE 500
GRASTFCGTP DYIAPEILQG LKYSFSVDWW SFGVLLYEML IGQSPFHGDD 550
EDELFESIRV DTPHYPRWIT KESKDIMEKL FERDPDKRLG VTGNIRIHPF 600
FKTINWSLLE KRKVEPPFKP KVKSPSDYSN FDPEFLNEKP QLSFSDKNLI 650
DSMDQEAFHG FSFVNPKFEQ FLDI
```

Sites of interest: A147, pseudosubstrate; H159, C172, C175, C189, C192, H197, C200, C208, H231, C244, C247, C261, C264, H269, C272, C280, conserved sites in Cys-rich sequences

nPKCε

```
MVVFNGLLKI KICEAVSLKP TAWSLRHAVG PRPQTFLLDP YIALNVDDSR  50
IGQTATKQKT NSPAWHDEFV TDVCNGRKIE LAVFHDAPIG YDDFVANCTI 100
QFEELLQNGS RHFEDWIDLE PEGKVYVIID LSGSSGEAPK DNEERVFRER 150
MRPRKRQGAV RRRVHQVNGH KFMATYLRQP TYCSHCRDFI WGVIGKQGYQ 200
CQVCTCVVHK RCHELIITKC AGLKKQETPD EVGSQRFSVN MPHKFGIHNY 250
KVPTFCDHCG SLLWGLLRQG LQCKVCKMNV HRRCETNVAP NCGVDARGIA 300
KVLADLGVTP DKITNSGQRR KKLAAGAESP QPASGNSPSE DDRSKSAPTS 350
PCDQELKELE NNIRKALSFD NRGEEHRASS ATDGQLASPG ENGEVRPGQA 400
KRLGLDEFNF IKVLGKGSFG KVMLAELKGK DEVYAVKVLK KDVILQDDDV 450
DCTMTEKRIL ALARKHPYLT QLYCCFQTKD RLFFVMEYVN GGDLMFQIQR 500
SRKFDEPRSR FYAAEVTSAL MFLHQHGVIY RDLKLDNILL DAEGHCKLAD 550
FGMCKEGIMN GVTTTTFCGT PDYIAPEILQ ELEYGPSVDW WALGVLMYEM 600
MAGQPPFEAD NEDDLFESIL HDDVLYPVWL SKEAVSILKA FMTKNPHKRL 650
GCVAAQNGED AIKQHPFFKE IDWVLLEQKK IKPPFKPRIK TKRDVNNFDQ 700
DFTREEPILT LVDEAIIKQI NQEEFKGFSY FGEDLMP
```

Sites of interest: A159, pseudosubstrate; H170, C183, C186, C201, C204, H209, C212, C220, H243, C256, C259, C273, C276, H281, C284, C292, conserved sites in Cys-rich sequences

nPKCη

```
MSSGTMKFNG YLRVRIGEAV GLQPTRWSLR HSLFKKGHQL LDPYLTVSVD  50
QVRVGQTSTK QKTNKPTYNE EFCANVTDGG HLELAVFHET PLGYDHFVAN 100
CTLQFQELLR TAGTSDTFEG WVDLEPEGKV FVVITLTGSF TEATLQRDRI 150
FKHFTRKRQR AMRRRVHQVN GHKFMATYLR QPTYCSHCRE FIWGVFGKQG 200
YQCQVCTCVV HKRCHHLIVT ACTCQNNINK VDAKIAEQRF GINIPHKFNV 250
HNYKVPTFCD HCGSLLWGIM RQGLQCKICK MNVHIRCQAN VAPNCGVNAV 300
ELAKTLAGMG LQPGNISPTS KLISRSTLRR QGKEGSKEGN GIGVNSSSRF 350
GIDNFEFIRV LGKGSFGKVM LARIKETGEL YAVKVLKKDV ILQDDDVECT 400
MTEKRILSLA RNHPFLTQLF CCFQTPDRLF FVMEFVNGGD LMFHIQKSRR 450
FDEARARFYA AEIISALMFL HEKGIIYRDL KLDNVLLDHE GHCKLADFGM 500
CKEGICNGVT TATFCGTPDY IAPEILQEML YGPAVDWWAM GVLLYEMLCG 550
HAPFEAENED DLFEAILNDE VVYPTWLHED ARGILKSFMT KNPTMRLGSL 600
TQGGEHEILR HPFFKEIDWA QLNHRQLEPP FRPRIKSRED VSNFDPDFIK 650
EEPVLTPIDE GHLPMINQDE FRNFSYVSPE LQL
```

Sites of interest: A161, pseudosubstrate; H172, C185, C188, C203, C206, H211, C214, C222, H246, C259, C262, C276, C279, H284, C287, C295, conserved sites in Cys-rich sequences

nPKCθ

```
MSPFLRIGLS NFDCGTCQAC QGEAVNPYCA VLVKEYVESE NGQMYIQKKP  50
TMYPPWDSTF DAHINKGRVM QIIVKGKNVD LISETTVELY SLAERCRKNN 100
GRTEIWLELK PQGRMLMNAR YFLEMSDTKD MSEFENEGFF ALHQRRGAIK 150
QAKVHHVKCH EFTATFFPQP TFCSVCHEFV WGLNKQGYQC RQCNAAIHKK 200
CIDKVIAKCT GSAINSRETM FHKERFKIDM PHRFKVYNYK SPTFCEHCGT 250
LLWGLARQGL KCDACGMNVH HRCQTKVANL CGINQKLMAE ALAMIESTQQ 300
ARSLRDSEHI FREGPVEIGL PCSTKNETRP PCVPTPGKRE PQGISWDSPL 350
DGSNKSAGPP EPEVSMRRTS LQLKLKIDDF ILHKMLGKGS FGKVFLAEFK 400
RTNQFFAIKA LKKDVVLMDD DVECTMVEKR VLSLAWEHPF LTHMFCTFQT 450
KENLFFVMEY LNGGDLMYHI QSCHKFDLSR ATFYAAEVIL GLQFLHSKGI 500
VYRDLKLDNI LLDRDGHIKI ADFGMCKENM LGDAKTNTFC GTPDYIAPEI 550
LLGQKYNHSV DWWSFGVLVY EMLIGQSPFH GQDEEELFHS IRMDNPFYPR 600
WLEREAKDLL VKLFVREPEK RLGVRGDIRQ HPLFREINWE ELERKEIDPP 650
FRPKVKSPYD CSNFDKEFLS EKPRLSFADR ALINSMDQNM FSNFSFINPG 700
METLICS
```

Sites of interest: A148, pseudosubstrate; H160, C173, C176, C190, C193, H198, C201, C209, H232, C245, C248, C262, C265, H270, C273, C281, conserved sites in Cys-rich sequences

aPKCζ

```
MPSRTDPKMD RSGGRVRLKA HYGGDILITS VDAMTTFKDL CEEVRDMCGL  50
HQQHPLTLKW VDSEGDPCTV SSQMELEEAF RLVCQGRDEV LIIHVFPSIP 100
EQPGMPCPGE DKSIYRRGAR RWRKLYRANG HLFQAKRFNR GAYCGQCSER 150
IWGLSRQGYR CINCKLLVHK RCHVLVPLTC RRHMDSVMPS QEPPVDDKND 200
GVDLPSEETD GIAYISSSRK HDNIKDDSED LKPVIDGVDG IKISQGLGLQ 250
DFDLIRVIGR GSYAKVLLVR LKKNDQIYAM KVVKKELVHD DEDIDWVQTE 300
```

aPKCζ *continued*

```
KHVFEQASSN PFLVGLHSCF QTTSRLFLVI EYVNGGDLMF HMQRQRKLPE 350
EHARFYAAEI CIALNFLHER GIIYRDLKLD NVLLDADGHI KLTDYGMCKE 400
GLGPGDTTST FCGTPNYIAP EILRGEEYGF SVDWWALGVL MFEMMAGRSP 450
FDIITDNPDM NTEDYLFQVI LEKPIRIPRF LSVKASHVLK GFLNKDPKER 500
LGCRPQTGFS DIKSHAFFRS IDWDLLEKKQ TLPPFQPQIT DDYGLDNFDT 550
QFTSEPVQLT PDDEDVIKRI DQSEFEGFEY INPLLLSAEE SV
```

Sites of interest: A119, pseudosubstrate; H131, C144, C147, C161, C164, H169, C172, C180, conserved sites in Cys-rich sequence

Homologues in other species

See **Database accession numbers** for vertebrate homologues. Homologues are also found in *D. melanogaster* (**DmPKC**), *S. cerevisiae* (**ScPKC**) and *C. elegans* (**CePKC**).

Physiological substrates and specificity determinants

The consensus sequence for phosphorylation by cPKC[30] is:

$$K/R_{1-3}X_{0-2}S/TX_{0-2}R/K_{1-3} > S/TX_{0-2}R/K_{1-3} >= R/K_{1-3}X_{0-2}S/T$$

The list below shows a selection of sequences around sites believed to be phosphorylated *in vivo* by cPKC (and/or nPKC):

MARCKS	KKRFSFKKSFKLSGFSFKK
EGF receptor	VRKRTLRRLLQ
Neurogranin/p17	KIQASFRGHMA
Neuromodulin/GAP43/F1	KIQASFRGHIT
Transferrin receptor	YTRFSLAR
Insulin receptor	GRVLTLPRS
Acetylcholine receptor	LRRSSSVGYIS
c-src	PKDPSQRRRS
Ribosomal protein S6	RRLSSLRA
Glycogen synthase (site N7)	SRTLSVSS
Glycogen synthase (site C87)	PRRASCTS
Lipocortin I/p35	EYVQTVKSSKPYTN
Lipocortin II/p36	SAYGSVKPYTNFD

Assay

cPKC: cPKC activity is assayed as a Ca^{2+}- and phospholipid-dependent and diacylglycerol/phorbol ester-stimulated transfer of radioactivity from $[\gamma^{-32}P]ATP$ to a variety of protein and peptide substrates. Histone Type III and myelin basic protein are usually used as protein substrates, whereas the EGF receptor peptide, VRKRTLRRL is usually used as a peptide substrate.

nPKC: nPKC activity is assayed as a Ca^{2+}-independent, phospholipid-dependent and diacylglycerol/phorbol ester-stimulated transfer of radioactivity from $[\gamma^{-32}P]ATP$ to a variety of protein and peptide substrates.

aPKC: aPKC activity is assayed as a Ca^{2+}-independent, diacylglycerol/phorbol ester-independent, phospholipid-dependent transfer of radioactivity from $[\gamma\text{-}^{32}P]ATP$ to a variety of protein and peptide substrates.

Enzyme activators and inhibitors

	ACTIVATORS	INHIBITORS	SPECIFICITY
cPKC	Diacylglycerols		
	Ca^{2+}		
	Phospholipids		
	Phorbol esters		Good
		H7	Poor
		Staurosporine	Poor
nPKC	Diacylglycerols		
	Phospholipids		
	Phorbol esters		Good
		H7	Poor
		Staurosporine	Poor
aPKC	Phospholipids		

Pattern of expression

cPKCs generally have a cytosolic distribution. On stimulation of cells with phorbol esters such as TPA or phorbol dibutyrate, cPKCs can be seen to translocate to cytoplasmic membranes. Some physiological agents which are believed to stimulate PKC cause similar translocation of cPKC (and nPKC). Some reports suggest that cPKC enter the nucleus on stimulation. Tissue distribution is clearly different among cPKC members; α is expressed in almost all tissues and cells examined. $\beta I/II$ are expressed in cells and tissues of neuronal and haematopoietic lineage.

nPKCs generally have a cytosolic distribution. However, depending on the solubilizing condition, some part of nPKC can be recovered in particulate fractions. Translocation is similar to cPKC. Tissue distribution is clearly different among the members; δ is expressed in almost all tissues and cells examined; ϵ is expressed in tissues and cells of neuronal and haematopoietic lineage; η is expressed in tissues and cells of epithelial origin; θ is expressed predominantly in skeletal muscle and also in haematopoietic cells.

aPKC members also generally have a cytosolic distribution. However, depending on the solubilizing condition, some part of aPKC can be recovered in particulate fractions. ζ and λ are expressed in almost all the tissues and cells examined.

References
1 Nishizuka, Y. (1992) Science 258, 607–614.
2 Finkenzeller, G. et al. (1990) Nucl. Acids Res. 18, 2183.
3 Ono, Y. et al. (1988) Nucl. Acids Res. 16, 5199–5200.
4 Rose-John, S. et al. (1988) Gene 74, 465–471.
5 Parker, P.J. et al. (1986) Science 233, 853–859.

6   Ohno, S. et al. (1987) Nature 325, 161–166.
7   Kubo, K. et al. (1987) FEBS Lett. 223, 138–142.
8   Ono, Y. et al. (1987) FEBS Lett. 206, 347–352.
9   Housey, G.M. (1988) Cell 52, 343–354.
10 Coussens, L. et al. (1986) Science 233, 859–866.
11 Knopf, J.L. et al. Cell (1986) 46, 491–502.
12 Ashendel, C.L. (1990) Nucl. Acids Res. 18, 5310.
13 Bowers, B.J. et al., unpublished.
14 Ohno, S. et al. (1988) Biochemistry 27, 2083–2087.
15 Ono, Y. et al. (1988) J. Biol. Chem. 263, 6927–6932.
16 Mizuno, K. et al. (1991) Eur. J. Biochem. 202, 931–940.
17 Basta, P.V. et al. (1992) Biochim. Biophys. Acta 1132, 154–160.
18 Schaap, D. et al. (1989) FEBS Lett. 243, 351–357.
19 Ohno, S. et al. (1988) Cell 53, 731–741.
20 Bacher, N. et al. (1992) Mol. Cell. Biol. 12, 1404.
21 Osada, S. et al. (1990) J. Biol. Chem. 265, 22434–22440.
22 Baier, G. et al. (1993) J. Biol. Chem. 268, 4997–5004.
23 Ware, J.A. and Chang, J. D. unpublished.
24 Osada, S. (1992) Mol. Cell. Biol. 12, 3930–3938.
25 Hug, H.P. unpublished.
26 Ono, Y. et al. (1989) Proc. Natl. Acad. Sci. USA. 86, 3099–3103.
27 Goodnight, J. et al. (1992) Gene 122, 305–311.
28 Selbie, L.A. et al. (1993) J. Biol. Chem. 268, 24296–24302.
29 Alkimoto, K. et al. (1994) J. Biol. Chem. 269, 12677–12683.
30 Kennelly, P.J. and Krebs, E. G. (1991) J. Biol. Chem. 266, 15555–15558.

DmPKC

Protein kinase C (*D. melanogaster*) (eye-PKC, DPKC53 (ey), inaC)

Javier Vinós and Charles Zuker (University of California San Diego, USA)

D. melanogaster has three known PKCs: DPKC53E(br), a PKCα homologue expressed in the adult brain; DPKC98F, a δ homologue expressed throughout development; and eye-PKC, a PKCα homologue expressed exclusively in photoreceptor cells of the visual system. Since this PKC is the best characterized of the *D. melanogaster* kinases, this summary will focus on this enzyme. Eye-PKC is a negative regulator of the visual transduction cascade and has been shown to be required for photoreceptor cell inactivation[1] and light adaptation[2].

Subunit structure and isoforms

SUBUNIT	AMINO ACIDS	MOL. WT	SDS-PAGE
eye-PKC	700	79 848	82 Da

eye-PKC behaves as a monomer.

Genetics

Chromosomal location: II, right arm.
Map position: 53E6–10, 25 kb from DPKC53(br).
Several mutants affecting the light response of photoreceptors which map to this locus have been isolated. P element mediated germline transformation of mutant lines with cloned eye-PKC DNA demonstrated that the *inaC* gene encoded eye-PKC[1]. Eight *inaC* alleles[3] have been reported in the literature, three of them have been sequenced: *inaC*[P209] [W93STOP], *inaC*[P207] [V201D], *US*[3741] [W139STOP][1].

Domain structure

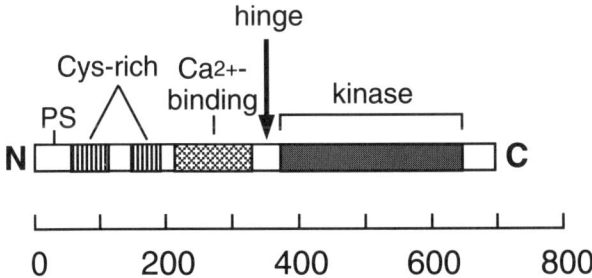

Assignment of the domains above is based on sequence similarity with other Ca^{2+}-dependent PKCs.

Database accession numbers

	PIR	SWISSPROT	EMBL/GENBANK	REF
DPKC53E(ey)	A32392	P13677	J04845	4
DPKC53E(br)	A32545	P05130	X05076	5
DPKC98F	B32392	P13678	J04848	4

Amino acid sequence of *D. melanogaster* eye-PKC

```
MAAAAVATPG ATVLPPSVPS AAPGAKAPAA GAGKGPGNLL EITGEANIVN  50
YMKNRLRKGA MKRKGLEMVN GHRFGVRFFK NPTYCGHCKD FIWGFGKQGF 100
QCEECRFNIH QKCCKFVVFK CPGKDTDFDA DCAKVKHGWI STTYTTPTFC 150
DECGLLLHGV AHQGVKCENC NLNVHHACQE TVPPMCGADI SEVRGKLLLY 200
VELKGNNLKV DIKEAANLIP MDTNGFSDPY IAVQMHPDRS GRTKKKTKTI 250
QKNLNPVFNE TFTFELQPQD RDKRLLIEVW DWDRTSRNDF MGSFSFSLEE 300
LQKEPVDGWY KFLSQVEGEH YNIPCVDAFN DIARLRDEVR HDRRPNEKRR 350
MDNKDMPHNM SKRDMIRAAD FNFVKVIGKG SFGKVLLAER RGTDELYAVK 400
VLRKDVIIQT DDMELPMNEK KILALSGRPP FLVSMHSCFQ TMDRLFFVME 450
YCKGGDLMYH MQQYGRFKES VAIFYAVEVA IALFFLHERD IIYRDLKLDN 500
ILLDGEGHVK LVDFGLSKEG VTERQTTRTF CGTPNYMAPE IVSYDPYSIA 500
ADWWSFGVLL FEFMAGQAPF EGDDETTVFR NIKDKKAVFP KHFSVEAMDI 600
ITSFLTKKPN NRLGAGRYAR QEITTHPFFR NVDWDKAEAC EMEPPIKPMI 650
KHRKDISNFD DAFTKEKTDL TPTDKLFMMN LDQNDFIGFS FMNPEFITII 700
```
```
                                  *3            *2
```

Sites of interest: R57–K64, pseudosubstrate

Other PKC species in *D. melanogaster*

Two other PKCs have been sequenced in *Drosophila*, a mammalian PKCδ homologue, dPKC98F[4] and a second mammalian PKCα homologue, dPKC53E(br)[5]. dPKC98F maps to the third chromosome at position 98E6–F2 and dPKC53E(br) maps to the second chromosome at position 53E4–7.

Pattern of expression

Eye-PKC is expressed in the visual system of the adult (retina and ocelli). In the photoreceptor cells, eye-PKC expression is restricted to the rhabdomeres; these are the specialized microvillar organelles containing rhodopsin and the molecules involved in the phototransduction cascade.

References

[1] Smith, D.P. et al. (1991) Science 254, 1478–1484.
[2] Hardie, R.C. et al. (1993) Nature 363, 634–637.
[3] Lindsley, D.L. and Zimm, G.G. (1992) in The Genome of *Drosophila melanogaster*, Academic Press, San Diego, p. 288.
[4] Schaeffer, E. et al. (1989) Cell 57, 403–412.
[5] Rosenthal, A. et al. (1987) EMBO J. 6, 433–441.

Johji Miwa, Yo Tabuse, Kiyoji Nishiwaki and Tohru Sano
(NEC Corporation, Japan)

The *tpa-1* gene mediates the action of tumour-promoting phorbol esters in the nematode *C. elegans*[1-4]. Analysis of cDNA clones for *tpa-1* predicts a protein (TPA-1) of 557 amino acids, which is homologous to protein kinase C in other animals.

Subunit stucture and isoforms

SUBUNIT	AMINO ACIDS	MOL. WT	SDS-PAGE
TPA-1	557		76 kDa

No evidence for isoforms at present.

Genetics

Chromosomal location: LGIV, left arm.
Mutations:
EMS-induced: k501; transposon Tc1-insertions: k529, k531, k532 (resistant to phorbol esters)

Domain structure

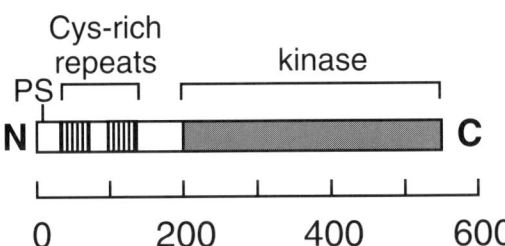

Like mammalian PKCs, TPA-1 has a C-terminal kinase domain and tandem Cys-rich repeats in the N-terminal region, which may represent the phorbol ester-binding region. Just N-terminal to the first Cys-rich repeat is a putative pseudosubstrate site.

Database accession numbers

	PIR	SWISSPROT	EMBL/GENBANK	REF
TPA-1		P34722	D14815	4

Partial amino acid sequence of *C. elegans* PKC

Note: the N-terminus of TPA-1 has not been determined

```
IQRRRGAIKH ARVHEIRGHQ FVATFFRQPH FCSLCSDFMW GLNKQGYQCQ  50
LCSAAVHKKC HEKVIMQCPG SAKNTKETMA LKERFKVDIP HRFKTYNFKS 100
PTFCDHCGSM LYGLFKQGLR CEVCNVACHH KCERLMSNLC GVNQKQLSEM 150
YHEIKRGTHA TASCPPNIAN LHLNGETSKN NGSLPNKLKN LFKSHQYSVE 200
EQKETDEYMD NIWGGGDGPV KKFALPHFNL LKVLGKGSFG KVMLVELKGK 250
```

Partial amino acid sequence of *C. elegans* PKC *continued*

```
NEFYAMKCLK KDVILEDDDT ECTYIERRVL ILASQCPFLC QLFCSFQTNE 300
YLFFVMEYLN GGDLMHHIQQ IKKFDEARTR FYACEIVVAL QFLHTNNIIY 350
RDLKLDNVLL DCDGHIKLAD FGMAKTEMNR ENGMASTFCG TPDYISPEII 400
KGQLYNEAVD FWSFGVLMYE MLVGQSPFHG EGEDELFDSI LNERPYFPKT 450
ISKEAAKCLS ALFDRNPNTR LGMPECPDGP IRQHCFFRGV DWKRFENRQV 500
PPPFKPNIKS NSDASNFDDD FTNEKAALTP VHDKNLLASI DPEAFLNFSY 550
TNPHFSK
```

Sites of interest: A7, putative pseudosubstrate site; C32, C35, C49, C52, C60, C68, cysteines in first Cys-rich repeat; C104, C107, C121, C124, C132, C140, cysteines in second Cys-rich repeat

Homologues in other species
TPA-1 is homologous to PKC isoforms in other species (**PKC, DmPKC, ScPKC**).

Assay
TPA-1 may be assayed using histone H1 as for mammalian PKC[5].

References
[1] Miwa, J. et al. (1980) Igaku no ayumi 114, 910–912.
[2] Miwa, J. et al. (1982) J. Cancer Res. Clin. Oncol. 104, 81–87.
[3] Tabuse, Y. et al. (1983) Carcinogenesis 4, 783–786.
[4] Tabuse, Y. et al. (1989) Science 243, 1713–1716.
[5] Sassa, T. and Miwa, J. (1992) Biochem. J. 282, 219–223.

David E. Levin (Johns Hopkins University, USA)

S. cerevisiae Pkc1 is related to the Ca^{2+}-dependent subtypes (α, β and γ) of mammalian protein kinase C[1]. Loss of Pkc1 function results in cell lysis due to a defect in cell wall construction[2,3]. Pkc1 has been proposed, based on genetic suppression and epistasis experiments, to mediate a protein kinase cascade[4] in which Pkc1 activates **Bck1**[5], which activates a redundant pair of MAP kinase kinases **Mkk1/2**[6], which activate the MAP kinase encoded by *MPK1*[7]. Pkc1 phosphorylates the Bck1 protein kinase *in vitro*. Conditional *pkc1* mutants are rescued under restrictive conditions by $CaCl_2$[2] and are hyper-sensitive to growth inhibition by the protein kinase inhibitor staurosporine[8].

Subunit structure

SUBUNIT	AMINO ACIDS	MOL. WT	SDS-PAGE
Pkc1	1151	131 524	150 kDa

Genetics

Chromosomal location: leftmost marker on chromosome II[1].
Mutations:
    K853R, inactive; N834I, Ca^{2+}-dependent; L887S, P1023L, temperature-sensitive; R398A, hyperactive.

Domain structure

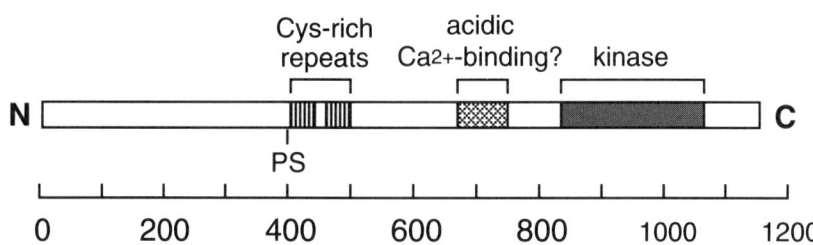

Database accession number

	PIR	SWISSPROT	EMBL/GENBANK	REF
Pkc1			M32491	1

Amino acid sequence of *S. cerevisiae* Pkc1

```
MSFSQLEQNI KKKIAVEENI IRGASALKKK TSNVMVIQKC NTNIREARQN   50
LEYLEDSLKK LRLKTAQQSQ GENGSEDNER CNSKEYGFLS TKSPNEHIFS  100
RLDLVKYDCP SLAQRIQYML QQLEFKLQVE KQYQEANTKL TKLYQIDGDQ  150
RSSSAAEGGA MESKYRIQML NKALKKYQAI NVDFDQFKHQ PNDIMDNQQP  200
KFRRKQLTGV LTIGITAARD VDHIQSPMFA RKPESYVTIK IDDSIKARTK  250
PSRNDRWSED FQIPVEKGNE IEITVYDKVN DSLIPVAIMW LLLSDIAEEI  300
RKKKAGQTNE QQGWVNASNI NGGSSLASEE GSTLTSTYSN SAIQSTSAKN  350
```

Amino acid sequence of *S. cerevisiae* Pkc1 *continued*

```
VQGENTSTSQ ISTNSWFVLE PSGQILLTLG FHKSSQIERK QLMGGLHRHG  400
AIINRKEEIF EQHGHHFVQK SFYNIMCCAY CGDFLRYTGF QCQDCKFLCH  450
KKCYTNVVTK CIAKTSTDTD PDEAKLNHRI PHRFLPTSNR GTKWCCHCGY  500
ILPWGRHKVR KCSECGIMCH AQCAHLVPDF CGMSMEMANK ILKTIQDTKR  550
NQEKKKRTVP SAQLGSSIGT ANGSDLSPSK LAERANAPLP PQPRKHDKTP  600
SPQKVERDSP TKQHDPIIDK KIPLQTHGRE KLNKFIDENE AYLNFTEGAQ  650
QTAEFSSPEK TLDPTSNRRS LGLTDLSIEH SQTWESKDDL MRDELELWKA  700
QREEMELEIK QDSGEIQEDL EVDHIDLETK QKLDWENKND FREADLTIDS  750
THTNPFRDMN SETFQIEQDH ASKEVLQETV SLAPTSTHAS RTTDQQSPQK  800
SQTSTSAKHK KRAAKRRKVS LDNFVLLKVL GKGNFGKVIL SKSKNTDRLC  850
AIKVLKKDNI IQNHDIESAR AEKKVFLLAT KTKHPFLTNL YCSFQTENRI  900
YFAMEFIGGG DLMWHVQNQR LSVRRAKFYA AEVLLALKYF HDNGVIYRDL  950
KLENILLTPE GHIKIADYGL CKDEMWYGNR TSTFCGTPEF MAPEILKEQE 1000
YTKAVDWWAF GVLLYQMLLC QSPFSGDDED EVFNAILTDE PLYPIDMAGE 1050
IVQIFQGLLT KDPEKRLGAG PRDAAEVMEE PFFRNINFDD ILNLRVKPPY 1100
IPEIKSPEDT SYFEQEFTSA PPTLTPLPSV LTTSQQEEFR GFSFMPDDLD 1150
L
```

Physiological substrates

The physiological substrate is thought to be **Bck1**, which is phosphorylated on serine and threonine residues.

Assay
In vitro Pkc1 phosphorylates the pseudosubstrate site peptide (GGLHRHGS/TIINRKEE) and myelin basic protein.

References
[1] Levin, D.E. et al. (1990) Cell 62, 213–224.
[2] Levin, D.E. and Bartlett-Heubusch, E. (1992) J. Cell Biol. 116, 1221–1229.
[3] Paravicini, G. et al. (1992) Mol. Cell. Biol. 12, 4896–4905.
[4] Errede, B. and Levin, D.E. (1993) Curr. Opin. Cell Biol. 5, 254–260.
[5] Lee, K.S. and Levin, D.E. (1992) Mol. Cell. Biol. 12, 172–182.
[6] Irie, K. et al. (1993) Mol. Cell. Biol. 13, 3076–3083.
[7] Lee, K.S. et al. (1993) Mol. Cell. Biol. 13, 3067–3075.
[8] Yoshida, S. et al. (1992) Mol. Gen. Genet. 231, 337–344.

RAC

RAC PK (vertebrates and D. melanogaster) (akt, PKB)

Evan Ingley and Brian A. Hemmings
(Friedrich Miescher-Institut, Basel, Switzerland)

The RAC (Related to PKA and PKC) PK subfamily was initially identified by homology cloning[1]. Further work has established that an oncogenic form (v-akt) exists[2], resulting from the addition of a truncated *gag* protein 6 bp upstream of the initiator methionine. Sequence analysis of the RAC kinase domain reveals that it is related to the PKA and PKC subfamilies, and to Sgk kinase[3].

Subunit structure and isoforms

SUBUNIT	AMINO ACIDS	MOL. WT	SDS-PAGE
α	480	55716	59 kDa
β1	520	60241	
β2	481	55768	
DmRAC	530	59910	66/85 kDa
v-*akt*		86360	105 kDa

Three isoforms of RAC PK have been identified in vertebrates (α, β1 and β2), and a single form in *D. melanogaster* (DmRAC). The β2 isoform has a different C-terminus to β1, which may arise from alternative splicing. At present no regulatory subunits have been identified.

Genetics

Chromosomal locations:

Human α[2]:	14q32
Human β1/β2[4]:	19q13.1–13.2
D. melanogaster:	Chromosome 3 at 89B

Domain structure

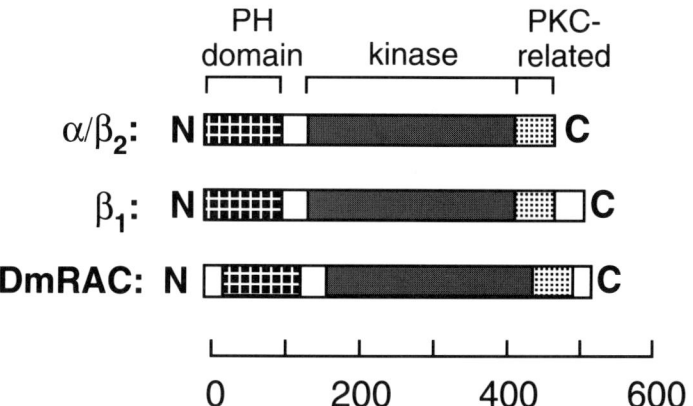

Three domains have been identified: (1) an N-terminal pleckstrin homology (PH) domain[5]; (2) a kinase domain closely related to that of PKA and PKC; (3) a C-terminal domain similar to the C-terminal region of the PKC subfamily. Pleckstrin, the major PKC substrate in platelets, contains N- and C-terminal regions of ~100 amino acids with very similar sequences. PH domains have been identified in a number of other proteins, and are proposed to function by promoting protein–protein interactions.

Database accession numbers

		PIR	SWISSPROT	EMBL/GENBANK	REF
α	Human			M63167	1
	Bovine			X61036	7
	Murine			M94335	
β1	Human			M77198	6
β2	Human			M95936	4
v-*akt*				M61767	2
C. elegans				M88917	8

Amino acid sequences of human RAC

α isoform

```
MSDVAIVKEG WLHKRGEYIK TWRPRYFLLK NDGTFIGYKE RPQDVDQREA  50
PLNNFSVAQC QLMKTERPRP NTFIIRCLQW TTVIERTFHV ETPEEREEWT 100
TAIQTVADGL KKQEEEEMDF RSGSPSDNSG AEEMEVSLAK PKHRVTMNEF 150
EYLKLLGKGT FGKVILVKEK ATGRYYAMKI LKKEVIVAKD EVAHTLTENR 200
VLQNSRHPFL TALKYSFQTH DRLCFVMEYA NGGELFFHLS RERVFSEDRA 250
RFYGAEIVSA LDYLHSEKNV VYRDLKLENL MLDKDGHIKI TDFGLCKEGI 300
KDGATMKTFC GTPEYLAPEV LEDNDYGRAV DWWGLGVVMY EMMCGRLPFY 350
NQDHEKLFEL ILMEEIRFPR TLGPEAKSLL SGLLKKDPKQ RLGGGSEDAK 400
EIMQHRFFAG IVWQHVYEKK LSPPFKPQVT SETDTRYFDE EFTAQMITIT 450
PPDQDDSMEC VDSERRPHFP QFSYSASSTA
```

β1 isoform

```
MNEVSVIKEG WLHKRGEYIK TWRPRYFLLK SDGSFIGYKE RPEAPDQTLP  50
PLNNFSVAEC QLMKTERPRP NTFVIRCLQW TTVIERTFHV DSPDEREEWM 100
RAIQMVANSL KQRAPGEDPM DYKCGSPSDS STTEEMEVAV SKARAKVTMN 150
DFDYLKLLGK GTFGKVILVR EKATGRYYAM KILRKEVIIA KDEVAHTVTE 200
SRVLQNTRHP FLTALKYAFQ THDRLCFVME YANGGELFFH LSRERVFTEE 250
RARFYGAEIV SALEYLHSRD VVYRDIKLEN LMLDKDGHIK ITDFGLCKEG 300
ISDGATMKTF CGTPEYLAPE VLEDNDYGRA VDWWGLGVVM YEMMCGRLPF 350
YNQDHERLFE LILMEEIRFP RTLSPEAKSL LAGLLKKDPK QRLGGGPSDA 400
KEVMEHRFFL SINWQDVVQK KLLPPFKPQV TSEVDTRYFD DEFTAQSITI 450
TPPDRYDSLG LLELDQRTHF PQFSYSAFRE EKDLLMSLFV SLILFSDFSS 500
LKSHSFSSNF ILLSFSSLKK
```

Sites of interest: F478–K520, in the β2 isoform, replaced by SIRE

Amino acid sequence of *D. melanogaster* RAC

```
MSINTTFDLS SPSVTSGHAL TEQTQVVKEG WLMKRGEHIK NWRQRYFVLH  50
SDGRLMGYRS KPADSASTPS DFLLNNFTVR GCQIMTVDRP KPFTFIIRGL 100
QWTTVIERTF AVESELERQQ WTEAIRNVSS RLIDVGEVAM TPSEQTDMTD 150
VDMATIAEDE LSEQFSVQGT TCNSSGVKKV TLENFEFLKV LGKGTFGKVI 200
LCREKATAKL YAIKILKKEV IIQKDEVAHT LTESRVLKST NHPFLISLKY 250
SFQTNDRLCF VMQYVNGGEL FWHLSHERIF TEDRTRFYGA EIISALGYLH 300
SQGIIYRDLK LENLLLDKDG HIKVADFGLC KEDITYGRTT KTFCGTPEYL 350
APEVLDDNDY GQAVDWWGTG VVMYEMICGR LPFYNRDHDV LFTLILVEEV 400
KFPRNITDEA KNLLAGLLAK DPKKRLGGGK DDVKEIQAHP FFASINWTDL 450
VLKKIPPPFK PQVTSDTDTR YFDKEFTGES VELTPPDPTG PLGSIAEEPL 500
FPQFSYQGDM ASTLGTSSHI STSTSLASMQ
```

Homologues in other species

RAC has been sequenced from the human (α isoform[1] MCF-7 and WI38 cells; β1 isoform[6] MCF-7 and WI38 cells; β2 isoform[4] thymus), mouse (α isoform, unpublished), bovine (α isoform[7]), pig (α and β isoforms LLC-PK$_1$ cells, unpublished), *D. melanogaster* (unpublished), *C. elegans*[8] and viral v-*akt*[2].

Physiological substrates and specificity determinants

The major phosphorylation site in myelin basic protein is S55, within the sequence RGSGKDGHHAAR. Data using model peptides indicate that basic residues are required on both the N- and C-terminal sides of the phosphorylated serine.

Assay

The enzyme has been assayed in immunoprecipitates by the transfer of radioactivity from [γ-^{32}P]ATP to myelin basic protein[1].

Pattern of expression

The mRNA for the α, β1 and β2 isoforms have been detected in all cells and tissues examined, though at different levels. The α isoform appears to be predominant.

References

1. Jones, P.F. et al. (1991) Proc. Natl Acad. Sci. USA 88, 4171–4175.
2. Bellacosa, A. et al. (1991) Science 254, 274–277.
3. Webster, M.K. et al. (1993) Mol. Cell. Biol. 13, 2031–2040.
4. Cheng, J.Q. et al. (1992) Proc. Natl Acad. Sci. USA 89, 9267–9271.
5. Haslam, R.J. et al. (1993) Nature 363, 309–310.
6. Jones, P.F. et al. (1991) Cell Regulation 2, 1001–1009.
7. Coffer, P.J. et al. (1991) Eur. J. Biochem. 201, 475–481.
8. Waterson, R. et al. (1992) Nature Genet. 1, 114–123.

 **β-Adrenergic receptor kinases (vertebrates)**

James Inglese, Richard T. Premont, Walter J. Koch, Julie Pitcher and
Robert J. Lefkowitz (Duke University Medical Center, USA)

Persistent stimulation of G protein-coupled receptors frequently leads to a period of diminished responsiveness. This process, termed desensitization, is mediated in part by βARK-mediated phosphorylation of the receptor. β-Adrenergic receptor kinases, two isoforms of which are currently known (βARK-1 and -2), phosphorylate the agonist-occupied forms of several G protein-coupled receptors. Phosphorylation is thought to be followed by the binding of another protein, β-arrestin, which diminishes the functional coupling of the receptor to the G protein. This mechanism thereby leads to homologous, or agonist-specific, desensitization[1].

Subunit structure and isoforms

SUBUNIT	AMINO ACIDS	MOL. WT	SDS-PAGE
βARK-1	689	79 555	80 kDa
βARK-2	688	79 712	76 kDa

Both isoforms are monomers.

Genetics

Chromosomal location
Human βARK-1:   between 11q13 and 11(cent)[2]
Mouse βARK-1:   19[3]
Mouse βARK-2:   5[3]

Domain structure

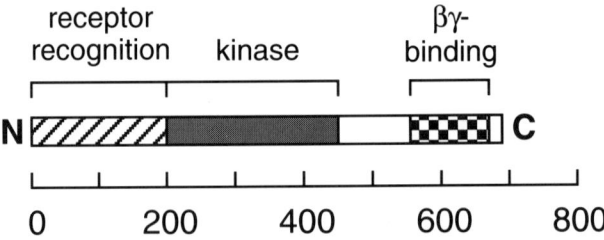

The kinase domain is in the centre, with an N-terminal region which is believed to contain recognition sites for activated receptor, and a C-terminal region which binds the βγ subunits of heterotrimeric G proteins[5]. In subdomain VII of the catalytic domain, L336 replaces the conserved phenylalanine of the DFG consensus found in the majority of serine/threonine protein kinases[4]. Such a substitution has been observed only among members of the G protein-coupled receptor kinases.

Database accession numbers

		PIR	SWISSPROT	EMBL/GENBANK	REF
1	Bovine	A40088	P21146	M34019	3
2	Bovine	A39336	P26818	M73216	4

Amino acid sequences of human βARK

βARK-1

```
MADLEAVLAD VSYLMAMEKS KATPAARASK KILLPEPSIR SVMQKYLEDR  50
GEVTFEKIFS QKLGYLLFRD FCLKHLEEAK PLVEFYEEIK KYEKLETEEE 100
RLVCSREIFD TYIMKELLAC SHPFSKSAIE HVQGHLVKKQ VPPDLFQPYI 150
EEICQNLRGD VFQKFIESDK FTRFCQWKNV ELNIHLTMND FSVHRIIGRG 200
GFGEVYGCRK ADTGKMYAMK CLDKKRIKMK QGETLALNER IMLSLVSTGD 250
CPFIVCMSYA FHTPDKLSFI LDLMNGGDLH YHLSQHGVFS EADMRFYAAE 300
IILGLEHMHN RFVVYRDLKP ANILLDEHGH VRISDLGLAC DFSKKKPHAS 350
VGTHGYMAPE VLQKGVAYDS SADWFSLGCM LFKLLRGHSP FRQHKTKDKH 400
EIDRMTLTMA VELPDSFSPE LRSLLEGLLQ RDVNRRLGCL GRGAQEVKES 450
PFFRSLDWQM VFLQKYPPPL IPPRGEVNAA DAFDIGSFDE EDTKGIKLLD 500
SDQELYRNFP LTISERWQQE VAETVFDTIN AETDRLEARK KTKNKQLGHE 550
EDYALGKDCI MHGYMSKMGN PFLTQWQRRY FYLFPNRLEW RGEGEAPQSL 600
LTMEEIQSVE ETQIKERKCL LLKIRGGKQF VLQCDSDPEL VQWKKELRDA 650
YREAQQLVQR VPKMKNKPRS PVVELSKVPL IQRGSANGL
```

βARK-2

```
MADLEAVLAD VSYLMAMEKS KATPAARASK KIVLPEPSIR SVMQKYLEER  50
HEITFDKIFN QRIGFLLFKD FCLNEINEAV PQVKFYEEIK EYEKLENEED 100
RLCRSRQIYD TYIMKELLSC SHPFSKQAVE HVQSHLSKKQ VTSTLFQPYI 150
EEICESLRCS IFQKFMESDK FTRFCQWKNV ELNIHLTMND FSVHRIICRC 200
GFGEVYGCRK ADTGKMYAMK CLDKKRIKMK QGETLALNER IMLSLVSTGD 250
CPFIVCMTYA FHTPDKLCFI LDLMNGGDLH YHLSQHGVFS EKEMRFYATE 300
IILGLEHMHN RFVVYRDLKP ANILLDEHGH VRISDLGLAC DFSKKKPHAS 350
VGTHGYMAPE VLQKGTAYDS SADWFSLGCM LFKLLRGHSP FRQHKTKDKH 400
EIDRMTLTMN VELPDVFSPE LKSLLEGLLQ RDVSKRLGCH GGSAQELKTH 450
DFFRGIDWQH VYLQKYPPPL IPPRGEVNAA DAFDIGSFDE EDTKGIKLLD 500
CDQELYKNFP LVISERWQQE VAETVYEAVN ADTDKIEARK RAKNKQLGHE 550
EDYALGRDCI VHGYMLKLGN PFLTQWQRRY FYLFPNRLEW RGEGESRQSL 600
LTMEQIVSVE ETQIKDKKCI LLRIKGGKQF VLQCESDPEF VQWKKELTET 650
FMEAQRLLRR APKFLNKSRS AVVELSKPPL CHRNSNGL
```

Homologues in other species

βARK-1 and -2 have been cloned from cow[3, 4], human[2, 7] and rat[6]. The GPRK-1 sequence cloned from *D. melanogaster*[8] is very similar to βARK-1/2.

Physiological substrates and specificity determinants

The β-adrenergic receptor kinases phosphorylate G protein-coupled receptors. One important feature of this phosphorylation event is that only the agonist-occupied, "active" form of the receptor acts as a substrate. *In vitro*, the βARK isoforms are capable of phosphorylating a number of purified reconstituted receptors (rhodopsin, $β_2$-adrenergic, $α_2$-adrenergic and muscarinic cholinergic receptors). Inhibition of these enzymes in permeabilized cell systems has demonstrated their role in regulating $β_2$-adrenergic and olfactory receptor function[9–11]. These observations, together with the relatively broad tissue distribution of the βARK isoforms (see below), suggest that they play a general role in phosphorylating and regulating the function of a wide range of G protein-coupled receptors. Utilizing synthetic peptide substrates it has been demonstrated that βARK preferentially phosphorylates serine and threonine residues located on the C-terminal side of acidic or phosphoserine residues.

However, peptides represent poor substrates for this enzyme, the V_{max}/K_m ratio being 10^4- to 10^7-fold lower than for receptor proteins. Interaction of βARK with activated (but not inactive) receptors significantly enhances the rate of peptide phosphorylation, a phenomenon observed even with activated receptors lacking the C-terminal phosphorylation sites. In addition, synthetic peptides derived from the first intracellular loop and the N-terminus of the third intracellular loop of the β_2-adrenergic receptor inhibit phosphorylation of the receptor[12]. Thus, although the nature of the environment immediately surrounding the phosphorylation site may be important, the major specificity determinants for the kinase appear to reside in the conformation of the activated receptor.

Assay
βARK activity is normally assayed as the agonist-dependent transfer of phosphate from $[\gamma\text{-}^{32}P]ATP$ to purified, reconstituted β_2AR. However the light-dependent transfer of phosphate to urea-washed, rod outer segment membranes can also be utilized as an assay[4].

Enzyme inhibitors and regulators[5,13]

INHIBITORS	IC_{50}
Heparin	0.77 µg/ml (0.15 µM)
Heparin sulphate	1.8 µg/ml
Dextran sulphate	0.76 µg/ml (0.15 µM)
Chondroitin sulphate C	6 µg/ml
Polyaspartic acid	14 µg/ml (1.3 µM)
REGULATORS	
G protein βγ subunits	50 nM

Pattern of expression
βARK isozymes are widely distributed as indicated by Northern blot analysis and RNAase protection assays. These methods have detected βARK-1 and -2 mRNA in most tissues examined, including the central nervous system[3,17]. Through in situ mRNA hybridization and immunohistochemistry, both βARK-1 and -2 were found to be primarily expressed in neurones, and were widely distributed throughout rat brain[6]. Overall, the βARK isozymes have a regional and subcellular distribution consistent with a general role in the desensitization of synaptic receptors. Evidence for tissue specificity of isozymes is limited to the recent finding that βARK-2 is enriched in olfactory cells, and appears to be the kinase involved in odorant signal transduction[10,11]. βARK-1 and -2 are cytosolic enzymes which rapidly distribute to the plasma membrane following stimulus-mediated receptor activation[14]. This membrane translocation event involves a specific

interaction between *β*ARK and the *βγ* subunits of heterotrimeric G proteins[5]. A geranylgeranyl moiety on the *γ* subunit is required for this agonist-dependent membrane translocation[5, 15]. The specific *βγ*-binding site on *β*ARK has been localized to a 125 amino acid region (Q546–S670) within the C-terminal domain[16].

References

1 Hausdorff, W.P. et al. (1990) FASEB J. 4, 2881–2889.
2 Benovic, J.L. et al. (1991) FEBS Lett. 283, 122–126.
3 Benovic, J.L. et al. (1991) J. Biol. Chem. 266, 14939–14946.
4 Benovic, J.L. et al. (1989) Science 246, 235–240.
5 Pitcher, J.A. et al. (1992) Science 257, 1264–1267.
6 Arriza, J.L. et al. (1992) J. Neurosci. 12, 4045–4055.
7 Parruti, G. et al. (1993) Biochem. Biophys. Res. Commun. 190, 475–481.
8 Cassill, J.A. et al. (1991) Proc. Natl Acad. Sci. USA 88, 11067–11070.
9 Lohse, M.J. et al. (1989) Proc. Natl Acad. Sci. USA 86, 3011–3015.
10 Schleicher, S. et al. (1993) Proc. Natl Acad. Sci. USA 90, 1420–1424.
11 Dawson, T.M. et al. (1993) Science 259, 825–829.
12 Chen, C.-Y. et al. (1993) J. Biol. Chem. 268, 7825–7831.
13 Benovic, J.L. et al. (1989) J. Biol. Chem. 264, 6707–6710.
14 Strasser, R.H. et al. (1986) Proc. Natl Acad. Sci. USA 83, 6362–6366.
15 Inglese, J. et al. (1992) Nature 359, 147–150.
16 Koch, W.J. et al. (1993) J. Biol. Chem. 268, 8256–8260.
17 Benovic, J.L. et al. (1986) Proc. Natl Acad. Sci. USA 83, 2797–2801.

James Inglese, Richard T. Premont, Walter J. Koch, Julie Pitcher and Robert J. Lefkowitz (Duke University Medical Center, USA)

The visual signalling cascade proceeds from the interaction of a photon with rhodopsin, to activation of transducin, then cGMP phosphodiesterase, and ultimately the gating of a cGMP-sensitive Na^+ channel. Signalling is quenched by the actions of a serine/threonine protein kinase, rhodopsin kinase, which phosphorylates the light bleached, or activated, conformation of rhodopsin (metarhodopsin II or Rho*). Subsequently, arrestin (48 kDa protein) binds to the phosphorylated rhodopsin and sterically interdicts its interactions with transducin[1].

Subunit structure and isoforms

SUBUNIT	AMINO ACIDS	MOL. WT	SDS-PAGE
RhK	558	62 862	63 kDa

RhK is a monomer.

Domain structure

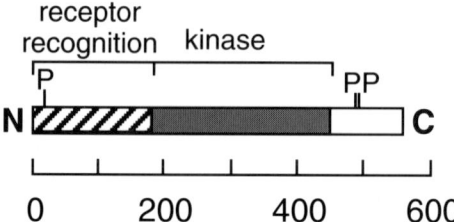

The catalytic domain is centrally located between the N-terminal domain, believed to contain the recognition sites for Rho*, and the C-terminal domain. In subdomain VII of the catalytic domain L333 replaces the conserved F of the DFG motif found in the majority of serine/threonine protein kinases. Such a substitution has been observed only among members of the G protein-coupled receptor kinases[2]. Three sites of autophosphorylation have been identified in RhK, two major sites in the C-terminus and one minor site in the N-terminus[3]. The C-terminus of the newly translated protein ends in the tetrapeptide sequence CVLS. This 'CAAX' motif signals a series of post-translational modifications which include farnesylation of the cysteine, proteolysis of the VLS sequence, and carboxymethylation of the newly exposed S-farnesyl-Cys C-terminus[4].

Database accession numbers

	PIR	SWISS-PROT	EMBL/GENBANK	REF
RhK Bovine	A41365	P28327	M73836	2

Amino acid sequence of bovine RhK

```
MDFGSLETVV ANSAFIAARG SFDASSGPAS RDRKYLARLK LPPLSKCEAL   50
RESLDLGFEG MCLEQPIGKR LFQQFLRTHE QHGPALQLWK DIEDYDTADD  100
ALRPQKAQAL RAAYLEPQAQ LFCSFLDAET VARARAGAGD GLFQPLLRAV  150
LAHLGQAPFQ EFLDSLYFLR FLQWKWLEAQ PMGEDWFLDF RVLGRGGFGE  200
VFACQMKATG KLYACKKLNK KRLKKRKGYQ GAMVEKKILA KVHSRFIVSL  250
AYAFETKTDL CLVMTIMNGG DIRYHIYNVD EDNPGFQEPR AIFYTAQIVS  300
GLEHLHQRNI IYRDLKPENV LLDDDGNVRI SDLGLAVELK AGQTKTKGYA  350
GTPGFMAPEL LLGEEYDFSV DYFALGVTLY EMIAARGPFR ARGEKVENKE  400
LKQRVLEQAV TYPDKFSPAS KDFCEALLQK DPEKRLGFRD GSCDGLRTHP  450
LFRDISWRQL EAGMLTPPFV PDSRTVYAKN IQDVGAFSTV KGVAFEKADT  500
EFFQEFASGT CPIPWQEEMI ETGVFGDLNV WRPDGQMPDD MKGVSGQEAA  550
PSSKSGMCVLS
```

Sites of interest: M1, blocked; S21, S488, T489, autophosphorylation; C558, farnesylation/carboxymethylation; C558–S561, 'CAAX' box; V559–S561, proteolytically removed after farnesylation.

Homologues in other species

Only the bovine sequence is determined to date[2].

Physiological substrates and specificity determinants

RhK catalyses phosphorylation of photolysed rhodopsin (Rho*), presumably the metarhodopsin II conformer, at multiple serine and threonine residues located exclusively in the C-terminal region. In addition to phosphorylating rhodopsin, RhK phosphorylates *in vitro* the red pigment, iodopsin. The substrate specificity of RhK appears, therefore not to be dictated by amino acid sequence but by the conformation of photolysed rhodopsin. RhK preferentially phosphorylates serine and threonine residues located in an acidic environment. Synthetic peptide substrates containing acidic residues C-terminal to the phosphorylated serine represent the most favourable substrates. However, all synthetic peptide substrates are poorly phosphorylated in comparison to Rho*. RhK has been shown to interact only weakly with the C-terminal region of Rho*. Rather, the enzyme appears to bind to the 3rd intracellular loop of Rho*, an interaction which results in RhK activation[5].

Assay

RhK is assayed as a light-dependent transfer of radioactivity from [γ-^{32}P]ATP to urea-washed, rod outer segment membranes[6].

Enzyme activators and inhibitors[7, 8]

Activators:
Polycations such as spermine or spermidine activate the kinase. At 5 mM spermidine activates to 210%. Spermine doubles activity at 3 mM with decreasing activation at higher concentrations.

INHIBITORS	K_i	IC_{50}
Sangivamycin	180 nM	
Poly(aspartic acid)	300 μM	
Dextran sulphate		200 μg/ml
Poly(adenylic acid)		40 μM
Heparin		200 μM
Chondroitin sulphate C		6 mg/ml

Pattern of expression

The localization of RhK mRNA is highly specific to the retina[2]. It is also present in lower amounts in the pineal gland, consistent with other biochemical similarities between the mammalian pineal gland and retina[2]. This mRNA specificity suggests that rhodopsin is the primary *in vivo* substrate of the kinase. Immunocytochemistry has shown that RhK is present in both rods and cones[11]. RhK is distributed primarily in the cytoplasm although, upon rhodopsin activation by light the kinase rapidly distributes to the plasma membrane[9]. The light-induced membrane trans-location of RhK has been shown to be dependent on the farnesyl isoprenoid which is covalently attached to the C-terminus of the enzyme[4]. Mutagenesis studies have revealed that the translocation of RhK is farnesyl-dependent, as non-isoprenylated or geranylgeranylated forms of the kinase do not undergo light-dependent translocation to rhodopsin-enriched rod outer disc membranes[10].

References

1 Hargrave, P.A. and McDowell, J.H. (1992) FASEB J. 6, 2323–2331.
2 Lorenz, W. et al. (1991) Proc. Natl Acad. Sci. USA 88, 8715–8719.
3 Palczewski, K. et al. (1992) J. Biol. Chem. 267, 18991–18998.
4 Inglese, J. et al. (1992) J. Biol. Chem. 267, 1422–1425.
5 Palczewski, K. et al. (1991) J. Biol. Chem. 266, 12949–12955.
6 Wilden, U. and Kuhn, H. (1982) Biochemistry 21, 3014–3022.
7 Palczewski, K. et al. (1990) Biochemistry 29, 6276–6282.
8 Palczewski, K. et al. (1989) Biochemistry 28, 8764–8770.
9 Kuhn, H. (1978) Biochemistry 21, 4389–4395.
10 Inglese, J. et al. (1992) Nature 359, 147–150.
11 Palczewski, K. et al. (1993) J. Biol. Chem. 268, 6004–6013.

Stevan Marcus (University of Texas M.D. Anderson Cancer Center, USA)

The *SCH9* gene was cloned from *S. cerevisiae* as a suppressor of a *cdc25* temperature-sensitive mutant[1]. Disruption of the *SCH9* gene results in a slow growth defect which can be suppressed by activation of the Ras-regulated cAMP pathway. When overexpressed, *SCH9* is capable of suppressing loss of the yeast **ScPKA**-encoding genes, *TPK1, 2* and *3*, as well as loss of function of *RAS* and *CYR1* (which encodes adenylyl cyclase). Sch9 may be a component of a growth control pathway that is partially redundant with the RAS-regulated cAMP pathway in *S. cerevisiae*.

Subunit structure and isoforms

SUBUNIT	AMINO ACIDS	MOL. WT	SDS-PAGE
Sch9	824	91 799	

Domain structure

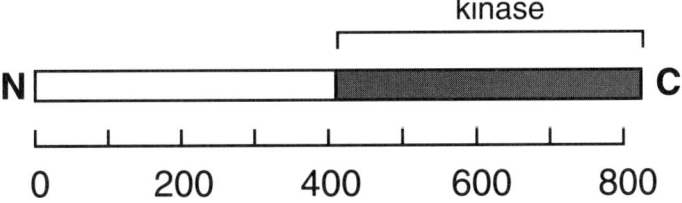

The kinase domain is homologous to mammalian **PKA** and yeast **ScPKA**[1]. The regulatory domain does not exhibit homology to any known protein.

Database accession numbers

	PIR	SWISSPROT	EMBL/GENBANK	REF
Sch9	A28429	P11792	X12560	1

Amino acid sequence of *S. cerevisiae* Sch9

```
MMNFFTSKSS NQDTGFSSQH QHPNGQNNGN NNSSTAGNDN GYPCKLVSSG  50
PCASSNNGAL FTNFTLQTAT PTTAISQDLY AMGTTGITSE NALFQMKSMN 100
NGISSVNNNN SNTPTIITTS QEETNAGNVH GDTGGNSLQN SEDDNFSSSS 150
TTKCLLSSTS SLSINQREAA AAAYGPDTDI PRGKLEVTII EARDLVTRSK 200
DSQPYVVCTF ESSEFISNGP ESLGAINNNN NNNNNNQHNQ NQHINNNNEN 250
TNPDAASQHH NNNSGWNGSQ LPSIKEHLKK KPLYTHRSSS QLDQLNSCSS 300
VTDPSKRSSN SSSGSSNGPK NDSSHPIWHH KTTFDVLGSH SELDISVYDA 350
AHDHMFLGQV RLYPMSHNLA HASQHQWHSL KPRVIDEVVS GDILIKWTYK 400
QTKKRHYGPQ DFEVLRLLGK GTFGQVYQVK KKDTQRIYAM KVLSKKVIVK 450
KNEIAHTIGE RNILVTTASK SSPFIVGLKF SFQTPTDLYL VTDYMSGGEL 500
FWHLQKEGRF SEDRAKFYIA ELVLALEHLH DNDIVYRDLK PENILLDANG 550
NIALCDFGLS KADLKDRTNT FCGTTEYLAP ELLLDETGYT KMVDFWSLGV 600
LIFEMCCGWS PFFAENNQKM YQKIAFGKVK FPRDVLSQEG RSFVKGLLNR 650
```

Amino acid sequence of *S. cerevisiae* Sch9 *continued*

```
NPKHRLGAID DGRELRAHPF FADIDWEALK QKKIPPPFKP HLVSETDTSN 700
FDPEFTTAST SYMNKHQPMM TATPLSPAMQ AKFAGFTFVD ESAIDEHVNN 750
KRKFLQNSYF MEPGSFIPGN PNLPPDEDVI DDDGDEDIND GFNQEKNMNN 800
SHSQMDFDGD QHMDDEFVSG RFEI
```

Reference
[1] Toda, T. et al. (1988) Genes Devel. 2, 517–527.

Ykr2 Ykr2 PK (*S. cerevisiae*)

Shigeo Ohno (Yokohama City University, Japan) and
Koichi Suzuki (Tokyo University, Japan)

The *YKR2* gene was identified by homology probing using mammalian PKC cDNA, and the protein product is not characterized. The kinase domain shows the highest sequence identity to PKC and PKA. The *YKR2* gene is not essential for growth and sporulation. Disruption of the gene had no effect on the growth of haploid or diploid cells, and homozygous diploid cells showed normal sporulation (Matsumoto, S., Kubo, K., Ohno, S., Suzuki, K. et al., unpublished results).

Subunit structure and isoforms

SUBUNIT	AMINO ACIDS	MOL. WT	SDS-PAGE
Ykr2	677	76577	

Domain structure

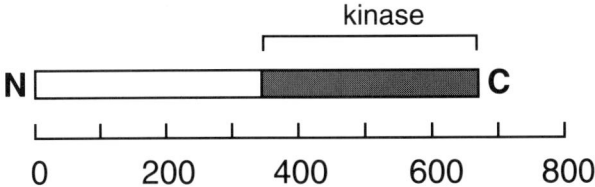

Database accession numbers

	PIR	SWISSPROT	EMBL/GENBANK	REF
Ykr2	JS0178	P18961	M24929	*1*

Amino acid sequence of *S. cerevisiae* Ykr2

```
MHSWRISKFK LGRSKEDDGS SEDENEKSWG NGLFHFHHGE KHHDGSPKNH  50
NHEHEHHIRK INTNETLPSS LSSPKLRNDA SFKNPSGIGN DNSKASERKA 100
SQSSTETQGP SSESGLMTVK VYSGKDFTLP FPITSNSTIL QKLLSSGILT 150
SSSNDASEVA AIMRQLPRYK RVDQDSAGEG LIDRAFATKF IPSSILLPGS 200
TNSSPLLYFT IEFDNSITTI SPDMGTMEQP VFNKISTFDV TRKLRFLKID 250
VFARIPSLLL PSKNWQQEIG EQDEVLKEIL KKINTNQDIH LDSFHLPLNL 300
KIDSAAQIRL YNHHWISLER GYGKLNITVD YKPSKNKPLS IDDFDLLKVI 350
GKGSFGKVMQ VRKKDTQKIY ALKALRKAYI VSKCEVTHTL AERTVLARVD 400
CPFIVPLKFS FQSPEKLYLV LAFINGGELF YHLQHEGRFS LARSRFYIAE 450
LLCALDSLHK LDVIYRDLKP ENILLDYQGH IALCDFGLCK LNMKDNDKTD 500
TFCGTPEYLA PEILLGQGYT KTVDWWTLGI LLYEMMTGLP PYYDENVPVM 550
YKKILQQPLL FPDGFDPAAK DLLIGLLSRD PSRRLGVNGT DEIRNHPFFK 600
DISWKKLLLK GYIPPYKPIV KSEIDTANFD QEFTKEKPID SVVDEYLSAS 650
IQKQFGGWTY IGDEQLGDSP SQGRSIS
```

Reference
1 Kubo, K. et al. (1989) Gene 76, 177–180.

Ypk1 (*S. cerevisiae*)

Richard Maurer (Oregon Health Sciences University, USA)

The *YPK1* gene was cloned by low stringency hybridization of a yeast genomic library, using as probe a cDNA encoding the catalytic subunit of bovine **PKA**. The presence of the gene has been deduced solely by identification of a long open reading frame, and no information is available on the expression, substrates, physiological role or possible regulators of this putative protein kinase. Ypk1 is closely related to, but distinct from, *S. cerevisiae* **Ykr2**.

Subunit structure and isoforms

SUBUNIT	AMINO ACIDS	MOL. WT	SDS-PAGE
YPK2	680	76477	

Domain structure

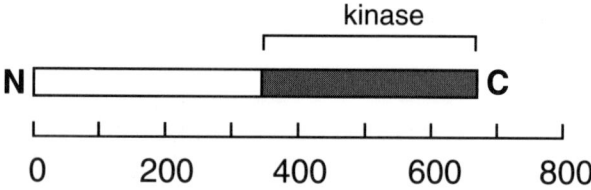

Database accession numbers

	PIR	SWISSPROT	EMBL/GENBANK	REF
Ypk1		P12688		*1*

Amino acid sequence of *S. cerevisiae* Ypk1

```
MYSWKSKFKF GKSKEEKEAK HSGFFHSSKK EEQQNNQATA GEHDASITRS  50
SLDRKGTINP SNSSVVPVRV SYDASSSTST VRDSNGGNSE NTNSSQNLDE 100
TANIGSTGTP NDATSSSGMM TIKVYNGDDF ILPFPITSSE QILNKLLASG 150
VPPPHKEISK EVDALIAQLS RVQIKNQGPA DEDLISSESA AKFIPSTIML 200
LGSSTLNPLL YFTIEFDNTV ATIEAEYGTI AKPGFNKIST FDVTRKLPYL 250
KIDVFARIPS ILLPSKTWQQ EMGLQDEKLQ TIFDKINSNQ DIHLDSFHLP 300
INLSFDSAAS IRLYNHHWIT LDNGLGKINI SIDYKPSRNK PLSIDDFDLL 350
KVIGKGSFGK VMQVRKKDTQ KVYALKAIRK SYIVSKSEVT HTLAERTVLA 400
RVDCPFIVPL KFSFQSPEKL YFVLAFINGG ELFYHLQKEG RFDLSRARFY 450
TAELLCALDN LHKLDVVYRD LKPENILLDY QGHIALCDFG LCKLNMKDDD 500
KTDTFCGTPE YLAPELLLGL GYTKAVDWWT LGVLLYEMLT GLPPYYDEDV 550
PKIYKKILQE PLVFPDGFDR DAKDLLIGLL SRDPTRRLGY NGADEIRNHP 600
FFSQLSWKRL LMKGYIPPYK PAVSNSMDTS NFDEEFTREK PIDSVVDEYL 650
SESVQKQFGG WTYVGNEQLG SSMVQGRSIR
```

References

1   Maurer, R.A. (1988) DNA 7, 469–474.

S6K

Ribosomal protein S6 kinase (vertebrates)
p70S6 kinase αI and αII (p70^{S6K}p85^{S6K})

Sara C. Kozma and George Thomas
(Friedrich Miescher Institute, Switzerland)

Two isoforms of S6K are activated by phosphorylation in cells stimulated to proliferate[1]. The kinase phosphorylates five serines near the C-terminus of ribosomal protein S6. S6 phosphorylation is implicated in the obligatory increase in protein synthesis required for the G_0/G_1 transition of the cell cycle.

Subunit structure and isoforms

SUBUNIT	AMINO ACIDS	MOL. WT	SDS-PAGE
p70	502	56 160	70 kDa (phospho)
p85	525	59 186	85 kDa (phospho)

Two isoforms (p70, p85) have been characterized[2–4]: both are monomers.

Domain structure

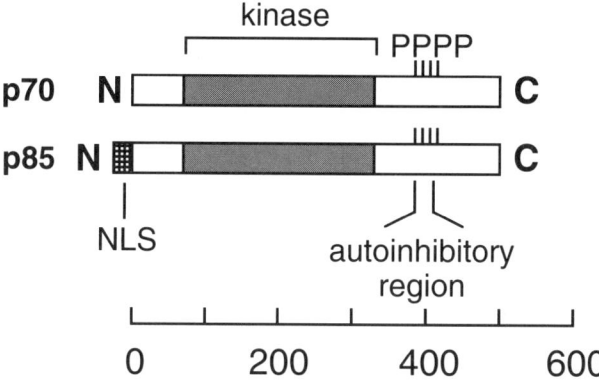

The only difference between the isoforms is that p85 contains 23 extra residues at the N-terminus, containing a nuclear localization signal (NLS). A single gene gives rise to four transcripts, two of which have been cloned. One clone encodes p70, the second encodes both forms[4]. In the regulatory domain are four SP/TP motifs, which are phosphorylated following mitogenic stimulation of quiescent cells[5]. These sites lie close to one another, in a putative autoinhibitory domain which has homology to the substrate recognition motif of S6[3, 6].

Database accession numbers

		PIR	SWISSPROT	EMBL/GENBANK	REF
p70	Human	B41687	P23443	M60725	7
	Rat	A38279	P21425	M35864	2
p85	Human	A41687		M60724	7
	Rat	TVRTK6		M37777	3
	Rabbit	S12906		X54415	8
	Xenopus	S20964			9

Amino acid sequence of rat liver p70 and p85

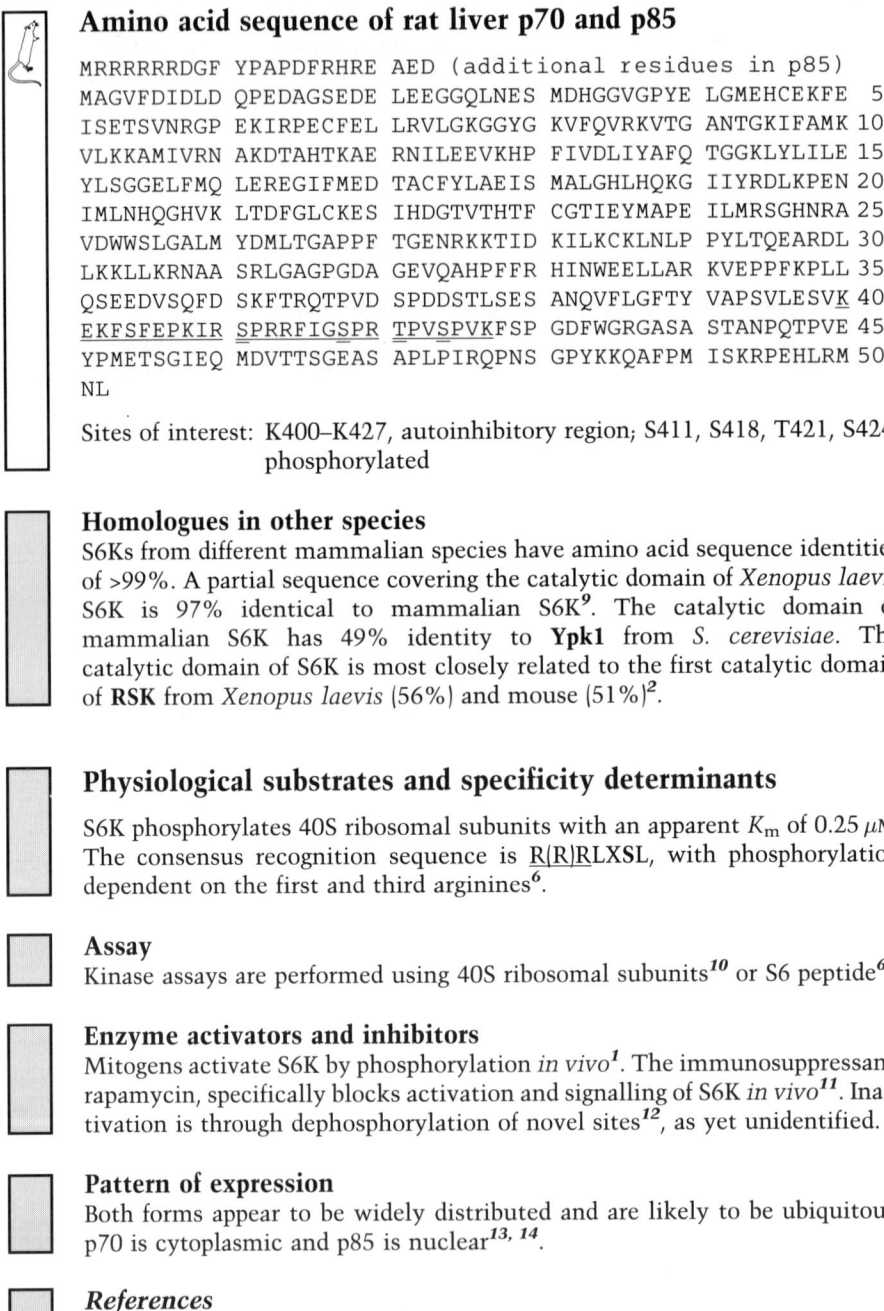

```
MRRRRRRDGF YPAPDFRHRE AED (additional residues in p85)
MAGVFDIDLD QPEDAGSEDE LEEGGQLNES MDHGGVGPYE LGMEHCEKFE  50
ISETSVNRGP EKIRPECFEL LRVLGKGGYG KVFQVRKVTG ANTGKIFAMK 100
VLKKAMIVRN AKDTAHTKAE RNILEEVKHP FIVDLIYAFQ TGGKLYLILE 150
YLSGGELFMQ LEREGIFMED TACFYLAEIS MALGHLHQKG IIYRDLKPEN 200
IMLNHQGHVK LTDFGLCKES IHDGTVTHTF CGTIEYMAPE ILMRSGHNRA 250
VDWWSLGALM YDMLTGAPPF TGENRKKTID KILKCKLNLP PYLTQEARDL 300
LKKLLKRNAA SRLGAGPGDA GEVQAHPFFR HINWEELLAR KVEPPFKPLL 350
QSEEDVSQFD SKFTRQTPVD SPDDSTLSES ANQVFLGFTY VAPSVLESVK 400
EKFSFEPKIR SPRRFIGSPR TPVSPVKFSP GDFWGRGASA STANPQTPVE 450
YPMETSGIEQ MDVTTSGEAS APLPIRQPNS GPYKKQAFPM ISKRPEHLRM 500
NL
```

Sites of interest: K400–K427, autoinhibitory region; S411, S418, T421, S424, phosphorylated

Homologues in other species

S6Ks from different mammalian species have amino acid sequence identities of >99%. A partial sequence covering the catalytic domain of *Xenopus laevis* S6K is 97% identical to mammalian S6K[9]. The catalytic domain of mammalian S6K has 49% identity to **Ypk1** from *S. cerevisiae*. The catalytic domain of S6K is most closely related to the first catalytic domain of **RSK** from *Xenopus laevis* (56%) and mouse (51%)[2].

Physiological substrates and specificity determinants

S6K phosphorylates 40S ribosomal subunits with an apparent K_m of 0.25 μM. The consensus recognition sequence is R(R)RLXSL, with phosphorylation dependent on the first and third arginines[6].

Assay

Kinase assays are performed using 40S ribosomal subunits[10] or S6 peptide[6].

Enzyme activators and inhibitors

Mitogens activate S6K by phosphorylation *in vivo*[1]. The immunosuppressant, rapamycin, specifically blocks activation and signalling of S6K *in vivo*[11]. Inactivation is through dephosphorylation of novel sites[12], as yet unidentified.

Pattern of expression

Both forms appear to be widely distributed and are likely to be ubiquitous. p70 is cytoplasmic and p85 is nuclear[13, 14].

References

1   Ferrari S. and Thomas G. (1994) CRC Press 29, 385–413.
2   Kozma S.C. et al. (1990) Proc. Natl Acad. Sci. USA 87, 7365–7369.
3   Banerjee P. et al. (1990) Proc. Natl Acad. Sci. USA 87, 8550–8554.
4   Reinhard C. et al. (1992) Proc. Natl Acad.Sci. USA 89, 4052–4056.

5   Ferrari S. et al. (1992) Proc. Natl Acad. Sci. USA 89, 7282–7286.

6   Flotow H. and Thomas G. (1992) J. Biol. Chem. 267, 3074–3078.

7   Harmann B. and Kilimann M.W. (1990) FEBS Lett. 273, 248–252.

8   Grove J.R. et al. (1991) Mol. Cell. Biol. 11, 5541–5550.

9   Lane H.A. et al. (1992) EMBO J. 11, 1743–1749.

10 Lane, H.A. and Thomas, G. (1991) Methods Enzymol. 200, 268–291.

11 Schreiber S.L. (1992) Cell 70, 365–368.

12 Ferrari, S. et al. (1993) J. Biol. Chem. 268, 16091–16094.

13 Coffer, P.J. and Woodgett J.R. (1994) Biochem. Biophys. Res. Comm. 198, 780–786.

14 Reinhard, C. et al. (1994) EMBO J. 13, 1557–1567.

RSK
Ribosomal protein S6 kinase II (vertebrates)
(Insulin-stiulated PK-1, ISPK1, MAP kinase-activated PK1)

Raymond L. Erikson (Harvard University, USA)

The protein kinase encoded by *RSK* was first purified from extracts of unfertilized *Xenopus* eggs[1]. In these extracts it displays the major activity for the phosphorylation of protein S6 in the ribosomal 40S subunit.

Subunit structure and isoforms

SUBUNIT	AMINO ACIDS	MOL. WT	SDS-PAGE
RSK^mo-1	724	81 600	≈90 kDa
RSK^mo-2	740	83 688	≈90 kDa

The enzyme is a monomer, with two isoforms identified in the mouse[2].

Genetics
Chromosomal locations:

RSK^mo-1:   chromosome 4 near Cri (Cribriform degeneration)

RSK^mo-2:   X chromosome near DXPas18

(unpublished results, Nancy Jenkins and Neil Copeland, Frederick Cancer Research Center)

Domain structure

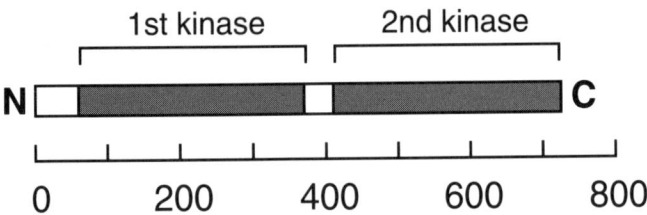

RSK has an unusual structure consisting of two apparent kinase domains. The N-terminal domain is related to the catalytic domains of **PKA** and **S6K**[3], while the C-terminal domain is related to the γ subunit of **PhK**.

Database accession numbers

	PIR	SWISSPROT	EMBL/GENBANK	REF
Xenopus			M20187	
Mouse			M28488	2

Amino acid sequences of mouse RSK

RSK^mo-1

```
MPLAQLKEPW PLMELVPLDP ENGQTSGEEA GLQPSKDEAI LKEISITHHV   50
KAGSEKADPS QFELLKVLGQ GSFGKVFLVR KVTRPDSGHL YAMKVLKKAT  100
LKVRDRVRTK MERDILADVN HPFVVKLHYA FQTEGKLYLI LDFLRGGDLF  150
TRLSKEVMFT EEDVKFYLAE LALGLDHLHS LGIIYRDLKP ENILLDEEGH  200
IKLTDFGLSK EAIDHEKKAY SFCGTVEYMA PEVVNRQGHT HSADWWSYGV  250
LMGKDRKETM TLILKAKLGM PQFLSTEAQS LLRALFKRNP ANRLGSGPDG  300
AEEIKRHIFY STIDWNKLYR REIKPPFKPA VAQPDDTFYF DTEFTSRTPR  350
DSPGIPPSAG AHQLFRGFSF VATGLMEDDG KPRTTQAPLH SVVQQLHGKN  400
LVFSDGYVVK ETIGVGSYSV CKRCVHKATN MEYAVKVIDK SKRDPSEEIE  450
ILLRYGQHPN IITLKDVYDD GKHVYLVTEL MRGGELLDKI LRQKFFSERE  500
ASFVLHTISK TVEYLHSQGV VHRDLKPSNI LYVDESGNPE CLRICDFGFA  550
KQLRAENGLL MTPCYTANFV APEVLKRQGY DEGCDIWSLG ILLYTMLAGY  600
TPFANGPSDT PEEILTRIGS GKFTLSGGNW NTVSETAKDL VSKMLHVDPH  650
QRLTAKQVLQ HPWITQKDKL PQSQLSHQDL QLVKGAMAAT YSALNSSKPT  700
PQLKPIESSI LAQRRVRKLP STTL
```

RSK^mo-2

```
MPLAHVADPW QKMAVESPSD SAENGQQIMD EPMGEEEINP QTEEGSIKEI   50
AITHHVKEGH EKADPSQFEL LKVLGQGSFG KVFLVKKISG SDARQLYAMK  100
VLKKATLKVR DRVRTKMERD ILVEVNHPFI VKLHYAFQTE GKLYLILDFL  150
RGGDLFTRLS KEVMFTEEDV KFYLAELALA LDHLHSLGII YRDLKPENIL  200
LDEEGHIKLT DFGLSKESID HEKKAYSFCG TVEYMAPEVV NRRGHTQSAD  250
WWSFGVLMFE MLTGTLPFQG KDRKETMTMI LKAKLGMPQF LSPEAQSLLR  300
MLFKRNPANR LGAGPDGVEE IKRHSFFSTI DWNKLYRREI HPPFKPATGR  350
PEDTFYFDPE FTAKTPKDSP GIPPSANAHQ LFRGFSFVAI TSDDESQAMQ  400
TVGVHSIVQQ LHRNSIQFTD GYEVKEDIGV GSYSVCKRCI HKATNMEFAV  450
KIIDKSKRDP TEEIEILLRY GQHPNIITLK DVYDDGKYVY VVTELMKGGE  500
LLDKILRQKF FSEREASAVL FTITKTVEYL HAQGVVHRDL KPSNILYVDE  550
SGNPESIRIC DFGFAKQLRA ENGLLMTPCY TANFVAPEVL KRQGYDAACD  600
IWSLGVLLYT MLTGYTPFAN GPDDTPEEIL ARIGSGKFSL SGGYWNSVSD  650
TAKDLVSKML HVDPHQRLTA ALVLRHPWIV HWDQLPQYQL NRQDAPHLVK  700
GAMAATYSAL NRNQSPVLEP VGRSTLAQRR GIKKITSTAL
```

Homologues in other species

RSK cDNAs have been sequenced from *Xenopus laevis*, mouse and avian sources. Unpublished sequences are available from human and *D. mela-nogaster*. All have the two domain structure. An MAP kinase-activated PK purified from rabbit skeletal muscle, termed ISPK1[4] or MAPKAP kinase-1[5], is likely to be the rabbit homologue.

Physiological substrates and specificity determinants

Although RSK was originally isolated as an S6 protein kinase, it has been argued that S6 is not a substrate *in vivo*. RSK^mo-1 is, however, highly expressed in proliferating tissue, whereas RSK^mo-2 is highly expressed in the brain and heart[2]. The N-terminus of rat RSK is 57% identical to the N-terminus of rat p70^{S6K}, which is regarded as an authentic S6 kinase *in vivo*. RSK shows a strong preference for serine residues C-terminal to arginine residues[3].

Assay
Purified 40S ribosomal subunits or kemptide are routinely used as substrates.

Enzyme activators and inhibitors
RSK activity is activated by phosphorylation by MAP kinase (**Erk1/2**) on serine residues.

Pattern of expression
RSK is activated after fertilization of *Xenopus laevis* eggs and upon mitogenic stimulation of quiescent fibroblasts. Activity in fibroblasts reaches a peak ≈15 min after stimulation, declines to about 20% of peak activity within 1 h, and then remains stable for several hours. Mouse intestine, thymus and lung show significant levels of RSK^{mo-1} and RSK^{mo-2}. Heart and brain yield low levels of RSK^{mo-1} mRNA, whereas RSK^{mo-2} is expressed at significantly higher levels in these tissues.

References
1   Erikson, E. and Maller, J. (1985) Proc. Natl Acad. Sci. USA 82, 742–746.
2   Alcorta, D. A. et al. (1989) Mol. Cell. Biol. 9, 3850–3859.
3   Erikson, R.L. (1991) J. Biol. Chem. 266, 6007–6010.
4   Lavoinne, A. et al. (1992) Eur. J. Biochem. 199, 723–728.
5   Stokoe, D. et al. (1992) EMBO J. 11, 3985–3994.

Jeremy H. Toyn and Leland H. Johnston
(National Institute for Medical Research, UK)

The *DBF2* gene encodes a protein kinase that is required for cell cycle progression at a late stage during nuclear division[1], and for efficient chromatid separation[2]. Temperature-sensitive mutations in *DBF2* cause arrest of cell division in which the cells have large buds, an elongated mitotic spindle, a divided nucleus, and high Cdc28 kinase activity. During incubation of *dbf2* mutants at the restrictive temperature, a high frequency of chromosomal non-disjunction occurs. In addition, *dbf2* mutants delay the onset of DNA replication, and can be suppressed by multiple copies of the *SIT4* protein phosphatase gene[3], consistent with a second, non-essential role for Dbf2 at the G1/S transition of the cell cycle. The Dbf20 and Dbf2 proteins are >80% identical in sequence[4]. Despite the fact that recessive conditional lethal mutations have been isolated in the *DBF2* gene[1], *DBF2* and *DBF20* form a functionally redundant gene pair: either gene can be deleted without loss of viability, but deletion of both is lethal. Genetic analysis suggests that the Spo12 protein is necessary for the activation of Dbf20. However, this activation is prevented by Dbf2, and only occurs either when no Dbf2 protein is present[2], or when Spo12 is over-expressed[3].

Subunit structure and isoforms

SUBUNIT	AMINO ACIDS	MOL. WT	SDS-PAGE
Dbf2	572	67 000	68 kDa
Dbf20	564	66 000	60 kDa

In SDS-PAGE, the Dbf2 protein runs as a doublet, with the upper band being due to phosphorylation.

Genetics
Chromosomal location:
　　Dbf2:　　7R
　　Dbf20:　　16R

Domain structure

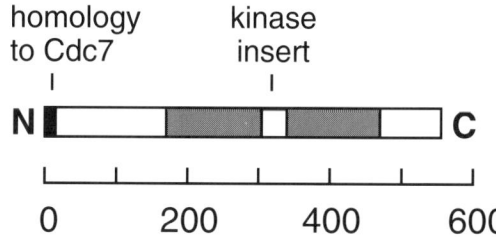

Dbf2 and Dbf20 have, like Cdc7, an insert of 54 amino acids between subdomains VII and VIII. Outside of the kinase domain, the sequences of Dbf2, Dbf20 and Cdc7 are closely related in the N-terminal 11 residues.

Database accession numbers

	PIR	SWISSPROT	EMBL/GENBANK	REF
Dbf2			M34146	1
Dbf20			M62506	4

Amino acid sequences of *S. cerevisiae* Dbf2/20

Dbf2
```
MLSKSEKNVD HMAGNMSNLS FDGHGTPGGT GLFPNQNITK RRTRPAGIND  50
SPSPVKPSFF PYEDTSNMDI DEVSQPDMDV SNSPKKLPPK FYERATSNKT 100
QRVVSVCKMY FLEYYCDMFD YVISRRQRTK QVLEYLQQQS QLPNSDQIKL 150
NEEWSSYLQR EHQVLRKRRL KPKNRDFEMI TQVGQGGYGQ VYLARKKDTK 200
EVCALKILNK KLLFKLNETK HVLTERDILT TTRSEWLVKL LYAFQELQSL 250
YLAMEFVPGG DFRTLLINTR CLKSGHARFY ISEMFCAVNA LHDLGYTHRD 300
LKPENFLIDA KGHIKLTDFG LAAGTISNER IESMKIRLEK IKDLEFPAFT 350
EKSIEDRRKM YNQLREKEIN YANSMVGSPD YMALEVLEGK KYDFTVDYWS 400
LGCMLFESLV GYTPFSGSST NETYDNLRRW KQTLRRPRQS DGRAAFSDRT 450
WDLITRLIAD PINRLRSFEH VKRMSYFADI NFSTLRSMIP PFTPQLDSET 500
DAGYFDDFTS EADMAKYADV FKRQDKLTAM VDDSAVSSKL VGFTFRHRNG 550
KQGSSGILFN GLEHSDPFST FY
```

Dbf20
```
MFSRSDREVD DLAGNMSHLG FYDLNIPKPT SPQAQYRPAR KSENGRLTPG  50
LPRSYKPCDS DDQDTFKNRI SLNHSPKKLP KDFHERASQS KTQRVVNVCQ 100
LYFLDYYCDM FDYVISRRQR TKQVLRYLEQ QRSVKNVSNK VLNEEWALYL 150
QREHEVLRKR RLKPKHKDFQ ILTQVGQGGY GQVYLAKKKD SDEICALKIL 200
NKKLLFKLNE TNHVLTERDI LTTTRSDWLV KLLYAFQDPE SLYLAMEFVP 250
GGDFRTLLIN TRILKSGHAR FYISEMFCAV NALHELGYTH RDLKPENFLI 300
DATGHIKLTD FGLAAGTVSN ERIESMKIRL EEVKNLQFPA FTERSIEDRS 350
KIYHNMRKTE INYANSMVGS PDYMALEVLE GKKYDFTVDY WSLGCMLFES 400
LVGYTPFSGS STNETYENLR YWKKTLRRPR TEDRRAAFSD RTWDLITRLI 450
ADPINRVRSF EQVRKMSYFA EINFETLRTS SPPFIPQLDD ETDAGYFDDF 500
TNEEDMAKYA DVFKRQNKLS AMVDDSAVDS KLVGFTFRHR DGKQGSSGIL 550
YNGSEHSDPF STFY
```

Physiological substrates

The physiological substrates of Dbf2/20 have not been identified. However, a number of high copy number suppressors of *ts dbf2* mutants have been isolated. Some or all of these may encode substrates of Dbf2. They include *SPO12* (a gene first characterized as essential for meiosis), *SIT4* (a protein phosphatase gene), *SDB23* (a gene with homology to a human small nuclear ribonucleoprotein), *SDB24* (which has a region of homology to coiled coil structures), and *SDB25* (encoding p40, a Cdc2-associated protein). Because of the functional redundancy between *DBF2* and *DBF20*, the same substrates may be involved in their vital cell cycle function. Some of the supressors of *dbf2* mutants may also be substrates for Dbf20.

Assay
Immunoprecipitates of Dbf2 and Dbf20 catalyse transfer of phosphate from ATP to histone H1.

Pattern of expression
The *DBF2* transcript is under cell cycle control, and peaks once per cell cycle during M phase, but distinctly later than the peak of expression of the B-type cyclin transcripts *CLB1* and *CLB2*. Furthermore, *SPO12* is expressed at precisely the same stage in the cell cycle as *DBF2*. This coordinate expression of *DBF2* and one of its suppressors suggests that the two genes may function together. As well as expression of *DBF2* late in the cell cycle, the transcript is also expressed transiently early in the first cell cycle after exit from stationary phase, and distinctly earlier than the G1/S phase transition transcripts, which are under the control of the Sbf and Dscl/Mbf transcription factors. Dbf2 kinase activity measured in immunoprecipitates is under cell cycle control, and maximum activity coincides with division of the chromatin in M phase, distinctly later in the cell cycle than maximum Cdc28 kinase activity[5]. *DBF20* mRNA is expressed at a low and constant level throughout the mitotic cell cycle. Thus, since the *DBF2* transcript is expressed periodically during M phase, the relative abundance of *DBF2* and *DBF20* transcripts varies in a cell cycle-dependent manner. Consequently, the ratio of Dbf2 to Dbf20 protein is thought to vary in a cell cycle-dependent manner also. Since the presence of Dbf2 prevents activation of Dbf20 by Spo12, it is possible that a cell cycle fluctuation in the Dbf2/Dbf20 ratio could allow periodic activation of Dbf20.

References
1   Johnston, L.H. et al. (1990) Mol. Cell. Biol. 10, 1358–1366.
2   Toyn, J.H. and Johnston, L.H. (1993), Genetics 135, 963–971.
3   Parkes, V. and Johnston, L.H. (1992) Nucl. Acids Res. 20, 5617–5623.
4   Toyn, J.H. et al. (1991) Gene 104, 63–70.
5   Toyn, J.H. and Johnston, L.H. (1994), EMBO J. 13, 1103–1113.

PvPK1 — PvPK1 (*Phaseolus vulgaris*)

Michael A. Lawton (Center for Agricultural Biotechnology, Rutgers University, USA)

PvPK1 was isolated by homology cloning using degenerate primers corresponding to conserved catalytic domain sequences. The kinase domain resides in the C-terminal region and is most highly related to **PKC** and **PKA**. The kinase domain is interrupted by a series of short repeats that are rich in cysteine and basic amino acids. The N-terminal domain is unrelated to proteins in the database. The function of this protein kinase is unknown.

Subunit structure and isoforms

SUBUNIT	AMINO ACIDS	MOL. WT	SDS-PAGE
PvPK1	609	68 100	

Domain structure

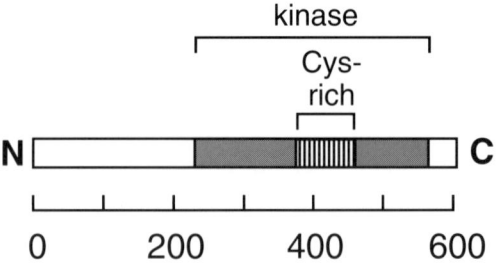

Database accession number

	PIR	SWISSPROT	EMBL/GENBANK	REF
PvPK1			J04555	1

Amino acid sequence of *Phaseolus vulgaris* PvPK1

```
MESSVNGVDS LSEVQNSVSG VHHHDPLPSG TPQPSRPPLR ASRNYDGGHQ  50
TKAIHHHNSH VINQKHSHQE GKTLKQEGLP TKLSSKQPPL DDSKGCEPNG 100
VLESEKKRVV DNHGKNYSQP DATFCASPQN SFYSATVYSE AKESFTNTEV 150
SECASVDKSC ESEVANSSDF NESRKTSICR ASTGSDASDE SSTSSLSSVL 200
YKPHKANDIR WEAIQAVRTR DGMLEMRHFR LLKKLGCGDI GSVYLAELSG 250
TRTSFAMKVM NKTELANRKK LLRAQTEREI LQSLDHPFLP TLYTHFETEI 300
FSCLVMEFCP GGDLHALRQR QPGKYFSEHA VRFYVAEVLL SLEYLHMLGI 350
IYRDLKPENV LVREDGHIML SDFDLSLRCS VSPTLVKSSN NLQTKSSGYC 400
VQPSCIEPTC VMQPDCIKPS CFTPRFLSGK SKKDKKSKPK NDMHNQVTPL 450
PELIAEPTNA RSMSFVGTHE YLAPEIIKGE GHGSAVDWWT FGIFLYELLF 500
GRTPFKGSAN RATLFNVIGQ PLRFPESPTV SFAARDLIRG LLVKEPQHRL 550
AYRRGATEIK QHPFFQNVNW ALIRCATPPE VPRQVINLPQ TEKDLGVKPS 600
GNYLDIDFF
```

Homologues in other species
The catalytic domain of **OsG11A** is homologous to the catalytic domain of PvPK1[1]. However, the N-terminal, non-catalytic regions are completely unrelated. A homologue is also present in *Arabidopsis thaliana* and again contains a unique N-terminal region (M. Lawton, unpublished data).

Reference
[1]   Lawton, M.A. et al. (1990) Proc. Natl Acad. Sci. USA 86, 3140–3144.

OsG11A G11A (*Oryzae sativa*)

Michael A. Lawton (Center for Agricultural Biotechnology, Rutgers University, USA)

G11A is an incomplete cDNA isolated by homology cloning using degenerate primers corresponding to conserved catalytic domain sequences. The kinase domain resides in the C-terminal region and is most highly related to **PKC** and **PKA**. The kinase domain is interrupted by a series of short repeats that are rich in cysteine and basic amino acids. The N-terminal domain is unrelated to proteins in the database. The function of this protein kinase is unknown.

Subunit structure and isoforms

SUBUNIT	AMINO ACIDS	MOL. WT	SDS-PAGE
G11A	535 (partial)	59 124 (partial)	

Domain structure

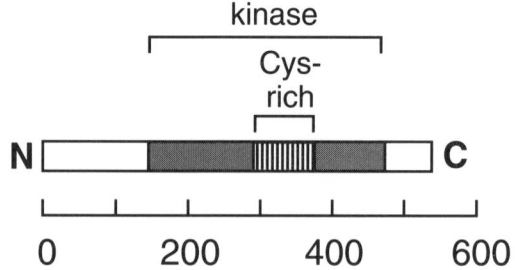

Database accession number

	PIR	SWISSPROT	EMBL/GENBANK	REF
G11A			J04556	*1*

Partial amino acid sequence of *Oryzae sativa* G11A

```
NSEVVQKEQK STQHQNESID LTGSNDPAEV KAEGNLVPKR LADEEKGVVE  50
DGIANGSLKS SSALGKEHGI ASASGSARLV GRSETGERGF SSSRCRPSTS 100
SDVSDESACS SISSVTKPHK ANDSRWEAIQ MIRTRDGILG LSHFKLLKKL 150
GCGDIGSVYL SELSGTESYF AMKVMDKASL ASRKKLLRAQ TEKEILQCLD 200
HPFLPTLYTH FETDKFSCLV MEFCPGGDLH TLRQRQRGKY FPEQAVKFYV 250
AEILLAMEYL HMLGIIYRDL KPENVLVRED GHIMLSDFDL SLRCAVSPTL 300
IRSSNPDAEA LRKNNQAYCV QPACVEPSCM IQPSCATPTT CFGPRFFSKS 350
KKDRKPKPEV VNQVSPWPEL IAEPSDARSM SFVGTHEYLA PEIIKGEGHG 400
SAVDWWTFGI FLYELLFGKT PFKGSGNRAT LFNVIGQPLR FPEYPVVSFS 450
ARDLIRGLLV KEPQQRLGCK RGATEIKQHP FFEGVNWALI RCASPPEVPR 500
PVEIERPPKQ PVSTSEPAAA PSDAAQKSSD SYLEF
```

Homologues in other species

The catalytic domain of G11A is homologous to the catalytic domain of

PvPK1[1]. The N-terminal, non-catalytic regions are completely unrelated. A homologue is also present in *A. thaliana* and this also contains a unique N-terminal region (M. Lawton, unpublished data).

Reference
[1]   Lawton, M.A. et al. (1990) Proc. Natl Acad. Sci. USA 86, 3140–3144.

DMPK   Myotonic dystrophy PK (vertebrates)

J. David Brook (University of Nottingham, UK)

The *DMPK* gene encodes a protein which, on the basis of sequence similarity, is a member of the protein kinase gene family[1-3]. The gene is implicated in the disorder myotonic dystrophy (dystrophia myotonica, DM), the most common form of muscular dystrophy affecting adults. The 3' untranslated region of this gene contains a repeated trinucleotide DNA sequence, CTG, that is greatly expanded in DM patients. In normal individuals 5–28 copies of this repeat are present. Myotonic dystrophy patients may have anything from 50 to several thousand copies of this repeat. Studies on the effect of the expanded repeat on *DMPK* RNA levels in myotonic dystrophy patients are contradictory. Three studies[4-6] indicate that the *DMPK* RNA level is reduced in DM patients and one study[7] indicates that the level is increased. Reduced protein levels have also been suggested in one report[4]. Recombinant DM protein kinase has been shown to phosphorylate itself and transphosphorylate histone HI. In both recombinant DM protein kinase and histone HI, threoine (predominantly) and serine were phosphorylated[11].

Subunit structure and isoforms

SUBUNIT	AMINO ACIDS	MOL. WT	SDS-PAGE
DMPK	624	69 013	

Genetics

Chromosomal location (human): 19q13.3.

Domain structure

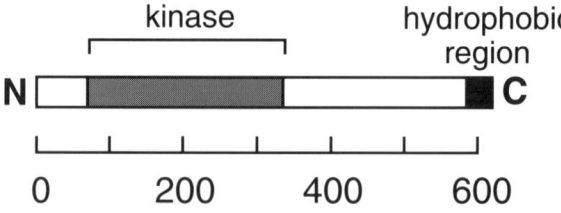

At the C-terminal end is a region of high hydrophobicity, which may be membrane bound. Alternatively spliced forms have been identified[4, 8-10] including one which would alter the reading frame 3' to residue 545.

Database accession numbers

	PIR	SWISSPROT	EMBL/GENBANK	REF
Human			L19266, L19267, L19268	10
			L08835	9
			L00727	4

Amino acid sequence of human DMPK

```
MSAEVRLRRL QQLVLDPGFL GLEPLLDLLL GVHQELGASE LAQDKYVADF    50
LQWAEPIVVR LKEVRLQRDD FEILKVIGRG AFSEVAVVKM KQTGQVYAMK 100
IMNKWDMLKR GEVSCFREER DVLVNGDRRW ITQLHFAFQD ENYLYLVMEY 150
YVGGDLLTLL SKFGERIPAE MARFYLAEIV MAIDSVHRLG YVHRDIKPDN 200
ILLDRCGHIR LADFGSCLKL RADGTVRSLV AVGTPDYLSP EILQAVGGGP 250
GTGSYGPECD WWALGVFAYE MFYGQTPFYA DSTAETYGKI VHYKEHLSLP 300
LVDEGVPEEA RDFIQRLLCP PETRLGRGGA GDFRTHPFFF GLDWDGLRDS 350
VPPFTPDFEG ATDTCNFDLV EDGLTAMETL SDIREGAPLG VHLPFVGYSY 400
SCMALRDSEV PGPTPMEVEA EQLLEPHVQA PSLEPSVSPQ DETAEVAVPA 450
AVPAAEAEAE VTLRELQEAL EEEVLTRQSL SREMEAIRTD NQNFASQLRE 500
AEARNRDLEA HVRQLQERME LLQAEGATAV TGVPSPRATD PPSHLDGPPA 550
VAVGQCPLVG PGPMHRRHLL LPARVPRPGL SEALSLLLFA VVLSRAAALG 600
CIGLVAHAGQ LTAVWRRPGA ARAP
```

Homologues in other species
The mouse homologue has been cloned and sequenced[8].

References
1 Brook, J.D. et al. (1992) Cell 68, 799–808.
2 Mahadevan, M.S. et al. (1992) Science 255, 1253–1255.
3 Fu, Y.H. et al. (1992) Science 255, 1256–1258.
4 Fu, Y.H. et al. (1993) Science 260, 235–238.
5 Hofmann-Radvanyi, H. et al. (1993) Human Mol. Genet. 2, 1263–1266.
6 Carango, P. et al. (1993) Genomics 18, 340–348.
7 Sabouri, L.A. et al. (1993) Nature Genet. 4, 233–238.
8 Jansen, G. et al. (1992) Nature Genet. 1, 261–266.
9 Mahadevan, M.S. et al. (1993) Human Mol. Genet. 2, 299–304.
10 Shaw, D.J. et al. (1993) Genomics 18, 673–679.
11 Dunne, P.W. et al. (1994) Biochemistry 33, 10809–10814.

Ddk2
Protein kinase 2 (*D. discoideum*)

**Bodduluri Haribabu (Duke University, USA) and
Robert P. Dottin (Hunter College, USA)**

Ddk2 represents one of the first serine/threonine protein kinases identified from the cellular slime mould *D. discoideum*[1]. It was first cloned as one of five kinase-like PCR fragments amplified from genomic DNA, using primers derived from the consensus sequences of serine/threonine kinases. It was subsequently isolated as a full length cDNA from a library made from cAMP-treated, developed cells.

Subunit structure

SUBUNIT	AMINO ACIDS	MOL. WT	SDS-PAGE
Ddk2	479	52 896	?

Domain structure

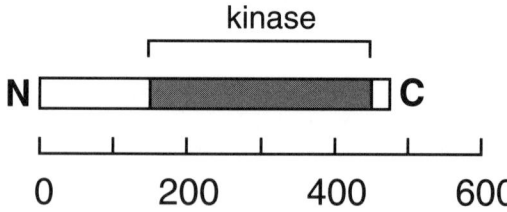

The deduced Ddk2 protein consists of a central kinase catalytic domain, a 150 amino acid highly basic N-terminal extension, and a short C-terminal extension.

Database accession numbers

	PIR	SWISSPROT	EMBL/GENBANK	REF
Ddk2		A38578	M59744	*1*

Amino acid sequence of *D. discoideum* Ddk2

```
MGKGQSKIKN GGSGKPAKAG KPKKGNKNDE TTPTSTPTPT PTPTQQNLDN  50
SAQQQQQQQQ TATAAVSLDN KEQQQQQNIP APATQTPITQ TGTPTIEESQ 100
KNTDNNNING ASNEASSSPD SPNGSGNGND DEDEGPEEVI FSKNKQSATK 150
DDFELLNVIG KGSFGKVMQV KKKGEDKIFA MKVLRKDAII ARKQVNHTKS 200
EKTILQCISH PFIVNLHYAF QTKDKLYMVL DFVNGGELFF HLKREGRFSE 250
PRVKIYAAEI VSALDHLHKQ DIVYRDLKPE NILLDSEGHI CITDFGLSKK 300
IETTDGTFTF CGTPEYLAPE VLNGHGHGCA VDWWSLGTLL YEMLTGLPPF 350
YSQNVSTMYQ KILNGELKIP TYISPEAKSL LEGLLTREVD KRLGTKGGGE 400
VKQHPWFKNI DWEKLDRKEV EVHFKPKVKS GTDISQIDPV FTQERPMDSL 450
VETSALGDAM GKDTSFEGFT YVADSILKD
```

Pattern of expression
Two differentially regulated mRNAs of Ddk2 are detected. While a 2.0 kb mRNA is constitutively expressed, a 2.2 kb mRNA appeared only during

development after cells established a cAMP-signalling system. The larger mRNA was also induced by agonists of the cell surface cAMP receptor[2] and is therefore regulated by transmembrane signal transduction. As only one copy of the gene was detected, the two mRNAs must originate from separate promoters, or by alternate splicing. The functional aspects of this gene are being investigated using molecular genetic approaches.

References
[1] Haribabu, B. and Dottin, R.P. (1991) Proc. Natl Acad. Sci. USA 88, 1115–1119.

[2] Haribabu, B. and Dottin, R.P. (1986) Mol. Cell. Biol. 6, 2402–2408.

ScSpk1   Spk1 (*S. cerevisiae*)

Pan Zheng, David Fay and David F. Stern
(Yale University School of Medicine)

SPK1 ("Spike1") was identified in an immunoscreen for protein-tyrosine kinases in *S. cerevisiae*[1]. It encodes one of the prototype dual-specificity (Ser/Thr/Tyr)-protein kinases. The Spk1 protein is associated with Ser/Thr and Tyr kinase activity, with the latter activity being weak relative to Ser and Thr activity[2].

Subunit structure

SUBUNIT	AMINO ACIDS	MOL. WT	SDS-PAGE
Spk1	821	92 051	90 kDa

Spk1 is a monomer[2].

Genetics

Chromosomal location: XVI, adjacent to *PEP4*[2].
Mutations: K227A, inactive[2].

Domain structure

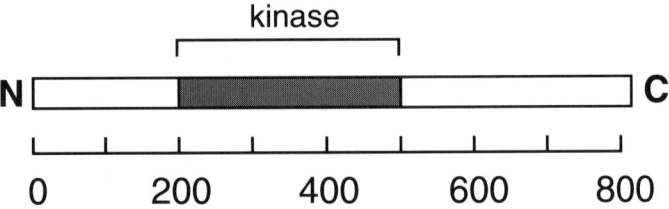

The kinase domain is similar to that of S/T-specific protein kinases. Although ScSpk1 resembles **PKG** most closely, this similarity does not extend outside the kinase domain to the cGMP-binding region[1].

Database accession numbers

	PIR	SWISS-PROT	EMBL/GENBANK	REF
ScSpk1			M55623	[1]

Amino acid sequence of *S. cerevisiae* Spk1

```
MENITQPTQQ STQATQRFLI EKFSQEQIGE NIVCRVICTT GQIPIRDLSA  50
DISQVLKEKR SIKKVWTFGR NPACDYHLGN ISRLSNKHFQ ILLGEDGNLL 100
LNDISTNGTW LNGQKVEKNS NQLLSQGDEI TVGVGVESDI LSLVIFINDK 150
FKQCLEQNKV DRIRSNLKNT SKIASPGLTS STASSMVANK TGIFKDFSII 200
DEVVGQGAFA TVKKAIERTT GKTFAVKIIS KRKVIGNMDG VTRELEVLQK 250
LNHPRIVRLK GFYEDTESYY MVMEFVSGGD LMDFVAAHGA VGEDAGREIS 300
RQILTAIKYI HSMGISHRDL KPDNILIEQD DPVLVKITDF GLAKVQGNGS 350
FMKTFCGTLA YVAPEVIRGK DTSVSPDEYE ERNEYSSLVD MWSMGCLVYV 400
ILTGHLPFSG STQDQLYKQI GRGSYHEGPL KDFRISEEAR DFIDSLLQVD 450
PNNRSTAAKA LNHPWIKMSP LGSQSYGDFS QISLSQSLSQ QKLLENMDDA 500
```

Amino acid sequence of *S. cerevisiae* Spk1 *continued*

```
QYEFVKAQRK LQMEQQLQEQ DQEDQDGKIQ GFKIPAHAPI RYTQPKSIEA 550
ETREQKLLHS NNTENVKSSK KKGNGRFLTL KPLPDSIIQE SLEIQQGVNP 600
FFIGRSEDCN CKIEDNRLSR VHCFIFKKRH AVGKSMYESP AQGLDDIWYC 650
HTGTNVSYLN NNRMIQGTKF LLQDGDEIKI IWDKNNKFVI GFKVEINDTT 700
GLFNEGLGML QEQRVVLKQT AEEKDLVKKL TQMMAAQRAN QPSASSSSMS 750
AKKPPVSDTN NNGNNSVLND LVESPINANT GNILKRIHSV SLSQSQIDPS 800
KKVKRAKLDQ TSKGPENLQF S
```

Assay

Spk1 can be assayed using enolase or poly(Glu/Tyr 5:1) as (poor) substrates, or by autophosphorylation[1,2].

Pattern of expression

Spk1 is a nuclear protein. The gene is transcriptionally regulated by an *Mlu*I cell cycle box that confers S phase-specific expression. This enhancer is also found upstream of a number of other S phase-specific genes, notably those encoding proteins that participate in DNA synthesis. *spk1* mutant cells are generally inviable, but those that survive (which may harbour suppressor mutations) are hypersensitive to ultraviolet light. The nuclear localization, transcriptional regulation and necessity for recovery from DNA damage all suggest that Spk1 plays an important role in replicative and repair synthesis of DNA[2].

References

[1] Stern, D.F. et al. (1991) Mol. Cell. Biol. 11, 987–1001.
[2] Zheng, P. et al. (1993) Mol. Cell. Biol. 13, 5829–5842.

CaMKI — CaM-dependent protein kinase I (vertebrates)

Arthur M. Edelman (State University of New York, Buffalo, USA)
and Angus C. Nairn (Rockefeller University, USA)

Calmodulin-dependent protein kinase I mediates some of the effects of increased intracellular Ca^{2+} via its dependence for activity on the $Ca^{2+}/$ CaM complex[1-3]. Site 1 of synapsins I[1-3] and II[1], cAMP response element-binding protein (CREB)[4] and the cystic fibrosis transmembrane conductance regulator (CFTR)[5] are *in vitro* substrates for CaM kinase I suggesting its potential involvement in the regulation of synaptic vesicle function, gene expression and ion channel activity. Based on a widespread tissue distribution of its activity[1] and mRNA[6-8], additional, as yet unidentified, physiologically relevant targets for this kinase are presumed to exist.

Subunit structure and isoforms

SUBUNIT	AMINO ACIDS	MOL. WT	SDS-PAGE
Bovine brain			37-42 kDa[1]
Rat brain			39[2], 43[3], 41[9] kDa
Rat brain	374[6,7]	41,636[6,7]	
Human HL-60 cells	370[8]	41,337[8]	

Preparations of CaM kinase I from bovine and rat brain are monomeric[1-3]. Partial amino acid sequence data from the bovine enzyme were used to clone CaMKI from a rat brain cDNA library[6]. The 39 and 43 kDa proteins, termed CaMKIb and CaMKIa, were purified separately from rat brain and found to have related properties[2,3]. However, these two apparent isoforms could be distinguished by differences in their kinetic and regulatory properties[2,3]. The 41 kDa protein, termed CaMKV, was purified from rat brain[9] and found to have properties very similar to that described for CaMKI from bovine brain[1] and CaMKIa rat brain[2,3]. The amino acid sequences of four peptides obtained from CaMKI[9] are found in the predicted sequence for rat brain CaMKI except for two changes (A37T and S239V). Whether the various molecular weight species are products of separate genes, alternate splice variants or are generated by post-translational modification (e.g. phosphorylation or proteolysis) remains to be determined.

Domain structure

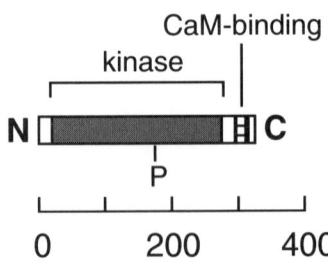

The domain structure above is based on the predicted amino acid sequence of the rat brain and human HL-60 cell enzymes[6-8]. CaMKI consists of a kinase domain with a short N-terminal extension. The slightly longer C-terminal extension contains a stretch of amino acids (K302-R314) that contains the CaM-binding domain of the enzyme[8, 10]. T177 has been identified as a site for autophosphorylation and phosphorylation by an activating kinase, CaMKI kinase[8, 11]. An autoinhibitory region has been identified that encompasses I294-A299[8, 10].

Database accession numbers

	PIR	SWISSPROT	EMBL/GENBANK	REF
Rat brain			L24907, L26288	6,7

Amino acid sequence of rat CaMKI

```
MPGAVEGPRW KQAEDIRDIY DFRDVLGTGA FSEVILAEDK RTQKLVAIKC
IAKKALEGKE GSMENEIAVL HKIKHPNIVA LDDIYESGGH LYLIMQLVSG
GELFDRIVEK GFYTERDASR LIFQVLDAVK YLHDLGIVHR DLKPENLLYY
SLDEDSKIMI SDFGLSKMED PGSVLSTACG TPGYVAPEVL AQKPYSKAVD
CWSIGVIAYI LLCGYPPFYD ENDAKLFEQI LKAEYEFDSP YWDDISDSAK
DFIRHLMEKD PEKRFTCEQA LQHPWIAGDT ALDKNIHQSV SEQIKKNFAK
SKWKQAFNAT AVVRHMRKLQ LGTSQEGQGQ TASHGELLTP TAGGPAAGCC
CRDCCVEPGS ELPPAPPPSS RAMD
```

Sites of interest: T177, phosphorylation site; I294-A299, autoinhibitory region; K302-R314, calmodulin-binding region.

Homologues in other species

CaMKI has been purified from rat and bovine brain[1-3, 9]. CaMKI activity has also been detected in nervous system from *Aplysia*[12], suggesting that it has a widespread species distribution.

Physiological substrates and specificity determinants

Hydrophobic and arginine residues in site 1 of synapsin I or II (LRRRLSDSNF) are of greatest importance for substrate recognition[11, 13, 20].

Assay

The enzyme is assayed by the transfer of radioactivity from $[\gamma\text{-}^{32}\text{P}]$ATP to a synthetic peptide based on site 1 of synapsin I or II (NYLRRRLSDSNF[2, 3] or YLRRRLSDSNF-amide[6]). Synapsin I may also be used as substrate: however, the use of a short peptide minimizes the interference from **CaMKII** activity.

Enzyme activators and inhibitors

Ca^{2+} and CaM are required for activity[1-3, 9]. The apparent K_a for CaM of CaMKIa, Ib and V are 11, 0.53 and 24.7 nM, respectively. A protein kinase (CaMKI kinase) has been purified[15, 16] that activates CaMKI through phosphorylation at T177[8]. CaM inhibitors, e.g. calmidozolium, trifluoperazine or W7 will inhibit activity. KN-62, previously described as a specific inhibitor of **CaMKII**[14], is also able to inhibit CaMKI with a K_i of 1 μM[9].

Each of the preparations of CaMKI have been shown to undergo autophosphorylation, however, the rate of phosphorylation is slow in comparison to the rate of phosphorylation by CaMKI kinase[8, 11, 15]. CaMKIb is unresponsive to CaMKI kinase, however, incubation with protein phosphatase-2A leads to a loss in kinase activity[3]. Therefore, CaMKIb may represent a phosphorylated CaMKI isoform.

Pattern of expression

CaMKI has a cytosolic distribution[1-3, 9, 18, 19]. The enzyme is present in all cell types and tissues examined as measured by enyme activity[1], mRNA distribution[6-8], or immunochemical methods[18, 19]. The highest level of CaMKI mRNA is found in the frontal cortex in brain. High levels of the kinase are also present in many tissues including, adrenal gland, lung, liver, kidney and peripheral blood leukocytes[6-8].

References

1 Nairn, A.C. and Greengard P. (1987) J. Biol. Chem. 262, 7273-7281.
2 DeRemer, M. F. et al (1992) J. Biol. Chem. 267, 13460-13465.
3 DeRemer, M.F. et al (1992) J. Biol. Chem. 267, 13466-13471.
4 Sheng, M. et al (1991) Science 252, 1427-1430.
5 Picciotto, M.R. et al (1992) J. Biol. Chem. 267, 12742-12752.
6 Picciotto, M.R. et al (1993) J. Biol. Chem. 268, 26512-26521.
7 Cho, F.S. et al. (1994) Biochim. Biophys. Acta 1224, 156-160.
8 Haribabu et al. (1995) Submitted.
9 Mochizuki, H. et al (1993) J. Biol. Chem. 268, 9143-9147
10 Yokokura, H. et al. (1995) Submitted.
11 Picciotto, M.R. et al. (1995) Submitted.
12 DeRiemer, S.A. et al (1984) J. Neurosci. 4, 1618-1625.
13 Lee, J.C. et al. (1994) Proc. Natl. Acad. Sci. USA 91, 6413-6417.
14 Tokumitsu, H. et al (1990) J. Biol. Chem. 265, 4315-4320.
15 Lee, J.C. and Edelman, A.M. (1994) J. Biol. Chem. 290, 2158-2164.
16 Edelman, A.M. and Lee, J.C. (1994) FASEB J. 8, 1235a.
17 Sugita, R. et al. (1994) Biochim. Biophys. Res. Commun. 203, 694-701.
18 Ito, T. et al. (1994) Arch. Biochem. Biophys. 312, 278-284.
19 Picciotto, M.R. et al. (1995) Synapse, In Press.
20 Dale, S. et al. (1995) FEBS Lett, In Press.

CaMKII	CaM-dependent PK II (vertebrates) (Type II CaM-dependent PK, multifunctional CaM-dependent PK, CaM-dependent multiprotein kinase)

Mary B. Kennedy (California Institute of Technology, USA)

Ca^{2+}/calmodulin-dependent protein kinase II is a multisubstrate protein kinase that is activated by increases in intracellular Ca^{2+} ion concentration. The kinase phosphorylates a wide variety of intracellular proteins including metabolic enzymes, transmitter synthesizing enzymes, ion channels, membrane receptor proteins, and transcription factors.

Subunit structure and isoforms

SUBUNIT	AMINO ACIDS	MOL. WT	SDS-PAGE
α	478	54 114	50 kDa
β	542	60 401	60 kDa
γ	527	59 038	59 kDa
δ	533	60 080	60 kDa

The holoenzyme is a heteromultimer composed of several catalytic subunits. Four homologous subunits, α, β, γ and δ, are encoded by distinct genes. The holoenzymes have molecular weights of 500 000 to 650 000 and are each composed of ten to twelve subunits mixed randomly in the proportion in which they are present in the cell [but see ref. 9]. mRNAs encoding the subunits are alternatively spliced, producing small variations in molecular weight.

Domain structure

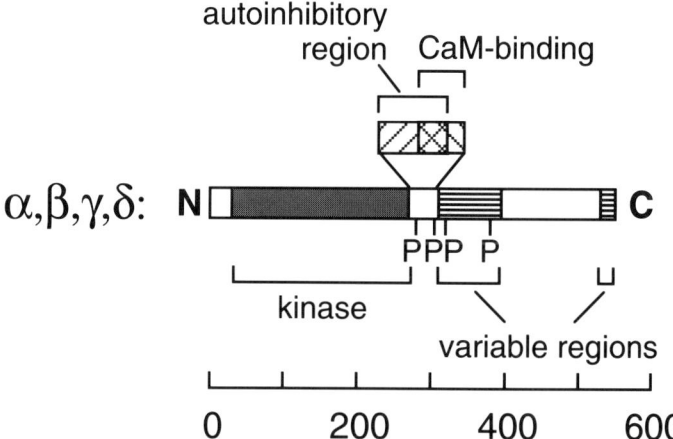

The α, β, γ and δ subunits contain an N-terminal kinase domain followed immediately by an autoinhibitory region (274–303) containing the autophosphorylation site at T286/287. The autoinhibitory region partially overlaps the calmodulin-binding domain (292–311). Immediately following the calmodulin-binding domain is a region of variable sequence (~315–392). Differences in the number of residues in this region account for much of the differences in molecular weights of the subunits. Alternative splicing of each subunit occurs at sites encoding residues in this region. The δ subunit contains a short 20 amino acid sequence at the C-terminus that is not present in the other three. Autophosphorylation of T286/287 partially relieves inhibition of catalytic activity and thus produces substantial Ca^{2+}/CaM-independent kinase activity. Autophosphorylation of T305 and T306 prevents binding of CaM. The functional consequences of autophosphorylation at S314, S315 and T382 are not known.

Database accession numbers

		PIR	SWISSPROT	EMBL/GENBANK	REF
α	Rat	A30355	P11275	J02942	1
	Mouse	S04365	P11798	X14836	5
β	Rat	A26464	P08413	M16112	2
	Mouse	S18915	P28652	X63615	
γ	Rat	A31908	P15791	J04063	3
δ	Rat	A34366	P11730	J05072	4

Amino acid sequences of rat CaMKII

α subunit

```
MATITCTRFT EEYQLFEELG KGAFSVVRRC VKVLAGQEYA AKIINTKKLS  50
ARDHQKLERE ARICRLLKHP NIVRLHDSIS EEGHHYLIFD LVTGGELFED 100
IVAREYYSEA DASHCIQQIL EAVLHCHQMG VVHRDLKPEN LLLASKLKGA 150
AVKLADFGLA IEVEGEQQAW FGFAGTPGYL SPEVLRKDPY GKPVDLWACG 200
VILYILLVGY PPFWDEDQHR LYQQIKAGAY DFPSPEWDTV TPEAKDLINK 250
MLTINPSKRI TAAEALKHPW ISHRSTVASC MHRQETVDCL KKFNARRKLK 300
GAILTTMLAT RNFSGGKSGG NKKNDGVKES SESTNTTIED EDTKVRKQEI 350
IKVTEQLIEA ISNGDFESYT KMCDPGMTAF EPEALGNLVE GLDFHRFYFE 400
NLWSRNSKPV HTTILNPHIH LMGDESACIA YIRITQYLDA GGIPRTAQSE 450
ETRVWHRRDG KWQIVHFHRS GAPSVLPH
```

Sites of interest: T286, T305, S314, autophosphorylation

β subunit

```
MATTVTCTRF TDEYQLYEDI GKGAFSVVRR CVKLCTGHEY AAKIINTKKL  50
SARDHQKLER EARICRLLKH SNIVRLHDSI SEEGFHYLVF DLVTGGELFE 100
DIVAREYYSE ADASHCIQQI LEAVLHCHQM GVVHRDLKPE NLLLASKCKG 150
AAVKLADFGL AIEVQGDQQA WFGFAGTPGY LSPEVLRKEA YGKPVDIWAC 200
GVILYILLVG YPPFWDEDQH KLYQQIKAGA YDFPSPEWDT VTPEAKNLIN 250
QMLTINPAKR ITAHEALKHP WVCQRSTVAS MMHRQETVEC LKKFNARRKL 300
KGAILTTMLA TRNFSVGRQT TAPATMSTAA SGTTMGLVEQ AKSLLNKKAD 350
GVKPQTNSTK NSSAITSPKG SLPPAALEPQ TTVIHNPVDG IKESSDSTNT 400
TIEDEDAKAR KQEIIKTTEQ LIEAVNNGDF EAYAKICDPG LTSFEPEALG 450
NLVEGMDFHR FYFENLLAKN SKPIHTTILN PHVHVIGEDA ACIAYIRLTQ 500
YIDGQGRPRT SQSEETRVWH RPDGKWQNVH FHCSGAPVAP LQ
```

Sites of interest: T287, T306, S315, T382, autophosphorylation; E378–K392, residues deleted by splicing in the β' isoform

γ subunit

```
MATTATCTRF TDDYQLFEEL GKGAFSVVRR CVKKTSTQEY AAKIINTKKL  50
SARDHQKLER EARICRLLKH PNIVRLHDSI SEEGFHYLVF DLVTGGELFE 100
DIVAREYYSE ADASHCIHQI LESVNHIHQH DIVHRDLKPE NLLLASKCKG 150
AAVKLADFGL AIEVQGEQQA WFGFAGTPGY LSPEVLRKDP YGKPVDIWAC 200
GVILYILLVG YPPFWDEDQH KLYQQIKAGA YDFPSPEWDT VTPEAKNLIN 250
QMLTINPAKR ITADQALKHP WVCQRSTVAS MMHRQETVEC LRKFNARRKL 300
KGAILTTMLV SRNFSVGRQS SAPASPAASA AGLAGQAAKS LLNKKSDGGV 350
KKRKSSSSVH LMEPQTTVVH NATDGIKGST ESCNTTTEDE DLKVRKQEII 400
KITEQLIEAI NNGDFEAYTK ICDPGLTSFE PEALGNLVEG MDFHKFYFEN 450
LLSKNSKPIH TTILNPHVHV IGEDAACIAY IRLTQYIDGQ GRPRTSQSEE 500
TRVWHRRDGK WLNVHYHCSG APAAPLQ
```

Sites of interest: T287, autophosphorylation

δ subunit

```
MASTTTCTRF TDEYQLFEEL GKGAFSVVRR CMKIPTGQEY AAKIINTKKL  50
SARDHQKLER EARICRLLKH PNIVRLHDSI SEEGFHYLVF DLVTGGELFE 100
DIVAREYYSE ADASHCIQQI LESVNHCHLN GIVHRDLKPE NLLLASKSKG 150
AAVKLADFGL AIEVQGDQQA WFGFAGTPGY LSPEVLRKDP YGKPVDMWAC 200
GVILYILLVG YPPFWDEDQH RLYQQIKAGA YDFPSPEWDT VTPEAKDLIN 250
KMLTINPAKR ITASEALKHP WICQRSTVAS MMHRQETVDC LKKFNARRKL 300
KGAILTTMLA TRNFSAAKSL LKKPDGVKIN NKANVVTSPK ENIPTPALEP 350
QTTVIHNPDG NKESTESSNT TIEDEDVKAR KQEIIKVTEQ LIEAINNGDF 400
EAYTKICDPG LTAFEPEALG NLVEGMDFHR FYFENALPKI NKPIHTIILN 450
PHVHLVGDDA ACIAYIRLTQ YMDGNGMPKT MQSEETRVWH RRDGKWQNIH 500
FHRSGSPTVP IKPPCIPNGK ENFSGGTSLW QNI
```

Sites of interest: T287, autophosphorylation

Homologues in other species

The α and β subunits have been sequenced from mouse (brain).

Physiological substrates and specificity determinants

The consensus sequence for phosphorylation by CaM kinase II is RXXS/T. A selection of sequences around sites believed to be phosphorylated *in vivo* by the kinase is shown below:

CaMKII (autophosphorylation):	MHRQE**T**VD/EC
Synapsin I, Site 2:	ATRQA**S**ISG
Synapsin I, Site 3:	PIRQA**S**QAG
Tyrosine hydroxylase:	FRRAV**S**EQD
CREB transcription factor:	LSRRP**S**YRK

Assay

CaMKII activity can be assayed as Ca^{2+}- and CaM-dependent transfer of radioactivity from $[\gamma$-^{32}P]ATP to synapsin I[6], or to the peptides syntide (PLARTLSVAGLPGKK)[7] or autocamtide (KKALRQETVDAL)[8], which have sequences related to site 2 on glycogen synthase, or the autophosphorylation site of CaMKII, respectively.

Enzyme activators and inhibitors

	SPECIFICITY	K_i (μM)
Autoinhibitory peptide (CaMKIIα 273–302)	Excellent	1.0
KN-62	Good	0.9

Pattern of expression

CaMKII is present in most tissues and is expressed at high concentrations in the brain. The α and β subunits are expressed only in nervous tissue; they comprise approximately 1% of total protein in the forebrain. In contrast, the γ and δ subunits are expressed at relatively low levels in a wide variety of tissues.

Within neurones, CaMKII is present throughout the cytosol. It is concentrated in postsynaptic densities, specializations of the submembranous cytoskeleton attached to the cytosolic face of the postsynaptic membrane, where it is thought to comprise approximately 20% of total protein. It is also found tightly bound to presynaptic vesicles where it serves as a binding protein for one of its substrates, synapsin I.

References

1   Lin, C.R. et al. (1987) Proc. Natl Acad. Sci. USA 84, 5962–5966.
2   Bennett, M.K. and Kennedy, M.B. (1987) Proc. Natl Acad. Sci. USA 84, 1794–1798.
3   Tobimatsu, T. et al. (1988) J. Biol. Chem. 263, 16082–16086.
4   Tobimatsu, T. and Fujisawa, H. (1989) J. Biol. Chem. 264, 17907–17912.
5   Hanley, R.M. et al. (1989) Nucl. Acids Res. 17, 3992.
6   Bennett, M.K. et al. (1983) J. Biol. Chem. 258, 12735–12744.
7   Schworer, C.M. et al. (1986) J. Biol. Chem. 261, 8581–8584.
8   Hanson, P.I. et al. (1989) Neuron 3, 59–70.
9   Kaneseki, T. et al. (1991) J. Cell Biol. 115, 1049–1060.

EF2K — Elongation factor-2 kinase (vertebrates) (CaM-dependent PK III)

Angus C. Nairn (Rockefeller University, USA)

EF-2 kinase phosphorylates and inactivates elongation factor-2 (EF-2)[1-5], an essential factor necessary for the extension of the polypeptide chain on the ribosome. EF-2 kinase is transiently activated, and EF-2 phosphorylated, in cells treated with hormones, mitogens and growth factors which increase intracellular calcium[6]. The exact function of Ca^{2+}-dependent inhibition of protein synthesis is unclear, but it is likely that it is involved in Ca^{2+}-dependent regulation of cell proliferation.

Subunit structure and isoforms

SUBUNIT	AMINO ACIDS	MOL. WT	SDS-PAGE
EF2K	?	?	95 kDa[7]/103 kDa[8]

EF2K is a monomer. It requires Ca^{2+} and calmodulin (16.5 kDa) for activity. The differences in subunit size[7, 8] probably reflect minor differences in the SDS-PAGE systems used rather than different isoforms. Degradation products of 85 or 95 kDa are observed in some preparations.

Amino acid sequence of EF2K

Unknown. A report that EF-2 kinase was identical to heat shock protein hsp90[9] is incorrect[7, 8]. Partial amino acid sequence of the rabbit reticulocyte enzyme suggests it is a novel protein kinase[8].

Homologues in other species

EF-2 kinase has been purified from rabbit reticulocytes and rat pancreas[7, 8]. The reticulocyte enzyme phosphorylates mammalian and yeast EF-2 at similar rates and with the same affinity[8]. An EF-2 kinase with properties similar to the mammalian enzyme has been detected in yeast[8].

Physiological substrates and specificity determinants

EF-2 kinase phosphorylates EF-2 with high specificity[2, 8]. EF-2 kinase does not phosphorylate a large number of other proteins and EF-2 is not phosphorylated by other protein kinases. EF-2 kinase phosphorylates two threonine residues[4, 10] (T56 and T58, bold type) with T56 being phosphorylated more rapidly than T58[10]. The exact consensus sequence for phosphorylation by EF-2 kinase is not known, but basic amino acids (underlined) adjacent to the phosphorylated threonine residues are likely to be important. T56 and T58, but not T53, are conserved in EF-2 from yeast to mammals.

EF-2 (50–60): AGET**R**FT**D**T**RK**

Assay

The enzyme is assayed using EF-2 as substrate[7, 8]. A short synthetic peptide (EF-2$_{50-60}$) is poorly phosphorylated by the kinase. Ca^{2+} (μM) and calmodulin (apparent $K_d \approx 1$ nM) are required for activity at pH >6.5. Some activity ($\approx$30% of maximum) can be measured in the absence of Ca^{2+}/CaM at pH 6.0[8].

Enzyme activators and inhibitors

Calcium and calmodulin are required for normal activity[8]. Calmodulin inhibitors, e.g. calmidozolium, trifluoperazine or W7 will inhibit activity. EF-2 kinase is autophosphorylated on multiple seryl and threonyl residues by an intramolecular mechanism[7,8]. Autophosphorylation is associated with the generation of a partially Ca^{2+}/CaM-independent activity ($\approx$30% of maximum)[7,8] and a 2–3-fold increase in total enzyme activity[7]. Furthermore, EF-2 kinase is phosphorylated *in vitro* by cAMP-dependent protein kinase[8], and this is also asociated with the generation of Ca^{2+}/Cam-independent activity[16].

Pattern of expression

EF-2 kinase has a cytosolic distribution. The enzyme is present in all cell types and tissues examined and is particularly abundant in rapidly dividing cells or in tissues exhibiting high levels of protein synthesis[2]. EF-2 phosphorylation has been found to increase during M phase[11] and EF-2 kinase activity may be regulated at other stages of the cell cycle[12]. The enzyme is present at low levels in non-dividing cells[13,14]. For example, in PC12 cells stimulated to differentiate into a neurone-like phenotype by treatment with nerve growth factor or agents that raise intracellular cAMP levels, the levels of EF-2 kinase are downregulated in 2-6 h[13]. The mechanism of down-regulation is not known, however, in PC12 cells this appears to involve cAMP-dependent protein kinase[15].

References

1   Proud, C.G. (1992) Curr. Top. Cell. Regul. 32, 243–369.
2   Nairn, A.C. et al (1985) Proc. Natl Acad. Sci. USA 82, 7939–7943.
3   Ryazanov, A.G. (1987) FEBS Lett. 214, 331–334.
4   Nairn, A.C. and Palfrey, H.C. (1987) J. Biol. Chem. 262, 17299–17303.
5   Ryazanov, A.G. et al. (1988) Nature 334, 170–173.
6   Palfrey, H.C. et al. (1987) J. Biol. Chem. 262, 9785–9792.
7   Redpath, N.T. and Proud, C. G. (1993) Eur. J. Biochem. 212, 511–520.
8   Mitsui, K. et al. (1993) J. Biol. Chem. 268, 13422–13433.
9   Nygärd, O. et al. (1991) J. Biol. Chem. 266, 16425–16430.
10 Price, N.T. et al. (1991) FEBS Lett. 282, 253–258.
11 Celis, J.E. et al. (1990) Proc. Natl Acad. Sci. USA 87, 4231–4235.
12 Carlberg, U. et al. (1991) Biochem. Biophys. Res. Commun. 180, 1372–1376.
13 Nairn, A.C. et al. (1987) J. Biol. Chem. 262, 14265–14272.
14 Bagaglio, D. et al. (1993) Cancer Res. 53, 2260–2264.
15 Brady, M.J. et al. (1990) J. Neurochem. 54, 1034–1039.
16 Redpath, N.T. and Proud, C.G. (1993) Biochem. J. 293, 31–34.

CaMKIV   CaM-dependent PK IV (vertebrates)

John Glod and James M. Sikela
(University of Colorado Health Sciences Center, USA)

Ca^{2+}/calmodulin-dependent protein kinase IV is a multifunctional kinase that has been shown to have a number of *in vitro* substrates. Its *in vivo* actions, however, are not as well understood. Transcripts from the CaMKIV gene can undergo alternative splicing to produce either a catalytically active calmodulin-dependent kinase or a testis-specific calmodulin binding protein, calspermin[1, 2, 3].

Subunit structure and isoforms

SUBUNIT	AMINO ACIDS	MOL. WT	SDS-PAGE
Mouse	469	52 627	65 and 67 kDa

CaMKIV is a monomer. Two isoforms can be discerned by SDS-PAGE. The higher molecular weight isoform is found only in the cerebellum[4, 5]. The discrepancy between the predicted molecular weight of the kinase and its migration on SDS-PAGE may be accounted for by anomalous migration due to the extremely acidic nature of the protein (pI = 4.56).

Genetics
Chromosomal location:

Human CaMKIV:	5q21–q23	(6)
Mouse CaMKIV:	18	(7)

Domain structure

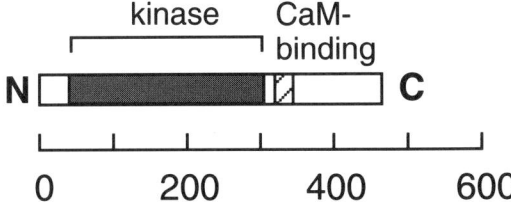

The CaM-binding domain has been localized to residues 317–334, 17 residues C-terminal to the catalytic domain[8]. The C-terminal end of the molecule contains an abundance of acidic residues which may target CaMKIV to the nucleus. The C-terminal end also contains a number of PEST sequences, suggesting that a possible cleavage site may be located here.

Database accession numbers

	PIR	SWISSPROT	EMBL/GENBANK	REF
Mouse			X58995	1
Rat			M74488	2
Rat			M64757	3

Amino acid sequence of mouse CaMKIV

```
MLKVTVPSCP SSPCSSVTAS TENLVPDYWI DGSNRDPLGD FFEVESELGR  50
GATSIVYRCK QKGTQKPYAL KVLKKTVDKK IVRTEIGVLL RLSHPNIIKL 100
KEIFETPTEI SLVLELVTGG ELFDRIVEKG YYSERDARDA VKQILEAVAY 150
LHENGIVHRD LKPENLLYAT PAPDAPLKIA DFGLSKIVEH QVLMKTVCGT 200
PGYCAPEILR GCAYGPEVDM WSVGIITYIL LCGFEPFYDE RGDQFMFRRI 250
LNCEYYFISP WWDEVSLNAK DLVKKLIVLD PKKRLTTFQA LQHPWVTGKA 300
ANFVHMDTAQ KKLQEFNARR KLKAAVKAVV ASSRLGSASS SHTSIQENHK 350
ASSDPPSTQD AKDSTDLLGK KMQEEDQEED QVEAEASADE MRKLQSEEVE 400
KDAGVKEEET SSMVPQDPED ELETDDPEMK RDSEEKLKSV EEEMDPMTEE 450
EAPDAGLGVP QQDAIQPEY
```

Homologues in other species

The sequences of mouse and rat CaMKIV have been reported. They are very similar in all areas of the protein except for a segment near the C-terminal end of the protein (residues 354–409 in the mouse) in which amino acid identity falls to approximately 45%.

Physiological substrates and specificity determinants

Information concerning *in vivo* substrates for CaMKIV is sparse, but based on colocalization, synapsin I and Rap-1b are potential candidates. A synthetic peptide comprising K345–S358 of CaMKIIγ has been found to be a specific substrate for CaMKIV[4]. The consensus sequence for phosphorylation by CaMKIV is not clear, but the following peptides can act as substrates:

Rap-1b: VPGKARKKSSCQLL   [9]
CaMKIIγ (345–358):   KSDGGVKKRKSSSS   [4]

Assay

CaMKIV is assayed by transfer of radioactivity from $[\gamma\text{-}^{32}P]ATP$ to synapsin I.

Enzyme activators and inhibitors

CaMKIV is activated by Ca^{2+}/CaM and through autophosphorylation[10]. The activity of the kinase is decreased by phosphorylation by PKA[11].

Pattern of expression

Immunocytochemical studies on cerebellar granule cells have found CaMKIV to be abundant in the nucleus (particularly regions of dispersed chromatin) and present to a lesser degree in cytoplasm[5]. Immunohistochemical and *in situ* hybridization show that CaMKIV is present in the brain, with highest levels in cerebellum. It is also present in testis, spleen, and thymus[1, 2]. The 67 kDa isoform appears to be present only in cerebellum. Studies in rat brain indicate that CaMKIV levels in cerebellar granule and Purkinje cells are modulated during early postnatal development[12].

References

[1] Jones, D.A. et al. (1991) FEBS Lett. 289, 105–109.

[2] Means, A.R. et al. (1991) Mol. Cell. Biol. 11, 3960–3971.

[3] Ohmstede, C.A. et al. (1991) Proc. Natl Acad. Sci. USA 88, 5784–5788.

[4] Miyano, O. et al. (1992) J. Biol. Chem. 267, 1198–1203.

[5] Ohmstede, C.A. et al. (1989) J.Biol.Chem. 264, 5866–5875.

[6] Sikela, J.M. and Hahn, W.E. (1987) Proc. Natl Acad. Sci. USA 84, 3038–3042.

[7] Sikela, J.M. et al. (1989) Genomics 4, 21–27.

[8] Sikela, J.M. et al. (1990) Genomics 8, 579–582.

[9] Sahyoun, N. et al. (1991) Proc. Natl Acad. Sci. USA 88, 2643–2647.

[10] Frangakis, M.V. et al. (1991) J. Biol. Chem. 266, 11309–11316.

[11] Kamashita, I. and Hitoshi, F. (1991) Biochem. Biophys. Res. Commun. 180, 191–196.

[12] Sakagami, H. et al. (1992) Mol. Brain Res. 16, 20–28.

Mary B. Kennedy (California Institute of Technology, USA)

Ca^{2+}/calmodulin-dependent protein kinase II is a multisubstrate protein kinase that is activated by increases in intracellular Ca^{2+} concentration. The mammalian kinase **CaMKII** phosphorylates a wide variety of intracellular proteins including metabolic enzymes, transmitter synthesizing enzymes, ion channels, membrane receptor proteins, and transcription factors.

Subunit structure and isoforms

SUBUNIT	AMINO ACIDS	MOL. WT	SDS-PAGE
490 aa	490	55 481	54 kDa
509 aa	509	57 568	58 kDa
516 aa	516	58 385	58 kDa
530 aa	530	59 919	60 kDa

Although the molecular weight of DmCaMKII has not been determined, by analogy with the mammalian enzyme it is thought to be a heteromultimer composed of several subunits, all of which are catalytic. At least four alternatively spliced catalytic subunits have been described. The smallest corresponds to the α subunit and the largest to the β and δ subunits of rat **CaMKII**.

Genetics
Chromosomal location: 4, between 102E and F.

Domain structure

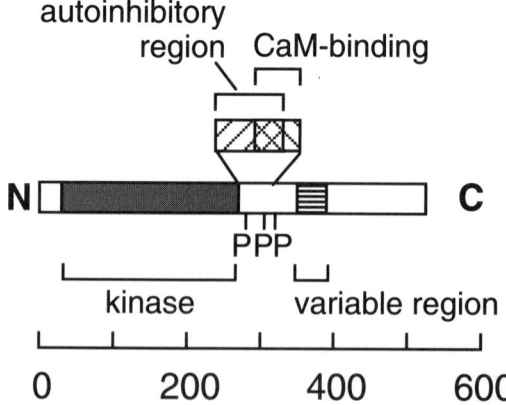

The four known subunits of DmCaMKII arise by alternative splicing of a single gene. Like rat **CaMKII**, each subunit contains an N-terminal kinase domain followed immediately by an autoinhibitory region (274–303)

containing the autophosphorylation site at T287. The inhibitory domain partially overlaps the calmodulin-binding domain (292–311). The variable sequences are found between residues 347 and 397. The autophosphorylation sites (T287, T306, S315) are conserved between rat and *D. melanogaster*.

Database accession numbers

	PIR	SWISSPROT	EMBL/GENBANK	REF
490 aa	JU0270		M74583/D13330	1, 2
509 aa			D13331	2
516 aa			D13332	2
530 aa			D13333	2

Amino acid sequence of *D. melanogaster* CaMKII

490 aa subunit

```
MAAPAACTRF SDNYDIKEEL GKGAFSIVKR CVQKSTGFEF AAKIINTKKL  50
TARDFQKLER EARICRKLHH PNIVRLHDSI QEENYHYLVF DLVTGGELFE 100
DIVAREFYSE ADASHCIQQI LESVNHCHQN GVVHRDLKPE NLLLASKAKG 150
AAVKLADFGL AIEVQGDHQA WFGFAGTPGY LSPEVLKKEP YGKSVDIWAC 200
GVILYILLVG YPPFWDEDQH RLYSQIKAGA YDYPSPEWDT VTPEAKNLIN 250
QMLTVNPNKR ITAAEALKHP WICQRERVAS VVHRQETVDC LKKFNARRKL 300
KGAILTTMLA TRNFSSRSMI TKKGEGSQVK ESTDSSSTTL EDDDIKAARR 350
QEIIKITEQL IEAINSGDFD GYTKICDPHL TAFEPEALGN LVEGIDFHKF 400
YFENVLGKNC KAINTTILNP HVHLLGEEAA CIAYVRLTQY IDKQGHAHTH 450
QSEETRVWHK RDNKWQNVHF HRSASAKISG ATTFDFIPQK
```

Sites of interest: T287, T306, S315, autophosphorylation; K346–A347, insertion sequences between these residues in alternatively spliced isoforms, 509 aa (EDKK GTVDRSTTVV SKEPEA), 516 aa (EDKK GTVDRSTTVV SKEPEVNLFT NKA), 530 aa (EDKK GTVDRSTTVV SKEPEDIRIL CPAKTYQQNI GNSQCSS)

Physiological substrates/Assay/Enzyme activators and inhibitors

See **CaMKII**.

Pattern of expression

During germ band extension, high levels of expression of DmCaMKII mRNA become apparent in the anterior and posterior midgut and in the neuroblasts of the central nervous system. After germ band retraction, expression increases in the central nervous system and decreases in the midgut. In later stages, expression is largely restricted to the central and peripheral nervous systems. In the adult, DmCaMKII subunits are much more highly concentrated in the head than in the body.

References

1   Cho, K.-O. et al. (1991) Neuron 7, 439–450.
2   Ohsako, S. et al. (1993) J. Biol. Chem. 268, 2052–2062.

Marc Melcher, Mark Pausch and Jeremy Thorner
(University of California, Berkeley, USA)

Protein kinase activities activated by Ca^{2+} and yeast or mammalian calmodulin were detected in yeast extracts[1, 2], purified[3, 4], and the corresponding genes isolated[3, 4]. The yeast enzymes are homologous to mammalian **CaMKII** and, like their mammalian counterparts, are able to phosphorylate a broad range of substrates *in vitro* when Ca^{2+}/CaM are present.

Subunit structure and isoforms

SUBUNIT	AMINO ACIDS	MOL. WT	SDS-PAGE
Cmk1	446	50 295	55–56 kDa
Cmk2	447	50 422	50 kDa

There are two known isoforms of CaMKII in yeast, products of the *CMK1* and *CMK2* genes. The native form of Cmk1 appears to be an oligomer, most likely a dimer[2, 5].

Genetics
Chromosomal locations[3,4]:
    Cmk1:   VI, tightly centromere-linked
    Cmk2:   XV, tightly centromere-linked

Site-directed mutations:
    Cmk1, Δ303–446: dominant growth inhibition, suggests constitutively active[6]
    Cmk2, Δ313–447: dominant growth inhibition, suggests constitutively active[6]
    Cmk2, Δ308–447: no detectable phenotype[5]

Strains carrying *cmk1Δ* or *cmk2Δ* null mutations, and even *cmk1Δ cmk2Δ* double mutants, display no discernible phenotype, suggesting that additional CaMKII genes may exist[3, 5].

Domain structure

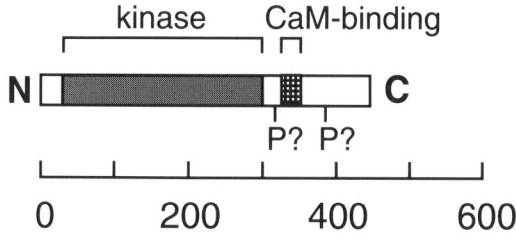

Cmk1 consists of a catalytic domain followed by a calmodulin-binding region[3] and a putative autoinhibitory region. Cmk2 has a very similar overall structure and is also a CaM-binding protein[3]. Binding of the $Ca^{2+}/$ CaM complex is thought to displace the autoinhibitory region thereby preventing occlusion of the active site of each enzyme. Both Cmk1 and Cmk2 become autophosphorylated in a Ca^{2+}/CaM-dependent manner, possibly at S417 in Cmk1 and T316 and S387 in Cmk2 (indicated above). However only Cmk2 becomes Ca^{2+}/CaM-independent after autophosphorylation[1, 3–5].

Database accession numbers

	PIR	SWISSPROT	EMBL/GENBANK	REF
Cmk1		P27466	X57782	3
Cmk1	A40896		D90375	4
Cmk2		P22517	X56961	3
Cmk2	B40896		D90376	4

Amino acid sequences of *S. cerevisiae* CaMKII

Cmk1[4]

```
MDDKVSEKES SPKQTEEDSE GKMAHVQPAS YVNKKKYVFG KTLGAGTFGV  50
VRQAKNTETG EDVAVKILIK KALKGNKVQL EALYDELDIL QRLHHPNIVA 100
FKDWFESKDK FYIITQLAKG GELFDRILKK GKFTEEDAVR ILVEILSAVK 150
YMHSQNIVHR DLKPENLLYI DKSDESPLVV ADFGIAKRLK SDEELLYKPA 200
GSLGYVAPEV LTQDGHGKPC DIWSIGVITY TLLCGYSAFR AERVQDFLDE 250
CTTGEYPVKF HRPYWDSVSN KAKQFILKAL NLDPSKRPTA AELLEDPWII 300
CTELKTHNLL PGLKEGLDAR QKFRNSVERV RLNMKIQKLR DLYLEQTESD 350
SDFDEGSQAN GSVPPLKATD TSQLSKKLSE EEQSKLKSEL TSKAFAQLVN 400
TVLAEKEKFL NINRVCSSDS DLPGSDIKSL DEAKEKPEGK DTKTEE.
```

Sites of interest: L313–R340, presumptive CaM-binding domain; S417, putative autophosphorylation site

The following unresolved differences exist between the sequences published by Pausch et al.[3] and Ohya et al.[4]: AGTA(199–202) vs. PAGS, insertion of S after Y236, TIDRK(238–242) vs. FRAER (239–243), and Q270 vs. K271.

Cmk2[3]

```
MPKESEVINS EFHVDVQDPE RLNGHPVAKF INKLSGQPES YVNRTNYIFG  50
RTLGAGSFGV VRQARKLSTN EDVAIKILLK KALQGNNVQL QMLYEELSIL 100
QKLSHPNIVS FKDWFESKDK FYIVTQLATG GELFDRILSR GKFTEVDAVE 150
IIVQILGAVE YMHSKNVVHR DLKPENVLYV DKSENSPLVI ADFGIAKQLK 200
GEEDLIYKAA GSLGYVAPEV LTQDGHGKPC DIWSIGVITY TLLCGYSPFI 250
AESVEGFMEE CTASRYPVTF HMPYWDNISI DAKRFILKAL RLNPADRPTA 300
TELLDDPWIT SKRVETSNIL PDVKKGFSLR KKLRDAIEIV KLNNRIKRLR 350
NMYSLGDDGD NDIEENSLNE SLLDGVTHSL DDLRLQSQKK GGELTEEQMK 400
LKSALTKDAF VQIVKAATKN KHKVLAGEEE DDSKKTLHDD RESKSED
```

Sites of interest: V323–R350, presumptive CaM-binding region, T316, S387, putative autophosphorylation sites

The published sequence of Ohya et al. has V rather than A at residue 282.

Homologues in other species

CaMKII has been identified in various mammals (**CaMKII**), in *D. mela-nogaster* (**DmCaMKII**), and in *Aspergillus nidulans*.

Physiological substrates and specificity determinants

Physiological substrates for Cmk1 and Cmk2 in yeast have not yet been characterized. Purified preparations of native or recombinant Cmk1 or Cmk2 will phosphorylate *in vitro* a wide variety of protein substrates, including myelin basic protein, synapsin I and casein, as well as synthetic peptides, such as kemptamide (based on a phosphorylation site in chicken gizzard myosin light chain) and autocamtide-2 (which contains the RQET sequence from the autophosphorylation site of rat brain CaMKII[2-5]).

Assay

Activity of Cmk1 and Cmk2 is most conveniently measured as the Ca^{2+}/CaM-dependent transfer of radioactivity from $[\gamma\text{-}^{32}P]ATP$ to autocamtide-2[3,5]. It can also be followed by autophosphorylation[2-5].

Enzyme activators and inhibitors

Phosphorylation of most of the proteins and peptides that serve as efficient phosphoacceptor substrates for Cmk1 and Cmk2, and autophosphorylation of Cmk1 and Cmk2, are strongly stimulated by Ca^{2+} and CaM.

Pattern of expression

CMK1 and *CMK2* are expressed at a similar level in all three yeast cell types (*MATa* haploids, *MATα* haploids, and *MATa/MATα* diploids) and are not pheromone-responsive genes[3]. Cmk1 and Cmk2 are localized primarily in the cytoplasm, as judged by immunofluorescence. Cmk1 is found almost exclusively in the soluble fraction of cell lysates, whereas a significant proportion of Cmk2 is associated with the particulate fraction[3,5].

References

[1] Miyakawa, T. et al. (1989) J. Bacteriol. 171, 1417–1422.
[2] Londesborough, J. (1989) J. Gen. Microbiol. 135, 3373–3383.
[3] **Pausch, M.H. et al. (1990) EMBO J. 10, 1511–1522.**
[4] Ohya, Y. et al. (1991) J. Biol. Chem. 266, 12784–12794.
[5] Melcher, M., Pausch, M.H. and Thorner, J., unpublished observations.
[6] Ohya, Y., personal communication.

Phosphorylase kinase (vertebrates)

Ludwig M.G. Heilmeyer Jr. and Manfred W. Kilimann
(Ruhr-Universität Bochum, Germany)

PhK regulates glycogen breakdown by phosphorylation of glycogen phosphorylase[1]. PhK is regulated, allosterically by Ca^{2+}, and by phosphorylation through PKA, in response to various neural and hormonal stimuli. PhK can phosphorylate various other proteins *in vitro*, but it is not known whether it has physiologically relevant protein substrates other than glycogen phosphorylase. PhK is isolated with multiple serine residues partially phosphorylated, but it is not clear what functional relevance these phosphate groups bear, and which kinases are responsible for their incorporation[2]. PhK is apparently found in all mammalian tissues. It has also been described in other organisms such as cartilaginous fish and insects.

Subunit structure and isoforms

SUBUNIT	AMINO ACIDS	MOL. WT	SDS-PAGE
α_M	1237	138 422	~133 kDa
α_L	1235	138 768	~138 kDa
β	1092	125 205	~125 kDa
γ_M	3P6	44 671	43 kDa
γ_T	406	46 677	43 kDa
δ	148	16 706	16 kDa

Subunit structure $(\alpha\beta\gamma\delta)_4$. γ is the catalytic subunit. α and β confer activation of the enzyme by phosphorylation. δ is calmodulin and mediates activation by Ca^{2+}; it is a stable constituent of the holoenzyme both in absence and presence of Ca^{2+}. Additionally, one molecule of calmodulin (δ') or skeletal muscle troponin C, can be bound per $\alpha\beta\gamma\delta$ protomer only in the presence of Ca^{2+}. There are many isoforms of PhK. Two known isoforms each of the subunits α (muscle and liver types, α_M and α_L) and γ (muscle and testis types, γ_M and γ_T) are encoded by distinct genes. In addition, the subunits α_M, α_L and β each have several subtypes arising through differential mRNA splicing. Differential splicing can affect both phosphorylation sites and calmodulin-binding sites.

Genetics

Chromosomal locations of human genes:

SUBUNIT	LOCUS	LOCATION	REF
α_M	PHKA1	Xq12–q13	3
α_L	PHKA2	Xp22.2–p22.1	4
β	PHKB	16q12–q13	3
β pseudogenes		unmapped	
γ_M	PHKG1	7p12–q21	5
pseudogenes or homologues		7q21 and 11p11–p14	5
γ_T	PHKG2	16p11.2–p12.1	

α_M, α_L and β genes have also been mapped in mouse and rabbit. The mouse and rat γ_M genes have been characterized. Heritable PhK deficiency, causing glycogen storage disease, is known in humans, mice and rats. There are multiple forms which differ in the range of tissues affected and in the mode of inheritance (X-chromosomal recessive or autosomal recessive). The genetic heterogeneity probably reflects the isoform diversity of PhK, and the location of PhK subunits on the X chromosome as well as on autosomes. Mutations in *PHKA1* are responsible for muscle PhK deficiency in mice and humans[7, 8].

Domain structure

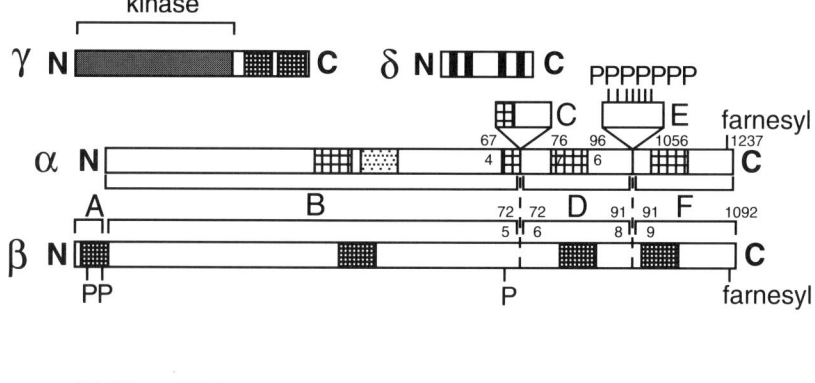

KEY:

▓ Calmodulin-binding, high affinity

▦ Calmodulin-binding, low affinity

▒ MgATP-binding

▌ Ca2+-binding EF hand

α and β are homologous over most of their length (regions B, D, F) but α has two inserts (C and E)[9]. Most phosphorylation sites lie in the α- or β-specific regions. The δ subunit binds to high-affinity calmodulin-binding domains on the γ subunit and possibly to two of the four calmodulin-binding domains on the α and β subunits. The remaining two calmodulin-binding domains on the α and β subunits are probably involved in binding δ[10]. Negative staining of PhK in electron micrographs visualizes a bilobal structure termed a "butterfly" which may complex four molecules of phosphorylase b[11].

Database accession numbers

		PIR	SWISSPROT	EMBL/GENBANK	REF
α_M	Rabbit muscle		P18688	J03247	12
	(slow-muscle)			M64656	13
	Mouse muscle			X74616	7
	Human muscle			X73874	14
α_L	Rabbit			X60421	4
β	Rabbit		P12798	J04120	9
	(liver)			M64657	13
	(brain)			M64658	13
γ_M	Rabbit		P00518	Y00684	15
γ_M	Mouse			J02731	16
γ_T	Rat		P31325	M73808	17

Amino acid sequences of rabbit PhK

α_M subunit

```
MRSRSNSGVR LDSYARLVQQ TILCHQNPVT GLLPASYDQK DAWVRDNVYS   50
ILAVWGLGLA YRKNADRDED KAKAYELEQS VVKLMRGLLH CMIRQVDKVE  100
SFKYSQSTKD SLHAKYNTKT CATVVGDDQW GHLQLDATSV YLLFLAQMTA  150
SGLHIIHSLD EVNFIQNLVF YIEAAYKTAD FGIWERGDKT NQGISELNAS  200
SVGMAKAALE ALDELDLFGV KGGPQSVIHV LADEVQHCQS ILNSLLPRAS  250
TSKEVDASLL SVISFPAFAV EDSKLVEITK QEIITKLQGR YGCCRFLRDG  300
YKTPKEDPNR LYYEPAELKL FENIECEWPL FWTYFILDGV FSGNAEQVQE  350
YREALEAVLI KGKNGVPLLP ELYSVPPDKV DEEYQNPHTV DRVPMGKLPH  400
MWGQSLYILG SLMAEGFLAP GEIDPLNRRF STVPKPDVVV QVSILAETEE  450
IKAILKDKGI NVETIAEVYP IRVQPARILS HIYSSLGCNN RMKLSGRPYR  500
HMGVLGTSKL YDIRKTIFTF TPQFIDQQQF YLALDNKMIV EMLRTDLSYL  550
CSRWRMTGQP TITFPISQTM LDEDGTSLNS SILAALRKMQ DGYFGGARIQ  600
TGKLSEFLTT SCCTHLSFMD PGPEGKLYSE DYDDNYDELE SGDWMDGYNS  650
TSTARCGDEV ARYLDHLLAH TAPHPKLAPA SQKGGLNRFR AAVQTTCDLM  700
SLVTKAKELH VQNVHMYLPT KLFQASRPSL NLLDSSHPSQ EDQVPTVRVE  750
VHLPRDQSGE VDFQALVLQL KETSSLQEQA DILYMLYTMK GPDWDTELYE  800
EGSATVRELL TELYGKVGKI RHWGLIRYIS GILRKKVEAL DEACTDLLSH  850
QKHLTVGLPP EPREKTISAP LPYEALTRLI EEACEGDMNI SILTQEIMVY  900
LAMYMRTQPG LFAEMFRLRI GLIIQVMATE LAHSLRCSAE EATEGLMNLS  950
PSAMKNLLHH ILSGKEFGVE RSVRPTDSNV SPAISIHEIG AVGATKTERT 1000
GIMQLKSEIK QVEFRRLSIS TESQPNGGHS LGADLMSPSF LSPGTSVTPS 1050
SGSFPGHHTS KDSRQGQWQR RRRLDGALNR VPIGFYQKVW KVLQKCHGLS 1100
VEGFVLPSST TREMTPGEIK FSVHVESVLN RVPQPEYRQL LVEAILVLTM 1150
LADIEIHSIG SIIAVEKIVH IANDLFLQEQ KTLGADDIML AKDPASGICT 1200
LLYDSAPSGR FGTMTYLSKA AATYVQEFLP HSICAMQ
```

Sites of interest: A654–V761, V1012–Q1024 and P1025–L1041 (underlined), these three sequences can be deleted in differentially spliced subtypes; S972, S985, S1007, autophosphorylation; S1018, PKA phosphorylation; S1020, S1023, S1030, endogenous phosphorylation; C1234, farnesylation

α_L subunit

```
MRSRSNSGVR LDGYARLVQQ TILCYQNPVT GLLSASHEQK DAWVRDNIYS   50
ILAVWGLGMA YRKNADRDED KAKAYELEQN VVKLMRGLLQ CMMRQVDKVE  100
KFKYTQSTKD SLHAKYNTAT CSTVVGDDQW GHLQVDATSL FLLFLAQMTA  150
SGLRIIFTLD EVAFIQNLVF YIEAAYKVAD YGMWERGDKT NQGIPELNAS  200
SVGMAKAALE AIDELDLFGA HGGRKSVIHV LPDEVEHCQS ILFSMLPRAS  250
TSKEIDAGLL SIISFPAFAV EDANLVNVTK SEIISKLQGR YGCCRFLRDG  300
YKTPREDPNR LHYDPAELKL FENIECEWPV FWTYFIIDGI FNGDALQVQE  350
YQEALEGILI RGKDGIRLVP ELYAIPPNKV DEEYKNPHTV DRVPLGKLPH  400
LWGQSLYILS SLLAEGFLAT GEIDPLNRRF STSVKPDVVV QVTVLAENSH  450
IKELLRKHGV DVQSIADIYP IRVQPGRILS HIYAKLGRNK NMKLSGRPYR  500
HIGVLGTSKL YVIRNQIFTF TPQFTDQHHF YLALDNEMIV EMLRIELAYL  550
CTCWRMTGRP TLTFPITHTM LTNDGSDIHS AVLSTIRKLE DGYFGGARVQ  600
LGNLSEFLTT SFYTYLTFLD PDCDEKLFDD ASEGSFSPDS DSDLGGYLEE  650
TYNQVTESQD ELDKYINHLL QSTYSKCHLP PLCKKMEDHN VFSAIHSTRD  700
ILSMMAKAKG LEVPFAPMTL PTKALSVHRK SLNLVDSPQP LLKRILKRLH  750
WPKDERGDVD CEKLVEQLKD CCTLQDQADI LYILYVLKGP SWDTALSGQH  800
GVTVHNLLSE LYGKAGLNQE WGLIRYISGL LRKKVEVLAE ACADLLSHQK  850
QLTVGLPPEP REKTISAPLP PEELTELIYE ASGEDISIAV LTQEIVVYLA  900
MYVRAQPALF VEMLRLRIGL IIQVMATELA RSLNCSGEEA SESLMNLSPF  950
DMKNLLHHIL SGKEFGVERS MRPIHSSASS PAISIHEVGH TGVTKTERSG 1000
INRLRSEMKQ MTRRFSADEQ FFPVSQTVSS SAYSKSVRSS TPSSPTGTSS 1050
SDSGGHHISW GERQGQWLRR RRLDGAINRV PVGFYQRVWK ILQKCHGLSI 1100
DGYVLPSSTT REMTPQEIKF AVHVESVLNR VSQPEYRQLL VEAIMVLTLL 1150
SDTEMESIGG IIHVDQIVQM ANQLFLQEQI STGAMDTLEK DQATGICHFF 1200
YDSAPSGAYG TMTYLTRAVA SHLQELLPSS GCQTQ
```

Sites of interest: M1011–Q1020, F1021–V1037, these two sequences may be deleted in differentially spliced subtypes

β subunit

```
AGATGLMAEV SWKVLERRAR TKRSGSVYEP LKSINLPRPD NETLWDKLDY   50
       MASS ADAVVSSPPA FLRS
YYKIVKSTLL LYQSPTTGLF PTKTCGGDQT AKIHDSLYCA AGAWALALAY  100
RRIDDDKGRT HELEHSAIKC MRGILYCYMR QADKVQQFKQ DPRPTTCLHS  150
LFNVHTGDEL LSYEEYGHLQ INAVSLYLLY LVEMISSGLQ IIYNTDEVSF  200
IQNLVFCVER VYRVPDFGVW ERGSKYNNGS TELHSSSVGL AKAALEAING  250
FNLFGNQGCS WSVIFVDLDA HNRNRQTLCS LLPRESRSHN TDAALLPCIS  300
YPAFALDDDV LYNQTLDKVI RKLKGKYGFK RFLRDGYRTS LEDPKRRYYK  350
PAEIKLFDGI ECEFPIFFLY MMIDGVFRGN PKQVKEYQDL LTPVLHQTTE  400
GYPVVPKYYY VPADFVEYEK RNPGSQKRFP SNCGRDGKLF LWGQALYIIA  450
KLLADELISP KDIDPVQRYV PLQNQRNVSM RYSNQGPLEN DLVVHVALVA  500
ESQRLQVFLN TYGIQTQTPQ QVEPIQIWPQ QELVKAYFHL GINEKLGLSG  550
RPDRPIGCLG TSKIYRILGK TVVCYPIIFD LSDFYMSQDV LLLIDDIKNA  600
LQFIKQYWKM HGRPLFLVLI REDNIRGSRF NPMLDMLAAL KNGMIGGVKV  650
HVDRLQTLIS GAVVEQLDFL RISDTEELPE FKSFEELEPP KHSKVKRQSS  700
TSNAPELEQQ PEVSVTEWRN KPTHEILQKL NDCSCLASQT ILLGILLKRE  750
GPNFITQEGT VSDHIERLYR RAGSKKLWLA VRYGAAFTQK FSSSIAPHIT  800
                              SV VRRAASLLNK VVDSLAPSIT
```

β subunit *continued*

```
TFLVHGKQVT LGAFGHEEEV ISNPLSPRVI KNIIYYKCNT HDEREAVIQQ 850
NVLVQG
ELVIHIGWII SNNPELFSGM LKIRIGWIIH AMEYELQIRS GDKPAKDLYQ 900
LSPSEVKQLL LDILQPQQNG RCWLNKRQID GSLNRTPTGF YDRVWQILER 950
TPNGIIVAGK HLPQQPTLSD MTMYEMNFSL LVEDMLGNID QPKYRQIVVE 1000
LLMVVSIVLE RNPELEFQDK VDLDKLVKEA FHEFQKDESR LKEIEKQDDM 1050
TSFYNTPPLG KRGTCSYLTK VVMNLLLEGE VKPSNEDSCL VS
```

Sites of interest: A1, acetylation; S11, S26, S700, phosphorylation; C1089, farnesylation. Sequences that are alternatively utilized in differentially spliced subtypes (1–24, 779–806) are aligned below their alternatively expressed counterpart

γ_M subunit

```
TRDAALPGSH STHGFYENYE PKEILGRGVS SVVRRCIHKP TCKEYAVKII  50
DVTGGGSFSA EEVQELREAT LKEVDILRKV SGHPNIIQLK DTYETNTFFF 100
LVFDLMKKGE LFDYLTEKVT LSEKETRKIM RALLEVICAL HKLNIVHRDL 150
KPENILLDDD MNIKLTDFGF SCQLDPGEKL REVCGTPSYL APEIIECSMN 200
DNHPGYGKEV DMWSTGVIMY TLLAGSPPFW HRKQMLMLRM IMSGNYQFGS 250
PEWDDYSDTV KDLVSRFLVV QPQKRYTAEE ALAHPFFQQY VVEEVRHFSP 300
RGKFKVICLT VLASVRIYYQ YRRVKPVTRE IVIRDPYALR PLRRLIDAYA 350
FRIYGHWVKK GQQQNRAALF ENTPKAVLFS LAEDDY
```

Sites of interest: T1, in the mature γ_M subunit, the initiator methionine has been replaced by an acetyl group

Amino acid sequences of rat PhK

γ_T subunit

```
MTLDVGPEDE LPDWAAAKEF YQKYDPKDII GRGVSSVVRR CVHRATGDEF  50
AVKIMEVSAE RLSLEQLEEV RDATRREMHI LRQVAGHPHI ITLIDSYESS 100
SFMFLVFDLM RKGELFDYLT EKVALSEKET RSIMRSLLEA VNFLHVNNIV 150
HRDLKPENIL LDDNMQIRLS DFGFSCHLEP GEKLRELCGT PGYLAPEILK 200
CSMDETHPGY GKEVDLWACG VILFTLLAGS PPFWHRRQIL MLRMIMEGQY 250
QFSSPEWDDR SNTVKDLIAK LLQVDPNARL TAEQALQHPF FERCKGSQPW 300
NLTPRQRFRV AVWTILAAGR VALSSHRLRP LTKNALLRDP YALRPVRRLI 350
DNCAFRLYGH WVKKGEQQNR AALFQHQPPR PFPIIATDLE GDSSAITEDE 400
VTLVRS
```

δ subunit (calmodulin)

```
ADQLTEEQIA EFKEAFSLFD KDGDGTITTK ELGTVMRSLG QNPTEAELAD  50
MINEVDADGN GTIDFPEFLT MMARKMKDTD SEEEIREAFR VFDKDGNGYI 100
SAAELRHVMT NLGEKLTDEE VDEMIREADI DGDGQVNYEE FVQMMTAK
```

Homologues in other species

Various PhK subunits have been sequenced from rabbit, human, mouse and rat tissues.

Physiological substrates and specificity determinants

The main physiological substrate is glycogen phosphorylase. Glycogen synthase and PhK (autophosphorylation) are phosphorylated *in vitro*, but it is not known if these are substrates *in vivo*. There appears to be a requirement for basic amino acids to be present on both the N- and C-terminal sides of the phosphorylatable serine or threonine.

Phosphorylase:	KNISVRG
Glycogen synthase:	RTLSVSS
PhK α subunit:	VERSVRP
PhK α subunit:	PAISIHE
PhK α subunit:	QLKSEIK
PhK β subunit:	AEVSWKV
Skeletal muscle troponin I:	RAITARR

Assay

The usual assay is based on the determination of phosphorylase *a*, which can be measured by liberation of inorganic phosphate from glucose-1-phosphate upon initiation of glycogen synthesis. The whole reaction sequence can conveniently be carried out in an AutoAnalyzer[18]. Alternatively, the incorporation of [^{32}P]phosphate into phosphorylase can be measured.

Enzyme activators and inhibitors

Physiologically, Ca^{2+} at micromolar concentrations is an important allosteric ligand. Ca^{2+} at millimolar concentrations inhibits. Millimolar Mg^{2+} is required for activity, in addition to MgATP. Stimulation of PhK by calmodulin, troponin C and glycogen may be physiologically relevant. Heparin and higher alcohols activate PhK *in vitro*.

Pattern of expression

PhK activity and mRNA is highest in skeletal muscle, heart and brain, but PhK is expressed in all mammalian tissues investigated. In most tissues several isoforms of each subunit coexist, although in some tissues one isoform clearly predominates (e.g. skeletal muscle). In skeletal muscle, $\approx$75% is soluble, $\approx$25% bound to glycogen particles, $\approx$1% to sarcoplasmic reticulum.

References

[1] Pickett-Gies, C.A. and Walsh, D.A. (1986) The Enzymes XVII, 395–459.
[2] Heilmeyer, L.M.G. (1991) Biochim. Biophys. Acta 1094, 168–174.
[3] Francke, U. et al. (1989) Am. J. Human Genet. 45, 276–282.
[4] Davidson, J.J. et al. (1992) Proc. Natl Acad. Sci. USA 89, 2096–2100.
[5] Jones, T.A. et al. (1990) Biochim. Biophys. Acta 1048, 24–29.
[6] Whitmare, S.A. et al. (1994) Genomics 20, 169–175.
[7] Schneider, A. et al. (1993) Nature Genet. 5, 381–385.
[8] Wehner, M. et al. (1994) Hum. Mol. Genet. 3, 1983–1987
[9] Kilimann, M.W. et al. (1988) Proc. Natl Acad. Sci. USA 85, 9381–9385.
[10] Heilmeyer, L.M.G. et al. (1993) Mol. Cell. Biochem. (in press).

[11] Edstrom, R.D. et al. (1990) Biophys. J. 58, 1437–1448.

[12] Zander, N.F. et al. (1988) Proc. Natl Acad. Sci. USA 85, 2929–2933.

[13] Harmann, B. et al. (1991) J. Biol. Chem. 266, 15631–15637.

[14] Wüllrich, A. et al. (1993) J. Biol. Chem. 268, 23208–23214.

[15] da Cruz e Silva, E.F. and Cohen, P.T.W. (1987) FEBS Lett. 220, 36–42.

[16] Bender, P.K. and Emerson, C.P. (1987) J. Biol. Chem. 262, 8799–8805.

[17] Calalb, M.B. et al. (1992) J. Biol. Chem. 267, 1455–1463

[18] Jennissen, H.P. and Heilmeyer, L.M.G. (1974) Anal. Biochem. 57, 118–126.

<table>
<tr><td>

SkMLCK
</td><td>

Skeletal muscle myosin light chain kinase (vertebrates)
</td></tr>
</table>

James T. Stull (University of Texas, USA)

SkMLCK is a highly specific Ca^{2+}/calmodulin-dependent protein kinase that phosphorylates the regulatory myosin light chain in skeletal muscle[1]. Phosphorylation moves the myosin head away from the backbone of the myosin thick filament, thereby increasing the rate at which the myosin cross bridge goes into the force-producing state upon attachment with actin[2]. Physiologically, this movement of the myosin head potentiates the contractile performance of skeletal muscle.

Subunit structure and isoforms

SUBUNIT	AMINO ACIDS	MOL. WT	SDS-PAGE
Rabbit	607	65 396	87 kDa
Chicken	825	87 116	150 kDa

SkMLCK is a monomer. The form found in skeletal muscle is structurally and genetically distinct from the form found in smooth and non-muscle cells.

Domain structure

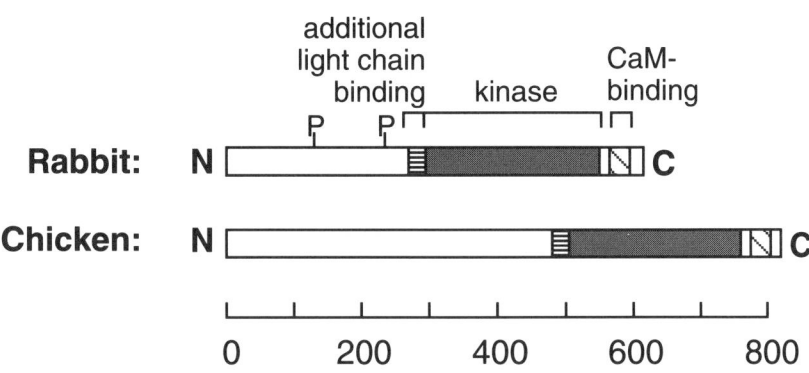

The Ca^{2+}/calmodulin (CaM)-binding and kinase domains form a globular structure with the N-terminal region having an extended conformation[3]. An acidic residue N-terminal to the kinase domain may bind to basic residues in the light chain substrate[4, 5]. The chicken skeletal muscle kinase is 83% identical to the rabbit enzyme in the catalytic core and calmodulin-binding domains, but there is no significant identity in the N-terminal regions. The function of the extended N-terminus is not known although it accounts for the size heterogeneity between species and the aberrant migration on SDS-PAGE. This region is rich in A, P, G, and charged residues. The autoinhibitory domain has not been defined in SkMLCK but, based on analogy with **SmMLCK**, most likely includes a portion of the calmodulin-binding domain and residues in the connecting peptide between the catalytic core and calmodulin-binding domain.

Database accession numbers

	PIR	SWISSPROT	EMBL/GENBANK	REF
Rabbit			J05194	4
Chicken			M81787	5

Amino acid sequences of rabbit SkMLCK

```
ATENGAVELG IQSLSTDEAS KGAASEESLA AEKDPAPPDP EKGPGPSDTK  50
QDPDPSTPKK DANTPAPEKG DVVPAQPSAG GSQGPAGEGG QVEAPAEGSA 100
GKPAALPQQT ATAEASEKKP EAEKGPSGHQ DPGEPTVGKK VAEGQAAARR 150
GSPAFLHSPS CPAIIASTEK LPAQKPLSEA SELIFEGVPA TPGPTEPGPA 200
KAEGGVDLLA ESQKEAGEKA PGQADQAKVQ GDTSRGIEFQ AVPSERPRPE 250
VGQALCLPAR EEDCFQILDD CPPPPAPFPH RIVELRTGNV SSEFSMNSKE 300
ALGGGKFGAV CTCTEKSTGL KLAAKVIKKQ TPKDKEMVML EIEVMNQLNH 350
RNLIQLYAAI ETPHEIVLFM EYIEGGELFE RIVDEDYHLT EVDTMVFVRQ 400
ICDGILFMHK MRVLHLDLKP ENILCVNTTG HLVKIIDFGL ARRYNPNEKL 450
KVNFGTPEFL SPEVVNYDQI SDKTDMWSLG VITYMLLSGL SPFLGDDDTE 500
TLNNVLSGNW YFDEETFEAV SDEAKDFVSN LIVKEQGARM SAAQCLAHPW 550
LNNLAEKAKR CNRRLKSQIL LKKYLMKRRW KKNFIAVSAA NRFKKISSSG 600
ALMALGV
```

Sites of interest: A1, acetylation; S160 and S234, intramolecular autophosphorylation; D270, binds light chain; K577–F593,
calmodulin-binding

Amino acid sequence of chicken SkMLCK

```
MEQPEVPGST ADTGGSGSTA PPGTESSTAQ PAPTAATNAV QAQNRGKEEP  50
MEKTVGAAGE PHGGTAETAQ EKVPTAAGKE EPGKEVAPGA AAEAQGAERE 100
KAVPEGGKAE PEQEKVPNAA PHWDGGEKGA AKDGASAAET NRGKEEPAKE 150
KAPTGESRGE TAEPTRGKGS AAAGAQEEPP KESAPGAPGA EAGGEAEPAK 200
KASAAAEAER GKAEPAGSVP AAAPAHGEGE PVKEQPAAVK AEDGGKNAAK 250
AKKSSAAKQP AKEKAPAAAK DGGKKPTKVE KIAAVTKDGG KETAKEKAMA 300
PAKEKATVAA QKPPAAVKAE DGGREPTEKA PAAAGAQGEG EKEASAEQTN 350
KEQPPRGSEP TRPTLMQSLS CPATCQREEQ PRQEVAAMET IPEETPPMAP 400
AGEEPTPAQG DLQPPEQPIP AGSTAVPQER TLPGAEEQPQ VGSEPPGHAE 450
QPGPGATGQP AATVAEPPPS PYLTPDFGKE DPFEILDDVP PPPAPFAHRI 500
ITLRTGSVSS QYNLSSKEIL GGGKFGEVHT CTEKQTGLKL AAKVIRKQGA 550
KDKEMVLLEI DVMNQLNHRN LIQLYDAIET PREIILFMEF VEGGELFERI 600
IDDDYHLTEV DCMVFVRQIC EGIRFMHHMR VLHLDLKPEN ILCVAATGHM 650
VKIIDFGLAR RYNPEEKLKV NFGTPEFLSP EVVNYEQVSY STDMWSMGVI 700
TYMLLSGLSP FLGDNDTETL NNVLAANWYF DEETFESVSD EAKDFVSNLI 750
IKEKSARMSA GQCLQHPWLT NLAEKAKRCN RRLKSQVMLK KYVMRRRWKK 800
NFIGVCAANR FKKITSSGSL TALGV
```

Sites of interest: unique repeat motifs in N-terminal region including EQP (7
repeats), GKE(A)EP (6 repeats) and PAAVKAEDGGK(R) (3
repeats); R795–F811, calmodulin-binding

Homologues in other species

SkMLCK has also been sequenced from rat and is very similar in structure to the rabbit enzyme[6].

Physiological substrates and specificity determinants

The consensus sequence for the phosphorylation of the myosin regulatory light chain from skeletal muscle is KKXKRRX$_{5-7}$SXYY where Y is a hydrophobic residue[7]. The variation in spacing reflects the differences in the chicken and rabbit skeletal muscle regulatory light chains, both of which have similar kinetic properties for phosphorylation by both rabbit and chicken kinases. Interestingly SkMLCK also phosphorylates the smooth muscle regulatory light chain with a different consensus sequence KKRXXRXXSXYY[7]. The binding of these two substrates to residues on the catalytic core of the skeletal muscle myosin light chain kinase is probably not identical.

Assay

The enzyme is assayed as the Ca^{2+}/calmodulin-dependent transfer of radioactivity from $[\gamma$-^{32}P]ATP to myosin regulatory light chain. For convenience these assays are usually performed in the presence of Ca^{2+} or EGTA. The synthetic peptide substrate based on the sequence of the skeletal muscle light chain can be used as a substrate, but the maximal rate of phosphorylation is about 10% of the rate obtained with light chain itself. The rate is better with the synthetic peptide substrate based upon the smooth muscle regulatory light chain.

Enzyme activators and inhibitors

A large number of calmodulin antagonists inhibit SkMLCK *in vitro*, but they are not specific. L-Thyroxine[8] and 1-(5-chloronaphthalenesulphonyl)-1H-hexahydro-1,4-diazepine (ML-9)[9] also inhibit, but the specificity of both is poor.

Pattern of expression

Most, if not all, SkMLCK can be easily separated from myofibrils. Thus, in contrast to **SmMLCK**, the skeletal muscle kinase does not bind tightly to contractile proteins. The same kinase is expressed in different adult skeletal muscle fibers although a greater amount is found in fast-twitch fibers compared to slow-twitch fibers.

References

1   Stull, J. T. et al. (1986) In The Enzymes, Krebs, E. G. and Boyer, P. D., eds, Academic Press, Orlando, pp. 113–166.
2   Sweeney, H.L. et al. (1993) Am. J. Physiol. 264, C108–C109.
3   Mayr, G.W. and Heilmeyer, L. M. G., Jr. (1983) Biochemistry 22, 4316–4326.
4   Herring, B. P. et al. (1992) J. Biol. Chem. 267, 25945–25950.

[5] Leachman, S. A. et al. (1992) J. Biol. Chem. 267, 4930–4938.
[6] Roush, C. L. et al. (1988) J. Biol. Chem. 263, 10510–10516.
[7] Michnoff, C.H. et al. (1986) J. Biol. Chem. 261, 8320–8326.
[8] Hagiwara, M. et al. (1989) J. Biol. Chem. 264, 40–44.
[9] Saitoh, M. et al. (1986) Biochem. Biophys. Res. Commun. 140, 280–287.

Smooth muscle myosin light chain kinase (vertebrates)

Anthony R. Means and Bruce E. Kemp (Duke University, USA, and St. Vincent's Institute of Medical Research, Australia)

Smooth muscle myosin light chain kinase is the key regulatory enzyme controlling the initiation of smooth muscle contraction[1,2]. It is associated with the thick filament and is regulated by Ca^{2+} and calmodulin. In response to increased intracellular concentrations of Ca^{2+}, myosin light chain kinase is activated by calmodulin and phosphorylates the 21 kDa myosin regulatory light chains at S19. This phosphorylation event permits actin to activate myosin ATPase and results in muscle contraction.

Subunit structure and isoforms

SUBUNIT	AMINO ACIDS	MOL. WT	SDS-PAGE
Rabbit	1147	126 000	152 kDa
Chicken	972	107 534	130 kDa

SmMLCK is monomeric. It is genetically distinct from the MLCK found in skeletal muscle fibres **SkMLCK**. Rabbit and chicken SmMLCKs are highly conserved (97% similarity) with the exception of a 175-residue insert in the N-terminal region of the rabbit sequence. Within this 175-residue insert sequence (residues 112–288) Gallagher *et al.*[3] found 15 copies of a 12 residue repeat of unknown function. This insert sequence accounts for the greater size of rabbit compared to chicken SmMLCK.

Genetics

A number of mutations have been introduced into the chicken SmMLCK cDNA. They are primarily in the autoregulatory domain. The currently available mutants are summarized in Bagchi *et al*[4].

Domain structure

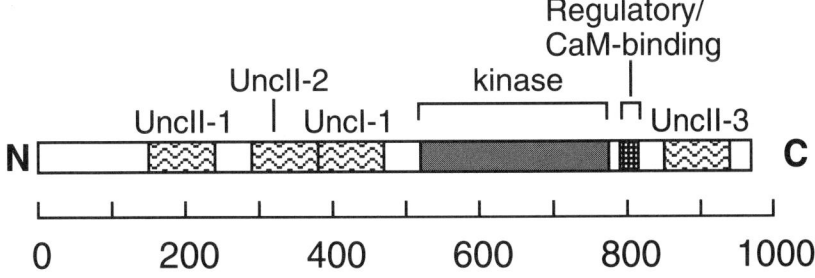

The unc domains comprise four motifs: uncII-1, uncII-2 fused to uncI-1 on the N-terminal side of the catalytic domain, and uncII-3 in the C-terminal domain. These motifs are named after similar extensive repeat motifs found in twitchin and titin, but their function is not known. The majority of the 157-residue C-terminal region, including uncII-3, is identical to telokin, a 23 kDa protein independently expressed from the same gene. In the N-terminal 60 residues of SmMLCK the sequence DFR is repeated three times: DFRAN, DFRSV and DFRSV. The function of this repeat is not known. On the C-terminal side of the catalytic domain are a connecting peptide (774–786) followed by overlapping pseudosubstrate (787–807) and calmodulin-binding (796–815) sequences. The connecting peptide contributes to the inhibitory potency of the pseudosubstrate peptide. The 3-dimensional crystal structure of the calmodulin-binding peptide complexed to Ca^{2+}/calmodulin has been solved to 2.4 Å[5].

Database accession numbers

	PIR	SWISSPROT	EMBL/GENBANK	REF
Chicken	A30593	P11799	M31048	2
Rabbit	A41674		M76233	3
Bovine			S57131	

Amino acid sequence of chicken gizzard MLCK

```
MDFRANLQRQ VKPKTLSEEE RKVHAPQQVD FRSVLAKKGT PKTPLPEKVP  50
PPKPAVTDFR SVLGAKKKPP AENGSASTPA PNARAGSEAQ NATPNSEAPA 100
PKPVVKKEEK NDRKCEHGCA VVDGGIIGKK AENKPAASKP TPPPSKGTAP 150
SFTEKLQDAK VADGEKLVLQ CRISSDPPAS VSWTLDSKAI KSSKSIVISQ 200
EGTLCSLTIE KVMPEDGGEY KCIAENAAGK AECACKVLVE DTSSTKAAKP 250
AEKKTKKPKT TLPPVLSTES SEATVKKKPA PKTPPKAATP PQITQFPEDR 300
KVRAGESVEL FAKVVGTAPI TCTWMKFRKQ IQENEYIKIE NAENSSKLTI 350
SSTKQEHCGC YTLVVENKLG SRQAQVNLTV VDKPDPPAGT PCASDIRSSS 400
LTLSWYGSSY DGGSAVQSYT VEIWNSVDNK WTDLTTCRST SFNVQDLQAD 450
REYKFRVRAA NVYGISEPSQ ESEVVKVGEK QEEELKEEEA ELSDDEGKET 500
EVNYQTVTIN TEQKVSDVYN IEERLGSGKF GQVFRLVEKK TGKVWAGKFF 550
KAYSAKEKEN IRDEISIMNC LHHPKLVQCV DAFEEKANIV MVLEMVSGGE 600
LFERIIDEDF ELTERECIKY MRQISEGVEY IHKQGIVHLD LKPENIMCVN 650
KTGTSIKLID FGLARRLESA GSLKVLFGTP EFVAPEVINY EPIGYETDMW 700
SIGVICYILV SGLSPFMGDN DNETLANVTS ATWDFDDEAF DEISDDAKDF 750
ISNLLKKDMK SRLNCTQCLQ HPWLQKDTKN MEAKKLSKDR MKKYMARRKW 800
QKTGHAVRAI GRLSSMAMIS GMSGRKASGS SPTSPINADK VENEDAFLEE 850
VAEEKPHVKP YFTKTILDME VVEGSAARFD CKIEGYPDPE VMWYKDDQPV 900
KESRHFQIDY DEEGNCSLTI SEVCGDDDAK YTCKAVNSLG EATCTAELLV 950
ETMGKEGEGE GEGEEDEEEE EE
```

Sites of interest: M1, acetylation; L774–L786, connecting peptide; A796–S815, calmodulin-binding site; S787–V807, pseudosubstrate peptide

Homologues in other species

SmMLCK has been cloned from the rabbit, and MLCKs have been cloned from skeletal muscle (**SkMLCK**) and *D. discoideum* (**DdMLCK**). SmMLCK is also homologous to the mammalian skeletal muscle protein titin[6], and the *C. elegans* protein twitchin (**Twn**). The non-muscle MLCK sequence reported by Shoemaker *et al.*[7] is in question, because the non-muscle enzyme has an identical size to the chicken gizzard enzyme on SDS-PAGE.

At residue 287 of the reported cDNA sequence of the non-muscle enzyme is the sequence MDFRANLQRQ, which corresponds exactly to the N-terminus of both mammalian and avian SmMLCKs, and the sequence continues with 97% homology to precisely the same C-terminus of chicken SmMLCK at E972.

Physiological substrates and specificity determinants

SmMLCK has a very restricted specificity, phosphorylating only smooth muscle myosin light chains with $V_{max} = 36\,\mu mol/min/mg$, $K_m = 7\,\mu M$. It does not phosphorylate other exogenous protein substrates, including skeletal muscle myosin light chains, at significant rates. An N-terminal peptide (1–23) from myosin light chain (SSKRTTKAKAKKRPQRATSNVFA) is phosphorylated with $V_{max} = 3.0\,\mu mol/min/mg$, $K_m = 2.7\,\mu M$. In this peptide, K12, R13 and R16 are important specificity determinants directing phosphorylation to S19. Substitution of the basic residues, or varying the distance between them and S19, alters the kinetics of phosphorylation. Residues on the C-terminal side of S19 are not essential for maintaining a low K_m[9].

Assay
The enzyme is usually assayed as the Ca^{2+}/calmodulin-dependent transfer of radioactivity from $[\gamma\text{-}^{32}P]ATP$ to the peptide KKRAARATSNVFA $(V_{max} = 1.42\,\mu mol/min/mg$, $K_m = 7.5\,\mu M)$[9].

Enzyme activators and inhibitors

	SPECIFICITY	K_i	REF
Pseudosubstrate (774–807)	Not known	0.3 nM	8
Pseudosubstrate (787–807)	Moderate	12 nM	8
ML-9	Moderate	4 mM	10
H-85	Moderate	28 mM	10
KT5926	Moderate	15 μM	11

Pattern of expression
SmMLCK is present in all smooth muscle and non-muscle cells of the adult animal[3]. In non-muscle cells the enzyme is a component of cytoplasmic stress fibres and is associated with the nuclear matrix[12].

References
1 Kamm, K.E. and Stull, J.T. (1985) Annu. Rev. Pharmacol. Toxicol. 25, 593–620.
2 Olson, N.J. et al. (1990) Proc. Natl Acad. Sci. USA 87, 2284–2288.
3 Gallagher, P.J. et al. (1991) J. Biol. Chem. 266, 23936–23944.
4 Bagchi, I.C. et al. (1992) J. Biol. Chem. 267, 3024–3029.
5 Meador, W.E. et al. (1992) Science 257, 1251–1255.
6 Benian, G.M. et al. (1989) Nature 342, 45–50.
7 Shoemaker, M.O. et al. (1990) J. Cell. Biol. 111, 1107–1125.
8 Knighton, D.R. et al. (1992) Science 258, 130–135.

9   Pearson, R.B. et al. (1994) in Protein Phosphorylation: A Practical Approach, Hardie, D.G., ed., Oxford University Press, pp. 265–291.

10 Hidaka H. et al. (1991) Methods Enzymol. 201, 328–339.

11 Nakanishi, S. et al. (1990) Mol. Pharmacol. 37, 482–488.

12 Guerriero, V. et al. (1981) Cell 27, 449–458.

Junyi Lei, Mario Valenzuela and Guy Benian
(Emory University School of Medicine, USA)

Twitchin, a remarkably large 750 kDa polypeptide encoded by the mutationally defined *unc-22* gene, is expressed in the muscle of the nematode, *C. elegans*, where it is localized to A-bands[1]. *Unc-22* mutants show a constant twitch of the body surface which originates in the underlying muscle, and have disorganized sarcomeres despite having normal numbers of thick and thin filaments[2]. Twitchin is believed to function both in regulating muscle contraction, and in the final stages of sarcomere assembly.

Subunit structure

SUBUNIT	AMINO ACIDS	MOL. WT	SDS-PAGE
Twn	6839	753 494	

Domain structure (scale only approximate)

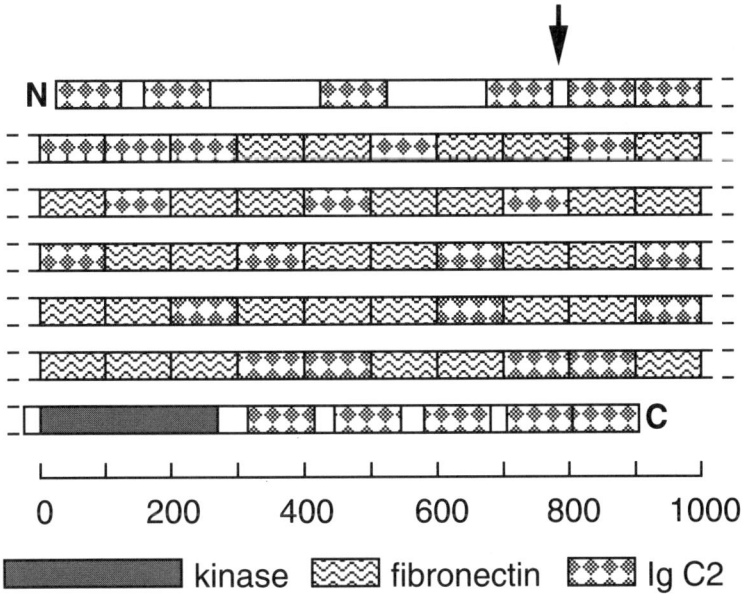

Twitchin consists of a single protein kinase domain, most related to myosin light chain kinases, plus 31 copies of a fibronectin type III-like domain and 30 copies of an immunoglobulin C2-like domain[3,4]. In the above diagram, the structure is shown in seven segments running from the N-terminus at the top to the C-terminus at the bottom. The arrow indicates a Gly-rich segment which could possibly form a flexible hinge.

Database accession numbers

	PIR	SWISSPROT	EMBL/GENBANK	REF
Twn		X15423	L10351	3,4

Amino acid sequence of *C. elegans* Twn

```
MVGAPRFTQK PSIQQTPTGD LLMECHLEAD PQPTIAWQHS GNLLEPSGRV   50
VQTLTPLGGS LYKATLVIKE PNAGDGGAYK CTARNQLGES NANINLNFAG  100
AGGDEAKSRG PSFVGKPRII PKDGGALIVM ECKVKSASTP VAKWMKDGVP  150
LSMGGLYHAI FSDLGDQTYL CQLEIRGPSS SDAGQYRCNI RNDQGETNAN  200
LALNFEEPDP SERQERKRST ASPRPSSRGP GSRPSSPKKS MKSREGTPKR  250
TLKPREGSPS KKLRSRTSTP VNEEVSQSES RRSSRTDKME VDQVSGASKR  300
KPDGLPPPGG DEKKLRAGSP STRKSPSRKS ASPTPSRKGS SAGGAASGTT  350
GASASATSAT SGGSASSDAS RDKYTRPPIV LEASRSQTGR IGGSVVLEVQ  400
WQCHSSTIIE WYRDGTLVRN SSEYSQSFNG SIAKLQVNKL TEEKSGLYKC  450
HAKCDYGEGQ SSAMVKIEQS DVEEELMKHR KDAEDEYQKE EQKSQTLQAE  500
TKKRVARRSK SKSKSPAPQA KKSTTSESGR QEASEVEHKR SSSVRPDPDE  550
ESQLDEIPSS GLTIPEERRR ELLGQVGESD DEVSESISEL PSFAGGKPRR  600
KTDDKPKKVS IAPVSTNKSS DDEPSTPRRR SSIDMRRESV QEILEKTSTP  650
LVPSGASGSA PKIVEVPENV TVVENETAIL TCKVSGSPAP TFRWFKGSRE  700
VISGGRFKHI TDGKEHTVAL ALLKCRSQDE GPYTLTIENV HGTDSADVKL  750
LVTSDNGLDF RAMLKHRESQ AGFQKDGEGG GAGGGGGEKK PMTEAERRQS  800
LFPGKKVEKW DIPLPEKTVQ QQVDKICEWK CTYSRPNAKI RWYKDRKEIF  850
SGGLKYKIVI EKNVCTLIIN NPEVDDTGKY TCEANGVPTH AQLTVLEPPM  900
KYSFLNPLPN TQEIYRTKQA VLTCKVNTPR APLVWYRGSK AIQEGDPRFI  950
IEKDAVGRCT LTIKEVEEDD QAEWTARITQ DVFSKVQVYV EEPRHTFVVP 1000
MKSQKVNESD LATLETDVND KDAEVVWWHD GKRIDIDGVK FKVESSNRKR 1050
RLIINGARIE DHGEYKCTTK DDRTMAQLIV DAKNKFIVAL KDTEVIEKDD 1100
VTLMCQTKDT KTPGIWFRNG KQISSMPGGK FETQSRNGTH TLKIGKIEMN 1150
EADVYEIDQA GLRGSCNVTV LEAEKRPILN WKPKKIEAKA GEPCVVKVPF 1200
QIKGTRRGDP KAQILKNGKP IDEEMRKLVE VIIKDDVAEI VFKNPQLADT 1250
GKWALELGNS AGTALAPFEL FVKDKPKPPK GPLETKNVTA EGLDLVWGTP 1300
DPDEGAPVKA YIIEMQEGRS GNWAKVGETK GTDFKVKDLK EHGEYKFRVK 1350
ALNECGLSDP LTGESVLAKN PYGVPGKPKN MDAIDVDKDH CTLAWEPPEE 1400
DGGAPITGYI IERREKSEKD WHQVGQTKPD CCELTDKKVV EDKEYLYRVK 1450
AVNKAGPGDP CDHGKPIKMK AKKASPEFTG GGIKDLRLKV GETIKYDVPI 1500
SGEPLPECLW VVNGKPLKAV GRVKMSSERG KHIMKIENAV RADSGKFTIT 1550
LKNSSGSCDS TATVTVVGRP TPPKGPLDIA DVCADGATLS WNPPDDDGGD 1600
PLTGYIVEAQ DMDNKGKYIE VGKVDPNTTT LKVNGLRNKG NYKFRVKAVN 1650
NEGESEPLSA DQYTQIKDPW DEPGKPGRPE ITDFDADRID IAWEPPHKDG 1700
GAPIEEYIVE VRDPDTKEWK EVKRVPDTNA SISGLKEGKE YQFRVRAVNK 1750
AGPGQPSEPS EKQLAKPKFI PAWLKHDNLK SITVKAGATV RWEVKIGGEP 1800
IPEVKWFKGN QQLENGIQLT IDTRKNEHTI LCIPSAMRSD VGEYRLTVKN 1850
SHGADEEKAN LTVLDRPSKP NGPLEVSDVF EDNLNLSWKP PDDDGGEPIE 1900
YYEVEKLDTA TGRWVPCAKV KDTKAHIDGL KKGQTYQFRV KAVNKEGASD 1950
ALSTDKDTKA KNPYDEPGKT GTPDVVDWDA DRVSLEWEPP KSDGGAPITQ 2000
YVIEKKGKHG RDWQECGKVS GDQTNAEILG LKEGEEYQFR VKAVNKAGPG 2050
EASDPSRKVV AKPRNLKPWI DREAMKTITI KVGNDVEFDV PVRGEPPPKK 2100
EWIFNEKPVD DQKIRIESED YKTRFVLRGA TRKHAGLYTL TATNASGSDK 2150
HSVEVIVLGK PSSPLGPLEV SNVYEDRADL EWKVPEDDGG APIDHYEIEK 2200
MDLATGRWVP CGRSETTKTT VPNLQPGHEY KFRVRAVNKE GESDPLTTNT 2250
AILAKNPYEV PGKVDKPELV DWDKDHVDLA WNAPDDGGAP IEAFVIEKKD 2300
KNGRWEEALV VPGDQKTATV PNLKEGEEYQ FRISARNKAG TGDPSDPSDR 2350
```

Amino acid sequence of *C. elegans* Twn *continued*

```
VVAKPRNLAP RIHREDLSDT TVKVGATLKF IVHIDGEPAP DVTWSFNGKG 2400
IGESKAQIEN EPYISRFALP KALRKQSGKY TITATNINGT DSVTINIKVK 2450
SKPTKPKGPI EVTDVFEDRA TLDWKPPEDD GGEPIEFYEI EKMNTKDGIW 2500
VPCGRSGDTH FTVDSLNKGD HYKFRVKAVN SEGPSDPLET ETDILAKNPF 2550
DRPDRPGRPE PTDWDSDHVD LKWDPPLSDG GAPIEEYQIE KRTKYGRWEP 2600
AITVPGGQTT ATVPDLTPNE EYEFRVVAVN KGGPSDPSDA SKAVIAKPRN 2650
LKPHIDRDAL KNLTIKAGQS ISFDVPVSGE PAPTVTWHWP DNREIRNGGR 2700
VKLDNPEYQS KLVVKQMERG DSGTFTIKAV NANGEDEATV KINVIDKPTS 2750
PNGPLDVSDV HGDHVTLNWR APDDDGGIPI ENYVIEKYDT ASGRWVPAAK 2800
VAGDKTTAVV DGLIPGHEYK FRVAAVNAEG ESDPLETFGT TLAKDPFDKP 2850
GKTNAPEITD WDKDHVDLEW KPPANDGGAP IEEYVVEMKD EFSPFWNDVA 2900
HVPAGQTNAT VGNLKEGSKY EFRIRAKNKA GLGDPSDSAS AVAKARNVPP 2950
VIDRNSIQEI KVKAGQDFSL NIPVSGEPTP TITWTFEGTP VESDDRMKLN 3000
NEDGKTKFHV KRALRSDTGT YIIKAENENG TDTAEVKVTV LDHPSSPRGP 3050
LDVTNIVKDG CDLAWKEPED DGGAEISHYV IEKQDAATGR WTACGESKDT 3100
NFHVDDLTQG HEYKFRVKAV NRHGDSDPLE AREAIIAKDP FDRADKPGTP 3150
EIVDWDKDHA DLKWTPPADD GGAPIEGYLV EMRTPSGDWV PAVTVGAGEL 3200
TATVDGLKPG QTYQFRVKAL NKAGESTPSD PSRTMVAKPR HLAPKINRDM 3250
FVAQRVKAGQ TLNFDVNVEG EPAPKIEWFL NGSPLSSGGN THIDNNTDNN 3300
TKLTTKSTAR ADSGKYKIVA TNESGKDEHE VDVNILDIPG APEGPLRHKD 3350
ITKESVVLKW DEPLDDGGSP ITNYVVEKQE DGGRWVPCGE TSDTSLKVNK 3400
LSEGHEYKFR VKAVNRQGTS APLTSDHAIV AKNPFDEPDA PTDVTPVDWD 3450
KDHVDLEWKP PANDGGAPID AYIVEKKDKF GDWVECARVD GKTTKATADN 3500
LTPGETYQFR VKAVNKAGPG KPSDPTGNVV AKPRRMAPKL NLAGLLDLRI 3550
KAGTPIKLDI AFEGEPAPVA KWKANDATID TGARADVTNT PTSSAIHIFS 3600
AVRGDTGVYK IIVENEHGKD TAQCNVTVLD VPGTPEGPLK IDEIHKEGCT 3650
LNWKPPTDNG GTDVLHYIVE KMDTSRGTWQ EVGTFPDCTA KVNKLVPGKE 3700
YAFRVKAVNL QGESKPLEAE EPIIAKNQFD VPDPVDKPEV TDWDKDRIDI 3750
KWNPTANNGG APVTGYIVEK KEKGSAIWTE AGKTPGTTFS ADNLKPGVEY 3800
EFRVIAVNAA GPSDPSDPTD PQITKARYLK PKILTASRKI KIKAGFTHNL 3850
EVDFIGAPDP TATWTVGDSG AALAPELLVD AKSSTTSIFF PSAKRADSGN 3900
YKLKVKNELG EDEAIFEVIV QDRPSAPEGP LEVSDVTKDS CVLNWKPPKD 3950
DGGAEISNYV VEKRDTKTNT WVPVSAFVTG TSITVPKLTE GHEYEFRVMA 4000
ENTFGRSDSL NTDEPVLAKD PFGTPGKPGR PEIVDTDNDH IDIKWDPPRD 4050
NGGSPVDHYD IERKDAKTGR WIKVNTSPVQ GTAFSDTRVQ KGHTYEYRVV 4100
AVNKAGPGQP SDSSAAATAK PMHEAPKFDL DLDGKEFRVK AGEPLVITIP 4150
FTASPQPDIS WTKEGGKPLA GVETTDSQTK LVIPSTRRSD SGPVKIKAVN 4200
PYGEAEANIK ITVIDKPGAP ENITYPAVSR HTCTLNWDAP KDDGGAEIAG 4250
YKIEYQEVGS QIWDKVPGLI SGTAYTVRGL EHGQQYRFRI RAENAVGLSD 4300
YCQGVPVVIK DPFDPPGAPS TPEITGYDTN QVSLAWNPPR DDGGSPILGY 4350
VVERFEKRGG GDWAPVKMPM VKGTECIVPG LHENETYQFR VRAVNAAGHG 4400
EPSNGSEPVT CRPYVEKPGA PDAPRVGKIT KNSAELTWNR PLRDGGAPID 4450
GYIVEKKKLG DNDWTRCNDK PVRDTAFEVK NLGEKEEYEF RVIAVNSAGE 4500
GEPSKPSDLV LIEEQPGRPI FDINNLKDIT VRAGETIQIR IPYAGGNPKP 4550
IIDLFNGNSP IFENERTVVD VNPGEIVITT TGSKRSDAGP YKISATNKYG 4600
KDTCKLNVFV LDAPGKPTGP IRATDIQADA MTLSWRPPKD NGGDAITNYV 4650
VEKRTPGGDW VTVGHPVGTT LRVRNLDANT PYEFRVRAEN QYGVGEPLET 4700
```

Amino acid sequence of *C. elegans* Twn *continued*

```
DDAIVAKNPF DTPGAPGQPE AVETSEEAIT LQWTRPTSDG GAPIQGYVIE 4750
KREVGSTEWT KAAFGNILDT KHRVTGLTPK KTYEFRVAAY NAAGQGEYSV 4800
NSVPITADNA PTRPKINMGM LTRDILAYAG ERAKILVPFA ASPAPKVTFS 4850
KGENKISPTD PRVKVEYSDF LATLTIEKSE LTDGGLYFVE LENSQGSDSA 4900
SIRLKVVDKP ASPQHIRVED IAPDCCTLYW MPPSSDGGSP ITNYIVEKLD 4950
LRHSDGKWEK VSSFVRNLNY TVGGLIKDNR YRFRVRAETQ YGVSEPCELA 5000
DVVVAKYQFE VPNQPEAPTV RDKDSTWAEL EWDPPRDGGS KIIGYQVQYR 5050
DTSSGRWINA KMDLSEQCHA RVTGLRQNGE FEFRIIAKNA AGFSKPSPPS 5100
ERCQLKSRFG PPGPPIHVGA KSIGRNHCTI TWMAPLEDGG SKITGYNVEI 5150
REYGSTLWTV ASDYNVREPE FTVDKLREFN DYEFRVVAIN AAGKGIPSLP 5200
SGPIKIQESG GSRPQIVVKP EDTAQPYNRR AVFTCEAVGR PEPTARWLRN 5250
GRELPESSRY RFEASDGVYK FTIKEVWDID AGEYTVEVSN PYGSDTATAN 5300
LVVQAPPVIE KDVPNTILPS GDLVRLKIYF SGTAPFRHSL VLNREEIDMD 5350
HPTIRIVEFD DHILITIPAL SVREAGRYEY TVSNDSGEAT TGFWLNVTGL 5400
PEAPQGPLHI SNIGPSTATL SWRPPVTDGG SKITSYVVEK RDLSKDEWVT 5450
VTSNVKDMNY IVTGLFENHE YEFRVSAQNE NGIGAPLVSE HPIIARLPFD 5500
PPTSPLNLEI VQVGGDYVTL SWQRPLSDGG GRLRGYIVEK QEEEHDEWFR 5550
CNQNPSPPNN YNVPNLIDGR KYRYRVFAVN DAGLSDLAEL DQTLFQASGS 5600
GEGPKIVSPL SDLNEEVGRC VTFECEISGS PRPEYRWFKG CKELVDTSKY 5650
TLINKGDKQV LIINDLTSDD ADEYTCRATN SSGTRSTRAN LRIKTKPRVF 5700
IPPKYHGGYE AQKGETIELK IPYKAYPQGE ARWTKDGEKI ENNSKFSITT 5750
DDKFATLRIS NASREDYGEY RVVVENSVGS DSGTVNVTVA DVPEPPRFPI 5800
IENILDEAVI LSWKPPALDG GSLVTNYTIE KREAMGGSWS PCAKSRYTYT 5850
TIEGLRAGKQ YEFRIIAENK HGQSKPCEPT APVLIPGDER KRRRGYDVDE 5900
QGKIVRGKGT VSSNYDNYVF DIWKQYYPQP VEIKHDHVLD HYDIHEELGT 5950
GAFGVVHRVT ERATGNNFAA KFVMTPHESD KETVRKEIQT MSVLRHPTLV 6000
NLHDAFEDDN EMVMIYEFMS GGELFEKVAD EHNKMSEDEA VEYMRQVCKG 6050
LCHMHENNYV HLDLKPENIM FTTKRSNELK LIDFGLTAHL DPKQSVKVTT 6100
GTAEFAAPEV AEGKPVGYYT DMWSVGVLSY ILLSGLSPFG GENDDETLRN 6150
VKSCDWNMDD SAFSGISEDG KDFIRKLLLA DPNTRMTIHQ ALEHPWLTPG 6200
NAPGRDSQIP SSRYTKIRDS IKTKYDAWPE PLPPLGRISN YSSLRKHRPQ 6250
EYSIRDAFWD RSEAQPRFIV KPYGTEVGEG QSANFYCRVI ASSPPVVTWH 6300
KDDRELKQSV KYMKRYNGND YGLTINRVKG DDKGEYTVRA KNSYGTKEEI 6350
VFLNVTRHSE PLKFEPLEPM KKAPSPPRVE EFKERRSAPF FTFHLRNRLI 6400
QKNHQCKLTC SLQGNPNPTI EWMKDGHPVD EDRVQVSFRS GVCSLEIFNA 6450
RVDDAGTYTV TATNDLGVDV SECVLTVQTK GGEPIPRVSS FRPRRAYDTL 6500
STGTDVERSH SYADMRRRSL IRDVSPDVRS AADDLKTKIT NELPSFTAQL 6550
SDSETEVGGS AEFSAAVSGQ PEPLIEWLHN GERISESDSR FRASYVAGKA 6600
TLRISDAKKS DEGQYLCRAS NSAGQEQTRA TLTVKGDQPL LNGHAGQAVE 6650
SELRVTKHLG GEIVNNGESV TFEARVQGTP EEVLWMRNGQ ELTNGDKTSI 6700
SQDGETLSFT INSADASDAG HYQLEVRSKG TNLVSVASLV VVGEKADPPV 6750
TRLPSSVSAP LGGSTAFTIE FENVEGLTVQ WFRGSEKIEK NERVKSVKTG 6800
NTFKLDIKNV EQDDDGIYVA KVVKEKKAIA KYAAALLLV
```

Sites of interest: G777–G787, Gly-rich segment; L6244–S6262, putative autoinhibitory region

Homologues in other species

The kinase domain of twitchin is most closely related to that of chicken **SmMLCK** (52% identity[3, 5]) and then to rabbit **SkMLCK** (44% identity[3, 6]): no gaps are needed for either of these alignments over a stretch of 250 amino acids. The third most similar protein kinase domain is from titin (39% identity with the introduction of one gap[7]). However, because the arrangement of 16 fibronectin and Ig domains on either side of the kinase domain is conserved between titin and twitchin, because both polypeptides are very long (titin is a 1 μm long, 3×10^6 Da polypeptide), and because both are probably associated with thick filaments (titin also extends into the I-band), the closest mammalian homologue is titin.

Physiological substrates and specificity determinants

The *in vivo* substrates for twitchin are presently unknown.

Assay

Expression and purification from bacteria of a portion of twitchin (5890–6262) comprising the catalytic core plus 47 residues N-terminal to it and 60 residues C-terminal to it, has allowed demonstration of phosphotransferase activity, both autophosphorylation and towards exogenous substrates[8]. Screening of a number of protein or peptide substrates *in vitro* showed that the best substrate was a chicken smooth muscle myosin light chain peptide (kMLC11–23), with a V_{max} close to 1 μmol/min/mg and a K_m of 98 μM. Autophosphorylation attained a maximum at 3 h with a stoichiometry of $\approx$1 mole/mole, probably via an intramolecular mechanism, and occurred predominantly N-terminal to the catalytic core, predominantly at T5910.

Enzyme activators and inhibitors

Twitchin (5890–6262) has very low activity, whereas twitchin (5890–6202) has the relatively high activity noted above, suggesting that the sequence 6202–6262 acts as an autoinhibitor[8]. A synthetic peptide corresponding to twitchin residues 6244–6268 acts as an inhibitor with an IC_{50} of 12.5 μM.

Pattern of expression

Twitchin is expressed in muscle cells of several types (body wall, vulval and anal muscles), and is localized to A-bands[1]. There is evidence for a second gene that is expressed in pharyngeal muscle.

References

1 Moerman, D.G. et al. (1988) Genes Devel. 2, 93–105.
2 Waterston, R.H. et al. (1980) Devel. Biol. 77, 271–302.
3 Benian, G.M. et al. (1989) Nature 342, 45–50.
4 Benian, G.M. et al. (1993) Genetics 134, 1097–1104.
5 Guerriero, V. et al. (1986) Biochemistry 25, 8372–8381.
6 Takio, K. et al. (1986) Biochemistry 25, 8049–8057.
7 Labeit, S. et al. (1992) EMBO J. 11, 1711–1716.
8 Lei, J. et al. (1995) submitted.

DdMLCK — Myosin light chain kinase-A (*D. discoideum*)

Janet L. Smith, Linda A. Silveira and James A. Spudich
(Stanford University and University of Redlands, USA)

Myosin light chain kinase-A (MLCK-A) activates myosin ATPase and contractile activity by phosphorylating the 18 kDa myosin regulatory light chain. In *D. discoideum*, myosin is required for cytokinesis and for fruiting body formation. Light chain phosphorylation is thought to regulate its activity during these processes.

Subunit structure and isoforms

SUBUNIT	AMINO ACIDS	MOL. WT	SDS-PAGE
MLCK-A	295	34 309	

DdMLCK is a monomer.

Genetics

MLCK-A is encoded by the *mlkA* locus. Disruption of this gene results in impaired ability to undergo cytokinesis. Substitution of the autophosphorylated Thr with Ala (T289A) decreases activity 7-fold *in vitro*.

Domain structure

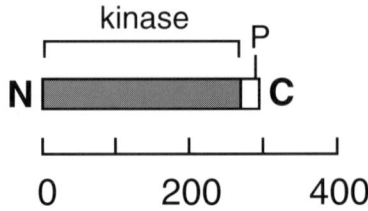

MLCK-A comprises a compact catalytic domain (1–266) and a short inhibitory domain (267–295). Autophosphorylation of T298 activates the enzyme.

Database accession numbers

	PIR	SWISSPROT	EMBL/GENBANK	REF
MLCK-A	A40811	P25323	M64176	1

Amino acid sequence of *D. discoideum* MLCK

```
MTEVEKIYEF KEELGRGAFS IVYLGENKQT KQRYAIKVIN KSELGKDYEK  50
NLKMEVDILK KVNHPNIIAL KELFDTPEKL YLVMELVTGG ELFDKIVEKG 100
SYSEADAANL VKKIVSAVGY LHGLNIVHRD LKPENLLLKS KENHLEVAIA 150
DFGLSKIIGQ TLVMQTACGT PSYVAPEVLN ATGYDKEVDM WSIGVITYIL 200
LCGFPPFYGD TVPEIFEQIM EANYEFPEEY WGGISKEAKD FIGKLLVVDV 250
SKRLNATNAL NHPWLKSNNS NNTIDTVKMK EYIVERQKTQ TKLVN
```

Sites of interest: T289, autophosphorylation

Homologues in other species

MLCKs have been characterized from several higher eukaryotes – **SmMLCK, SkMLCK, Twn**. These MLCKs have additional domains not found in MLCK-A. Unlike the other MLCKs, MLCK-A is not directly activated by Ca^{2+}/calmodulin.

Physiological substrates and specificity determinants

MLCK-A phosphorylates S13 of the myosin regulatory light chain from *D. discoideum*[2], and this results in an increase in actin-activated ATPase. S13 is homologous to the MLCK phosphorylation sites on myosins from higher eukaryotes.

Assay

MLCK-A can be assayed by transfer of radioactivity from $[\gamma\text{-}^{32}P]ATP$ to *D. discoideum* myosin[3] or to the synthetic peptide MASTKRRLNREESSVVL-$CONH_2$, corresponding to residues 1–17 of myosin regulatory light chain.

Pattern of expression

Cytosolic.

References

1 Tan, J.L. and Spudich, J.A. (1991) J. Biol. Chem. 266, 16044–16049.
2 Ostrow, B. and Chisholm, R., personal communication.
3 Tan, J.L. and Spudich, J.A. (1991) J. Biol. Chem. 265, 13818–13824.

CDPK

Ca²⁺-dependent protein kinase (soybean) (Calmodulin-like domain protein kinase)

Ca^{2+}-dependent protein kinase (soybean)
(Calmodulin-like domain protein kinase)

Alice C. Harmon (University of Florida, USA)

Ca^{2+}-dependent protein kinases (CDPKs) are activated by the direct binding of Ca^{2+} to regulatory domains, which contain four EF-hand Ca^{2+}-binding domains and are similar to calmodulin. This property distinguishes CDPKs from the Ca^{2+}/calmodulin-dependent protein kinases and protein kinase C. CDPKs are activated by micromolar concentrations of free Ca^{2+} and are thought to play a role in Ca^{2+}-mediated signal transduction pathways[1].

Subunit structure and isoforms

SUBUNIT	AMINO ACIDS	MOL. WT	SDS-PAGE
CDPKα	508	57 175	57 kDa

CDPK is a monomer. Biochemical, immunological and molecular genetic data indicate that there are at least four genes encoding CDPKs in soybean, with the enzymes ranging from 55 to 80 kDa. The complete sequence of one isoform (CDPKα encoded by cDNA SK5) has been determined. The predicted sequence of a second isoform encoded by a partial cDNA (SK2) is 70% identical to CDPKα.

Domain structure

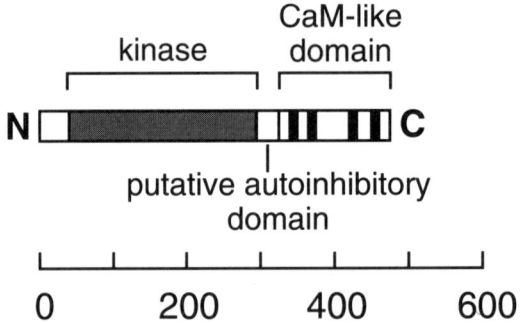

CDPKs have both catalytic and regulatory domains joined by a autoinhibitory domain. The regulatory domain is similar in sequence to calmodulin, especially within the four EF-hand Ca^{2+}-binding domains. The molecular organization of the three functional domains is similar in the four CDPKs for which sequence data are available. Soybean CDPKα has short N- and C-terminal sequences of unknown function. *Arabidopsis* AK1 is 70% identical to CDPKα in the 492 residue overlap but is 116 residues longer at the N-terminus and 16 residues shorter at the C-terminus.

Database accession numbers

	PIR	SWISSPROT	EMBL/GENBANK	REF
Soybean α			M64987	2
A. thaliana AK1			S58044	3
Rice Spk			D13436	4
Carrot			X56599	5
P. falciparum			X67288	6

Amino acid sequence of soybean CDPKα

```
MAAKSSSSST TTNVVTLKAA WVLPQRTQNI REVYEVGRKL GQGQFGTTFE  50
CTRRASGGKF ACKSIPKRKL LCKEDYEDVW REIQIMHHLS EHANVVRIEG 100
TYEDSTAVHL VMELCEGGEL FDRIVQKGHY SERQAARLIK TIVEVVEACH 150
SLGVMHRDLK PENFLFDTID EDAKLKATDF GLSVFYKPGE SFCDVVGSPY 200
YVAPEVLRKL YGPESDVWSA GVILYILLSG VPPFWAESEP GIFRQILLGK 250
LDFHSEPWPS ISDSAKDLIR KMLDQNPKTR LTAHEVLRHP WIVDDNIAPD 300
KPLDSAVLSR LKQFSAMNKL KKMALRVIAE RLSEEEIGGL KELFKMIDTD 350
NSGTITFDEL KDGLKRVGSE LMESEIKDLM DAADIDKSGT IDYGEFIAAT 400
VHLNKLEREE NLVSAFSYFD KDGSGYITLD EIQQACKDFG LDDIHIDDMI 450
KEIDQDNDGQ IDYGEFAAMM RKGNGGIGRR TMRKTLNLRD ALGLVDNGSN 500
QVIEGYFK
```

Homologues in other species

Sequences for CDPKs from carrot, *Arabidopsis thaliana*, alfalfa and *Plasmodium falciparum* (the protozoan which causes malaria) have been reported. The CDPK from alfalfa is the same size as CDPKα and is 95% identical[7]. CDPKs from other species have less identity (60–70%) with CDPKα, vary in size and may represent isoforms. *A. thaliana AK1* encodes a CDPK of predicted molecular weight 72 645 which is 70% identical to CDPKα in their overlapping regions[3]. This CDPK is 116 amino acids longer at the N-terminus. A partial carrot cDNA encodes a protein that is 70% identical to CDPKα and 60% identical to the *A. thaliana* enzyme[5]. A CDPK from *P. falciparum* is 36% identical to CDPKα[6].

Physiological substrates and specificity determinants

CDPKs phosphorylate synthetic peptides containing Basic–X–X–Thr/Ser, but not Basic–Basic–X–Thr/Ser[1]. CDPKs are multisubstrate kinases capable of phosphorylating numerous plant proteins *in vitro*. Nodulin-26, a membrane-bound protein in nitrogen-fixing root nodules, is likely to be a physiological substrate. Only one of three potential phosphorylation sites in this protein is phosphorylated by CDPK[8]. Other physiological substrates have not been identified[1].

Nodulin-26 (C-terminal sequence): KPLSEITK̲SASFLKGRAASK

Assay

CDPK may be assayed using syntide-2, PLARTLSVAGLPGKK[9] or histone IIIS. Assays are best performed with Ca^{2+} buffers which contain EGTA or EDTA, Mg^{2+}, and various amounts of Ca^{2+}[10].

Enzyme activators and inhibitors

Soybean CDPK is inhibited by agents that inhibit calmodulin activity; e.g. phenothiazines (trifluoperazine, chlorpromazine), and naphthalenesulfonamides (W series). *A. thaliana* CDPK is stimulated synergistically by Ca^{2+} and lipids[4].

Pattern of expression

CDPK is associated with F-actin, with the plasma membrane, with the symbiosome membrane of nitrogen-fixing root nodules, with chromatin, and with soluble fractions[1]. Specific expression of CDPKα has not yet been examined. Activity and immunological data show that CDPK is found in stems, roots, leaves, cotyledons and pollen[1]. Rice Spk is specifically expressed in developing seeds[4].

References

1 Roberts, D.M. and Harmon, A.C. (1992) Annu. Rev. Plant Physiol. Mol. Biol. 43, 375–414.
2 Harper, J.F. et al. (1991) Science 252, 951–954.
3 Harper, J.F. et al. (1994) Biochemistry 33, 7267–7277.
4 Kawasaki, T. et al. (1993) Gene 129, 183–189.
5 Suen K.-L. and Choi, J.H. (1991) Plant Molec. Biol. 17, 581–590.
6 Zhao, Y. et al. (1993) J. Biol. Chem. 268, 4347–4354.
7 Bögre, L et al. (1993) 12th Annual Missouri Plant Biochemistry, Molecular Biology and Physiology Symposium, Abstract 14.
8 Weaver, C.D. and Roberts, D.M. (1993) Biochemistry 31, 8954–8959.
9 Casnellie, J.E. (1991) Methods Enzymol. 200, 115–120.
10 Putnam-Evans, C.L. et al. (1990) Biochemistry 29, 2488–2495.

D. Grahame Hardie and David Carling (Dundee University and
MRC Unit for Molecular Medicine, UK)

The AMP-activated protein kinase was discovered as an activity that phosphorylated and inactivated HMG-CoA reductase[1]. However it was later found that it phosphorylated a number of other target proteins, and it was renamed after its physiological activator, 5'-AMP[2]. AMP activates AMPK both directly by an allosteric mechanism, and indirectly by promoting phosphorylation by a distinct "kinase kinase". The combination of these two effects means that elevation of AMP can give >100-fold activation. AMPK is now thought to exert a protective function, phosphorylating and inactivating biosynthetic enzymes, and other targets, when AMP is elevated due to cellular stress[3].

Subunit structure and isoforms

SUBUNIT	AMINO ACIDS	MOL. WT	SDS-PAGE
p63	552	62 250	63 kDa
p38	?	?	38 kDa
p35	?	?	35 kDa

AMPK has recently been shown to be most likely a heterotrimer (native molecular weight 190 000) comprising three distinct subunits[4]. Only the catalytic p63 subunit has been cloned and sequenced to date.

Domain structure

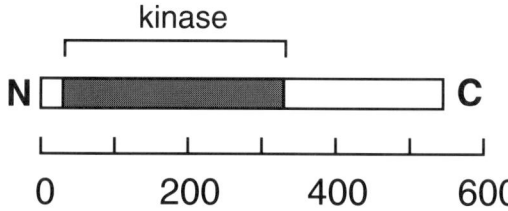

p63 is ≈46% identical in sequence (≈60% within the kinase domain) to *S. cerevisiae* **Snf1** and its higher plant homologues, **pSnf1**[5, 6]. All members of this subfamily have an N-terminal kinase domain and a less highly conserved C-terminal region that may have a regulatory function. The AMP-binding site on AMPK is located, at least in part, on the p63 subunit[7], and this may be one function of the C-terminal region. AMPK is phosphorylated (at unidentified sites) and activated by the upstream "kinase kinase", and also autophosphorylates slowly without detectable change in activity.

Database accession numbers

	PIR	SWISSPROT	EMBL/GENBANK	REF
p63			Z29486	5

Amino acid sequence of rat AMPK

```
MAEKQKHDGR VKIGHYVLGD TLGVGTFGKV KIGEHQLTGH KVAVKILNRQ  50
KIRSLDVVGK IKREIQNLKL FRHPHIIKLY QVISTPTDFF MVMEYVSGGE 100
LFDYICKHGR VEEVEARRLF QQILSAVDYC HRHMVVHRDL KPENVLLDAQ 150
MNAKIADFGL SNMMSDGEFL RTSCGSPNYA APEVISGRLY AGPEVDIWSC 200
GVILYALLCG TLPFDDEHVP TLFKKIRGGV FYIPEYLNRS IATLLMHMLQ 250
VDPLKRATIK DIREHEWFKQ DLPSYLFPED PSYDANVIDD EAVKEVCEKF 300
ECTESEVMNS LYSGDPQDQL AVAYHLIIDN RRIMNQASEF YLASSPPTGS 350
FMDDMAMHIP PGLKPHPERM PPLIADSPKA RCPLDALNTT KPKSLAVKKA 400
KWHLGIRSQS KPYDIMAEVY RAMKQLDFEW KVVNAYHLRV RRKNPVTGNY 450
VKMSLQLYLV DNRSYLLDFK SIDDEVVEQR SGSSTPQRSC SAAGLHRPRS 500
SVDSSTAENH SLSGSLTGSL TGSTLSSASP RLGSHTMDFF EMCASLITAL 550
AR
```

Homologues in other species

Although only well characterized from rat liver, it has recently been purified from pig liver[6], and activities with similar biochemical properties have been observed in human, rabbit and mouse cells. Similar biochemical activities in *S. cerevisiae* and higher plants[6, 8], now appear to be functions of the **Snf1** and **pSnf1** gene products. However neither the yeast nor the plant homologues appear to be regulated directly by AMP.

Physiological substrates and specificity determinants

In vitro substrates where the phosphorylation site have been identified are listed below. Evidence that these are substrates *in vivo* is particularly strong for HMG-CoA reductase and acetyl-CoA carboxylase. Studies using synthetic peptides based on the S79 site on acetyl-CoA carboxylase[9] have revealed that hydrophobic residues (M, V, L, I, F; bold, underlined) at $P-5/P-4$ and $P+4/P+5$, and a single basic residue (R, K or H; underlined) at $P-4$ or $P-3$ are critical to recognition. AMPK phosphorylates serine or threonine but not tyrosine.

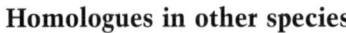

Rat HMG-CoA reductase (S871):	H**M**VHNRSKIN**L**QDL
Rat acetyl-CoA carboxylase (S79):	H**MR**SSMSGLH**L**VKQ
Rat acetyl-CoA carboxylase (S1200):	PT**L**N**R**MSFASN**L**NH
Rat acetyl-CoA carboxylase (S1215):	G**M**THVASVSD**V**LLD
Rat hormone-sensitive lipase:	P**M**RRSVSEAA**L**TQP
Rabbit muscle glycogen synthase (S7):	P**L**S**R**TLSVSS**L**PGL

Assay

Although originally assayed by ATP-dependent inactivation of HMG-CoA reductase or phosphorylation of acetyl-CoA carboxylase, AMPK is now routinely assayed by transfer of radioactivity from $[\gamma\text{-}^{32}P]$ATP to the synthetic peptide HMRSAMSGLHLVKRR[10].

Pattern of expression

AMPK is purified from the soluble fraction of the cell, but the subcellular location has not yet been addressed using antibodies. p63 mRNA and protein is expressed very widely in mammalian tissues, with particularly high levels in skeletal muscle. Using a biochemical kinase assay, liver contains the highest levels, and activity in skeletal muscle is low[10]. However this may be an artefact caused by variable elevation of AMP during cell harvesting.

References
1   Beg, Z.H. et al. (1973) Biochem. Biophys. Res. Commun. 54, 1362–1369.
2   Hardie, D.G. and MacKintosh, R.W. (1992) BioEssays 14, 699–704.
3   Corton, J.M. et al. (1994) Current Biol. 4, 315–324.
4   Davies, S.P. et al. (1994) Eur. J. Biochem 223, 351–357.
5   Carling, D. et al. (1994) J. Biol. Chem. 269, 11442–11448.
6   Mitchelhill, K.I. et al. (1994) J. Biol. Chem. 269, 2361–2364.
7   Carling, D. et al. (1989) Eur. J. Biochem. 186, 129–136.
8   MacKintosh, R.W. et al. (1992) Eur. J. Biochem. 209, 923–931.
9   Weekes, J. et al. (1993) FEBS Lett. 334, 335–339.
10 Davies, S.P. et al. (1989) Eur. J. Biochem. 186, 123–128.

Note added in proof
p38 has recently been shown to be related to Sip1/Sip2 from *S. cerevisiae*, and p35 to Snf4 (see Snf1 entry p.174) [Stapleton, D. et al. (1994) J. Biol. Chem. 269, 29343–29346.]

Snf1

Snf1 protein kinase (S. cerevisiae)
($CAT1^1$, $CCR1^{2,3}$)

E. Jane Albert Hubbard and Marian Carlson (Columbia University, USA)

Snf1 is required for expression of glucose-repressed genes in response to glucose deprivation[4,5]. Snf1 also affects other aspects of growth control such012 as carbohydrate storage and sporulation[6,7]. It is not clear whether the activity of Snf1 is regulated in response to glucose[8,9]. The Snf4 protein co-immunoprecipitates with Snf1 and is necessary for maximal Snf1 kinase activity[8]. A number of other proteins co-immunoprecipitate with Snf1 and are phosphorylated in immune-complex kinase assays[8]. One of these proteins, Sip1, has been characterized[10].

Subunit structure and isoforms

SUBUNIT	AMINO ACIDS	MOL. WT	SDS-PAGE
Snf1	633	72 045	72 kDa

On gel filtration, Snf1 migrates as a high molecular weight complex, apparently involving Snf4[9].

Genetics

Chromosomal location: right arm of chromosome IV[11].
Mutations: K84R, inactive; T210A, T210D, inactive; G53R, elevated kinase
activity.

Domain structure

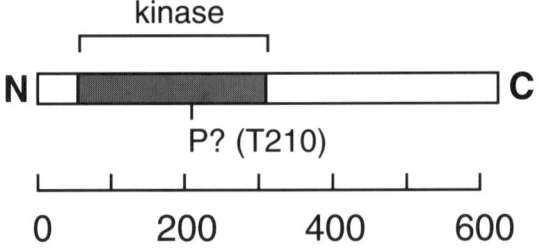

The domain structure of Snf1 has been analysed via specific deletions and mutations[8,9]. No glucose-responsive regulatory domain has been identified. The N-terminus includes a stretch of 13 consecutive histidines, but deletion of these residues has no effect. Deletion of most of the region N-terminal to the kinase domain (codons 5–52) partially impairs kinase activity but has no regulatory consequences; this region also does not mediate the stimulatory effects of Snf4. Deletion of the region C-terminal to the kinase domain (codons 309–628) yields a protein that is able to provide partial Snf1 function and allows regulated expression of glucose-repressed genes. Overexpression of this C-terminally deleted protein is deleterious to the cell. Two different mutations in a conserved putative phosphorylation site, analogous to T197 in **PKA**[12–14], i.e. T210A and T210D, abolish function.

Database accession numbers

	PIR	SWISSPROT	EMBL/GENBANK	REF
Snf1	A26030	P06782	M13971	4

Amino acid sequence of *S. cerevisiae* Snf1

```
MSSNNNTNTA PANANSSHHH HHHHHHHHHH GHGGSNSTLN NPKSSLADGA  50
HIGNYQIVKT LGEGSFGKVK LAYHTTTGQK VALKIINKKV LAKSDMQGRI 100
EREISYLRLL RHPHIIKLYD VIKSKDEIIM VIEYAGNELF DYIVQRDKMS 150
EQEARRFFQQ IISAVEYCHR HKIVHRDLKP ENLLLDEHLN VKIADFGLSN 200
IMTDGNFLKT SCGSPNYAAP EVISGKLYAG PEVDVWSCGV ILYVMLCRRL 250
PFDDESIPVL FKNISNGVYT LPKFLSPGAA GLIKRMLIVN PLNRISIHEI 300
MQDDWFKVDL PEYLLPPDLK PHPEEENENN DSKKDGSSPD NDEIDDNLVN 350
ILSSTMGYEK DEIYESLESS EDTPAFNEIR DAYMLIKENK SLIKDMKANK 400
SVSDELDTFL SQSPPTFQQQ SKSHQKSQVD HETAKQHARR MASAITQQRT 450
YHQSPFMDQY KEEDSTVSIL PTSLPQIHRA NMLAQGSPAA SKISPLVTKK 500
SKTRWHFGIR SRSYPLDVMG EIYIALKNLG AEWAKPSEED LWTIKLRWKY 550
DIGNKTNTNE KIPDLMKMVI QLFQIETNNY LVDFKFDGWE SSYGDDTTVS 600
NISEDEMSTF SAYPFLHLTT KLIMELAVNS QSN
```

Sites of interest: T210, putative phosphorylation

Homologues in other species

Snf1 has close relatives in plants (**pSnf1**), and is also closely related to vertebrate **AMPK**.

References

1. Schuller, H.-J. and Entian, K.-D. (1987) Mol. Gen. Genet. 209, 366–373.
2. Ciriacy, M. (1977) Mol. Gen. Genet. 154, 213–220.
3. Denis, C.L. (1984) Genetics 108, 833–844.
4. Celenza, J.L. and Carlson, M. (1986) Science 233, 1175–1180.
5. Johnston, M. and Carlson, M. (1992) In The Molecular and Cellular Biology of the Yeast Saccharomyces: Gene Expression, Jones, E.W., Pringle, J.R. and Broach, J.R., eds, Cold Spring Harbor Laboratory Press, Cold Spring Harbor, NY, pp. 193–281.
6. Thompson-Jaeger, S. et al. (1991) Genetics 129, 697–706.
7. Hubbard, E.J.A. et al. (1992) Genetics 130, 71–80.
8. Celenza, J.L. and Carlson, M. (1989) Mol. Cell Biol. 9, 5034–5044.
9. Estruch, F. et al. (1992) Genetics 132, 639–650.
10. Yang, X. et al. (1992) Science 257, 680–682.
11. Celenza, J.L. and Carlson, M. (1984) Mol. Cell Biol. 4, 49–53.
12. Shoji, S. et al. (1983) Biochemistry 22, 3702–3709.
13. Levin, L.R. et al. (1990) Science 240, 68–70.
14. Levin, L.R. and Zoller, M.J. (1990) Mol. Cell Biol. 10, 1066–1075.

Kin1/2 — Kin1 and Kin2 (*S. cerevisiae*)

David E. Levin (Johns Hopkins University, USA)

Kin1 and Kin2 are closely related gene products from *S. cerevisiae*. Deletion of both *KIN1* and *KIN2* results in no apparent phenotypic defect[1]. Kin1 autophosphorylates on serine and threonine, and phosphorylates α-casein on serine and threonine[2]. Kin1 and Kin2 are also closely related to the **SpKin1** gene product of *S. pombe*[3].

Subunit structure

SUBUNIT	AMINO ACIDS	MOL. WT	SDS-PAGE
Kin1	1064	117 000	145 kDa
Kin2	1148	126 000	

Genetics

Chromosomal location:
Kin1: VII, tightly linked to CLY8.

Domain structure

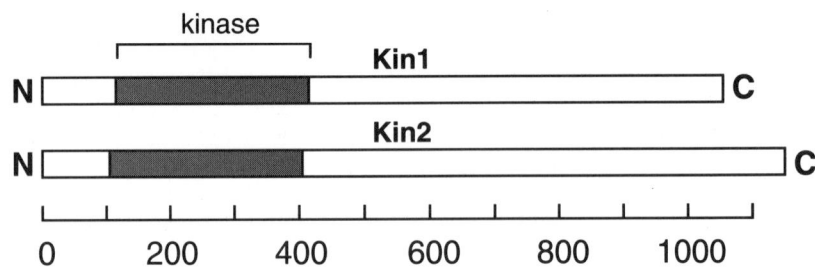

Database accession numbers

	PIR	SWISSPROT	EMBL/GENBANK	REF
Kin1			M69017	1
Kin2			M69018	1

Amino acid sequence of *S. cerevisiae* Kin1

```
MDDYHVNTAV SMGRGNQQDD GNSERNSMHT QPSTMAPATL RMMGKSPQQQ    50
QQQNTPLMPP ADIKYANNGN SHQAEQKERQ VELEGKSREN APKPNTTSQS   100
RVSSSQGMPK QFHRKSLGDW EFVETVGAGS MGKVKLAKHR YTNEVCAVKI   150
VNRATKAFLH KEQMLPPPKN EQDVLERQKK LEKEISRDKR TIREASLGQI   200
LYHPHICRLF EMCTLSNHFY MLFEYVSGGQ LLDYIIQHGS IREHQARKFA   250
RGIASALIYL HANNIVHRDL KIENIMISDS SEIKIIDFGL SNIYDSRKQL   300
HTFCGSLYFA APELLKANPY TGPEVDVWSF GVVLFVLVCG KVPFDDENSS   350
VLHEKIKQGK VEYPQHLSIE VISLLSKMLV VDPKRRATLK QVVEHHWMVR   400
GFNGPPPSYL PKRVPLTIEM LDINVLKEMY RLEFIDDVEE TRSVLVSIIT   450
DPHYGLLSRQ YWTLAAKMNA ESSDNGNAPN ITESFEDPTR AYHPMISIYY   500
LTSEMLDRKH AKIRNQQQRQ SHENIEKLSE IPESVKQRDV EVNTTAMKSE   550
PEATLATKDT SVPFTPKNSD GTEPPLHVLI PPRLAMPEQA HTSPTSRKSS   600
```

Amino acid sequence of *S. cerevisiae* Kin1   *continued*

```
DNQRREMEYA LSPTPQGNDY QQFRVPSTTG DPSEKAKFGN IFRKLSQRRK  650
KTIEQTSVNS NNSINKPVQK THSRAVSDFV PGFAKPSYDS NYTMNEPVKT  700
NDSRGGNKGD FPALPADAEN MVEKQREKQI EEDIMKLHDI NKQNNEVAKG  750
SGREAYAAQK FEGSDDDENH PLPPLNVAKG RKLHPSARAK SVGHARRESL  800
KYMRPPMPSS AYPQQELIDT GFLESSDDNK SDSLGNVTSQ TNDSVSVHSV  850
NAHINSPSVE KELTDEEILQ EASRAPAGSM PSIDFPRSLF LKGFFSVQTT  900
SSKPLPIVRY KIMFVLRKMN IEFKEVKGGF VCMQRFSSNN VAAKREGTPR  950
SIMPLSHHES IRRQGSNKYS PSSPLTTNSI HQRKTSITET YGDDKHSGTS 1000
LENIHQQGDG SEGMTTTEKE PIKFEIHIVK VRIVGLAGVH FKKISGNTWL 1050
YKELASSILK ELKL
```

Amino acid sequence of *S. cerevisiae* Kin2

```
MPNPNTADYL VNPNFRTSKG GSLSPTPEAF NDTRVAAPAT LRMMGKQSGP   50
RNDQQQAPLM PPADIKQGKE QAAQRQNDAS RPNGAVELRQ FHRRSLGDWE  100
FLETVGAGSM GKVKLVKHRQ TKEICVIKIV NRASKAYLHK QHSLPSPKNE  150
SEILERQKRL EKEIARDKRT VREASLGQIL YHPHICRLFE MCTMSNHFYM  200
LFEYVSGGQL LDYIIQHGSL KEHHARKFAR GIASALQYLH ANNIVHRDLK  250
IENIMISSSG EIKIIDFGLS NIFDYRKQLH TFCGSLYFAA PELLKAQPYT  300
GPEVDIWSFG IVLYVLVCGK VPFDDENSSI LHEKIKKGKV DYPSHLSIEV  350
ISLLTRMTVV DPLRRATLKN VVEHPWMNRG YDEKAPSYVP NRVPLTPEMT  400
DSQVLKEMYR LEFIDDIEDT RRSLIRLVTE KEYIQLSQEY WDKLSNAKGL  450
SSSLNNNYLN STAQQTLIQN HITSNPSQSG YNEPDSNFED PTLAYHPLLS  500
IYHLVSEMVA RKLAKLQRRQ ALALQAQAQQ RQQQQQVALG TKVALNNNSP  550
DIMTKMRSPQ KEVVPNPGIF QVPAIGTSGT SNNTNTSNKP PLHVMVPPKL  600
TIPEQAHTSP TSRKSSDIHT ELNGVLKSTP VPVSGEYQQR SASPVVGEHQ  650
EKNTIGGIFR RISQSGQSQH PTRQSGTYSS KENLQHICQN QMKFPSKYRK  700
AIVVLYQTYI PSARRYPSYV PNSVDVKQKP AKNTTIAPPI RSVSQKQNSD  750
LPALPQKRQL IVQKQRQKLL QENLDKLQIN DNDNNNVAV VDGINNDNSD  800
HYLSVPKGRK LHPSARAKSV GHARRESLKF TRPPIPAALP PSDMTNDNGF  850
LGEANKERYN PVSSNFSTVP EDSTTYSNDT NNRLTSVYSQ ELTEKQILEE  900
ASKAPPGSMP SIDYPKSMFL KGFFSVQTTS SKPLPIVRHN IISVLTRMNI  950
DFKEVKGGFI CVQQRPSIET AAVPVITTTG VGLDSGKAMD LQNSLDSQLS 1000
SSYHSTASSA SRNSSIKRQG SYKRGQNNIP LTPLATNTHQ RNSSIPMSPN 1050
YGNQSNGTSG ELSSMSLDYV QQQDDILTTS RAQNINNVNG QTEQTNTSGI 1100
KERPPIKFEI HIVKVRIVGL AGVHFKKVSG NTWLYKELAS YILKELNL
```

References
[1] Levin, D. E. et al. (1987) Proc. Natl Acad. Sci. USA 84, 6035–6039.
[2] Lamb, A. et al. (1991) Yeast 7, 219–228.
[3] Levin, D. E. and Bishop, J. M. (1990) Proc. Natl Acad. Sci. USA 87, 8272–8276.

SpKin1 — Kin1 (*S. pombe*)

David E. Levin (Johns Hopkins University, USA)

Loss of function of *S. pombe kin1⁺* results in morphological and growth defects[1]. *Kin1*-disrupted cells grow slowly, and as spheres, rather than rods. Differential sensitivity of *kin1*-disrupted cells to lysis by treatment with α- and β-glucanases suggests an alteration in either the composition or organization of their cell walls. SpKin1 is closely related to **Kin1/2** from *S. cerevisiae*[2].

Subunit structure

SUBUNIT	AMINO ACIDS	MOL. WT	SDS-PAGE
SpKin1	891	98 000	

Domain structure

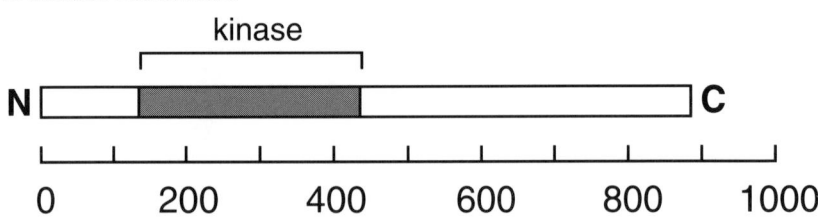

Database accession numbers

	PIR	SWISSPROT	EMBL/GENBANK	REF
SpKin1			M36060	*1*

Amino acid sequence of *S. pombe* Kin1

```
MEYRTNNVPV GNETKSAALN ALPKIKISDS PNRHHNLVDA FMQSPSYSTQ  50
PKSAVEPLGL SFSPGYISPS SQSPHHGPVR SPSSRKPLPA SPSRTRDHSL 100
RVPVSGHSYS ADEKPRERRK VIGNYVLGKT IGAGSMGKVK DAHHLKTGEQ 150
FAIKIVTRLH PDITKAKAAA SAEATKAAQS EKNKEIRTVR EAALSTLLRH 200
PYICEARDVY ITNSHYYMVF EFVDGGQMLD YIISHGKLKE KQARKFERQI 250
GSALSYLHQN SVVHRDLKIE NILISKTGDI KIIDFGLSNL YRRQSRLRTF 300
CGSLYFAAPE LLNAQPYIGP EVDVWSFGIV LYVLVCGKVP FDDQNMSALH 350
AKIKKGTVEY PRYLSSDCKG LLSRMLVTDP LKRATLEEVL NHPWMIRNYE 400
GPPASFAPER SPITLPLDPE IIREMNGFDF GPPEKIVREL TKVISSEAYQ 450
SLAKTGFYSG PNSADKKKSF FEFRIRHAAH DIENPILPSL SMNTDIYDAF 500
HPLISIYYLV SERRVYEKGG NWNRIAKTPV SSVPSSPVQP TSYNRTLPPM 550
PEVVAYKGDE ESPRVSRNTS LARRKPLPDT ESHSPSPSAT SSIKKNPSSI 600
FRRFSSRRKQ NKSSTSTLQI SAPLETSQSP PTPRTKPSHK PPVSYKNKLV 650
TQSAIGRSTS VREGRYAGIS SQMDSLNMDS TGPSASNMAN APPSVRNNRV 700
LNPRGASWGH GRMSTSTTNR QKQILNETMG NPVDKNSTSP SKSTDKLDPI 750
KPVFLKGLFS VSTTSTKSTE SIQRDLMLVM GMLDIEYKEI KGGYACLYKP 800
QGIRTPTKST SVHTRRKPSY GSNSTTDSYG SVPDTVPLDD NGESPASNLA 850
FEIYIVKVPI LSLRGVSFHR ISGNSWQYKT LASRILNELK L
```

References
[1] Levin, D.E. and Bishop, J.M. (1990) Proc. Natl Acad. Sci. USA 87, 8272–8276.
[2] Levin, D.E. et al. (1987) Proc. Natl Acad. Sci. USA 84, 6035–6039.

Nim1
Nim1 protein kinase (*S. pombe*)
(Cdr1)

Paul Russell (Scripps Research Institute, La Jolla, California, USA)

Nim1 functions as a mitotic inducer[1] by phosphorylating and inactivating Wee1[2-4]. The *nim1* gene was also genetically identified as *cdr1*[5].

Subunit structure and isoforms

SUBUNIT	AMINO ACIDS	MOL. WT	SDS-PAGE
Nim1	593	66 969	76 kDa

Nim1 is a monomer[2].

Domain structure

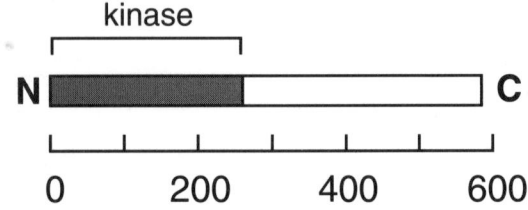

Database accession numbers

	PIR	SWISSPROT	EMBL/GENBANK	REF
Nim1		P07334	X57549	5

Amino acid sequence of *S. pombe* Nim1

```
MVKRHKNTIG VWRLGKTLGT GSTSCVRLAK HAKTGDLAAI KIIPIRYASI  50
GMEILMMRLL RHPNILRLYD VWTDHQHMYL ALEYVPDGEL FHYIRKHGPL 100
SEREAAHYLS QILDAVAHCH RFRFRHRDLK LENILIKVNE QQIKIADFGM 150
ATVEPNDSCL ENYCGSLHYL APEIVSHKPY RGAPADVWSC GVILYSLLSN 200
KLPFGGQNTD VIYNKIRHGA YDLPSSISSA AQDLLHRMLD VNPSTRITIP 250
EFFSHPFLMG CTSLSSMDST TPPTPSLSID EIDPLVVDCM CVLWKKSSSK 300
KVVRRLQQRD DNDEKYVYKV LSEILRDDML KKQRFDENKY LSLYDLIHDN 350
NLFTKASIST TSLVKSNVST NSRKSSNFED ELARRVSSPL SALNQMSQSP 400
IPIRVSSDKD YDSYACHEVV SNPSTLDDDY NYMFVCPPEE YTYSTDNVRT 450
DSLDLQSLPT PTLEQLESVP FNRYGYVRIF PSTTLSSTAS GYYTPDSLST 500
PEPSIDGLTN LDDVQVGGFV QGSGNQNRRP ISFPVISNMQ PNITNVRSAS 550
APLCSSPVPS RRYSQYATNI RYTPRKVSSG SVLRKISSFF RKD
```

Homologues in other species
None currently known.

Physiological substrates and specificity determinants

SpWee1 PK is the physiological substrate[2-4]. The phosphorylation site(s) are not known.

Assay

Assayed by transfer of radioactivity from $[\gamma\text{-}^{32}P]$ATP to **SpWee1**[2–4].

References

[1]   Russell, P. and Nurse, P. (1987) Cell 49, 569–576.

[2]   Coleman, T.R. et al. (1993) Cell 72, 919–929.

[3]   Parker, L.L. et al. (1993) Nature 363, 736–738.

[4]   Wu, L. and Russell, P. (1993) Nature 363, 738–741.

[5]   Feilotter, H. et al. (1991) Genetics 127, 309–318.

Nigel Halford (Long Ashton Research Station, UK)

Genes, cDNAs and PCR products related to *SNF1*, a yeast gene which encodes a protein kinase required for the derepression of glucose-repressible genes, have been isolated from rye (*Secale cereale: RKIN1*)[1], barley (*Hordeum vulgare: BKIN12*)[2] and *Arabidopsis thaliana* (*AKIN10*)[3]. Expression of the cDNA from rye in a yeast *snf1* mutant restored *SNF1* function[1], demonstrating a functional relationship between the yeast and plant enzymes. Although the exact function of these genes in plants has still to be demonstrated, the possibility that they play a role in regulating carbohydrate metabolism makes them of particular interest and a potential target for genetic manipulation.

Subunit structure and isoforms

SUBUNIT	AMINO ACIDS	MOL. WT	SDS-PAGE
Bkin12	513	58 715	58 kDa

A second related gene (*BKIN2*) is found in barley. By contrast, a single Snf1-related gene (*AKIN10*) is found in *A. thaliana*. It is not known whether Bkin2 represents an isoform or a different but closely-related protein kinase. The entire sequence of the Bkin2 type has not yet been determined, but there is considerable divergence in sequence outside of the catalytic domain.

Domain structure

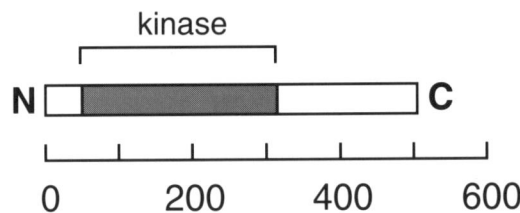

Database accession numbers

	PIR	SWISSPROT	EMBL/GENBANK	REF
Bkin12			X65606	2
Rkin1			M74113	1
Akin10			M93023	3

Amino acid sequence of barley Bkin12

```
MDGNNRGGGH SEVLKNYNLG KTLGLGTFGD VKVAEHKLTG QRVAIKILNR  50
RKMETMEMEE KANREIKIMR LFIDFIHPHI IRVYQVIETP KDIFIVMEYC 100
NNGELLDYII ENGRLQEDEA RRIFQQILAG VEYCHRIMVV HRDLKPENLL 150
LDSKYNVKLA DFGLSNVMRD GHFLKTSCGS LNYAAPEIIS SKLYAGPEVD 200
VVWSCGVILYA LLCGSVPFDD DNIPSLFRKI KGGTYILPSY LSDSARDLIP 250
KLLNIDPMKR ITIHEIRVHP WFKNHLPCYL AVPPPYKAPK AKMIDEDILR 300
DVVNLGYDKD HVCESLWNRL QNEETVAYYL LLDNRFRSTS GYLGADHQHL 350
MDRSFNEFTL SESASPSTRN YLPGINDSQG GGLRPYYPVQ RKWAIGLQSG 400
AHPRDIMIEV LKALKELNVC WKKNGLYNMK CRWCPGFPQV SAMLLDSNHN 450
FVDDSTIMDN GNADGRLPAV VKFEIQLYKT KDNKYLLDIQ RVTGPQLLFL 500
EFCGAFFTNL RVL
```

Homologues in other species

Snf1 *(S. cerevisiae)*[4], Rkin1 (rye), Akin10 *(A. thaliana)*, **AMPK** (vertebrates).

Pattern of expression

SNF1-related transcripts are detectable in all tissues of barley that we have analysed: endosperms and aleurones from seeds, leaves, roots, root tips, coleoptiles and internodes. The level of expression in the seed tissues is considerably higher than in the rest of the plant. The use of probes which distinguish between the *BKIN2* and *BKIN12* subtypes has shown that the *BKIN2* type is expressed in all tissues, although at very low levels in the seed, whereas the *BKIN12* type is expressed at high levels in the seed, with low or undetectable levels of expression in the non-seed tissues[5]. There is therefore a clear distinction between the expression patterns of the two subtypes of gene. In contrast, the single gene in *A. thaliana* is expressed in roots, shoots and leaves. Other tissues have not been analysed.

References

[1] Alderson, A. et al. (1991) Proc. Natl Acad. Sci. USA 88, 8602–8605.
[2] Halford, N.G. et al. (1992) The Plant Journal 2, 791–797.
[3] Le Guen, L. et al. (1992) Gene 120, 249–254.
[4] Celenza, J.L. and Carlson, M. (1986) Science 233, 1175–1180.
[5] Hannappel, U. et al. (1995), in preparation.

Cdc2

Cdc2 protein kinase (vertebrates)
(Maturation-promoting factor (MPF), M-phase kinase,
growth-associated H1 kinase)

Chris Norbury (ICRF Molecular Oncology Laboratory, Oxford, UK)

Activation of the Cdc2 protein kinase in the G2 phase of the mitotic cell cycle brings about the onset of mitosis in human cells, and probably in all other eukaryotes[1]. The human *CDC2* gene was isolated through its ability to substitute for a defective *cdc2* gene in *S. pombe*[2] (see **SpCdc2**). In addition to its mitotic role, human Cdc2 can perform the G1 (START) and meiotic functions of Cdc2 in *S. pombe*, and may also act during G1 and meiosis in human cells, though this has yet to be established.

Despite early reports of Cdc2 kinase activity associated in some systems with the monomeric protein, it is now generally held that the active kinase consists of a heterodimer containing a cyclin protein in addition to Cdc2. Association with a cyclin B subunit is essential for generation of the active M-phase form of the kinase, but this association generally results in phosphorylation of the Cdc2 subunit at T14 and Y15, blocking kinase activity, and at T161, phosphorylation of which may stabilize the heterodimer and is essential for its activation. Phosphorylation at Y15 is probably carried out by **Wee1**; T161 may be phosphorylated by a human equivalent of *Xenopus laevis* **Mo15**, also termed CAK. Final activation at the G2/M boundary results from Cdc25-mediated dephosphorylation of the Cdc2 subunit at T14 and Y15. The kinase is inactivated in late M-phase through a mechanism involving proteolytic degradation of the cyclin B subunit, probably as a result of its ubiquitination. In species closely related to humans two cyclin B isoforms (B1, B2) have been described, suggesting two may exist in humans, although only one sequence has so far been reported. The biological consequences of association between cyclin A and Cdc2 are not yet understood, but in this case the catalytic subunit may escape negative regulation by phosphorylation at T14 and Y15.

Subunit structure

SUBUNIT	AMINO ACIDS	MOL. WT	SDS-PAGE
Cdc2	297	34 073	34 kDa
Cyclin B	433	48 337	62 kDa

Cdc2 is a catalytic subunit barely larger than the canonical kinase domain. The majority of the active M-phase form of Cdc2 consists of heteromers with cyclin B, although some authors have reported physiological association between Cdc2 and cyclin A. The *CKShs1* and *2* genes encode small (9 kDa) Cdc2-binding proteins functionally homologous to p13^{suc1} of fission yeast, although the biological role of these proteins is not yet known.

Genetics

Chromosomal location:
 Cdc2: 10q21.1
 Cyclin B: 5q11

Mutations: a T14A, Y15F double mutant is constitutively activated, although still requiring association with cyclin and phosphorylation at T161.
 K20A, R22A, H23A, K24A, K56A, E57A, R59A, H60A, T161A/D/E, R170A,

E173A, R180A, D186A, E196A, K200A, K201A, R215A, R218A, D271A, K274A, R275A, inactive. P272S (mouse), *ts*
Deletion of the N-terminal ≈90 amino acids of B-type cyclins generates proteins that form stable complexes with Cdc2, are not subject to degradation and prevent exit from mitosis.

Domain structure

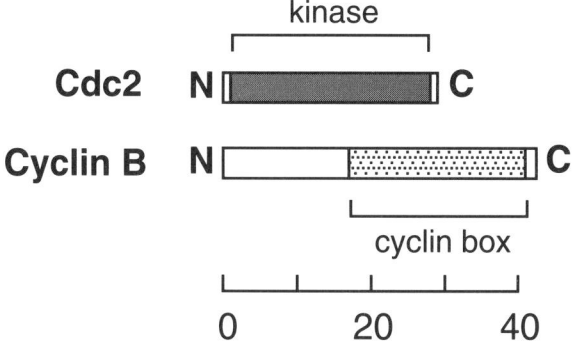

Database accession numbers

	PIR	SWISSPROT	EMBL/GENBANK	REF
Human Cdc2	A29539	P06493	X05360	2
Cyclin B	A32992	P14635	M25753	3

Amino acid sequence of human Cdc2

```
MEDYTKIEKI GEGTYGVVYK GRHKTTGQVV AMKKIRLESE EEGVPSTAIR  50
EISLLKELRH PNIVSLQDVL MQDSRLYLIF EFLSMDLKKY LDSIPPGQYM 100
DSSLVKSYLY QILQGIVFCH SRRVLHRDLK PQNLLIDDKG TIKLADFGLA 150
RAFGIPIRVY THEVVTLWYR SPEVLLGSAR YSTPVDIWSI GTIFAELATK 200
KPLFHGDSEI DQLFRIFRAL GTPNNEVWPE VESLQDYKNT FPKWKPGSLA 250
SHVKNLDENG LDLLSKMLIY DPAKRISGKM ALNHPYFNDL DNQIKKM
```

Sites of interest: T14, Y15, T161, phosphorylation

Amino acid sequence of human cyclin B

```
MALRVTRNSK INAENKAKIN MAGAKRVPTA PAATSKPGLR PRTALGDIGN  50
KVSEQLQAKM PMKKEAKPSA TGKVIDKKLP KPLEKVPMLV PVPVSEPVPE 100
PEPEPEPEPV KEEKLSPEPI LVDTASPSPM ETSGCAPAEE DLCQAFSDVI 150
LAVNDVDAED GADPNLCSEY VKDIYAYLRQ LEEEQAVRPK YLLGREVTGN 200
MRAILIDWLV QVQMKFRLLQ ETMYMTVSII DRFMQNNCVP KKMLQLVGVT 250
AMFIASKYEE MYPPEIGDFA FVTDNTYTKH QIRQMEMKIL RALNFGLGRP 300
LPLHFLRRAS KIGEVDVEQH TLAKYLMELT MLDYDMVHFP PSQIAAGAFC 350
LALKILDNGE WTPTLQHYLS YTEESLLPVM QHLAKNVVMV NQGLTKHMTV 400
KNKYATSKHA KISTLPQLNS ALVQDLAKAV AKV
```

Homologues in other species

The human Cdc2 sequence is from qm637 SV40-transformed fibroblasts. Cdc2 homologues have been identified in *S. pombe* **SpCdc2**, *S. cerevisiae* **Cdc28** and *D. melanogaster* **DmCdc2/DmCdc2c**, alfalfa (EMBL X70707), *Arabidopsis thaliana*[4], *C. elegans* (EMBL X68384), chicken[5], *D. discoideum*[6], maize[7], mouse[8], *Plasmodium falciparum* (PIR S22008), rat (PIR S24913), rice[9] and *Xenopus laevis*[10].

Physiological substrates and specificity determinants

The consensus sequence for phosphorylation is S/TPXK/R, frequently with additional basic residues on either side. The site of phosphorylation (bold type) has been unambiguously mapped for the following substrates:

c-Abl:	{ APD**T**PEL
	{ PAV**S**PLL
Elongation factor-1γ:	KKE**T**PKK
Glial fibrillary acidic protein:	RIT**S**AAR
High mobility group protein 1:	{ EVP**T**PKR
	{ TTT**T**PGR
Histone H1 (multiple sites):	K**S**/**T**PXK
Nuclear lamin B2*:	TPL**S**PTR
Nucleolin (multiple sites):	**T**PXK
Src:	{ ASQ**T**PNK
	{ THR**T**PSR
	{ TVT**S**PQR

*Site conserved in lamins A and C.

Vimentin and caldesmon are also thought to be substrates *in vivo*, although the sites of phosphorylation have not yet been reported. For a number of other *in vitro* substrates, phosphorylation *in vivo* by distinct, Cdc2-related kinases has not been ruled out[1].

Assay

Cdc2 activity is usually assayed by the cyclic nucleotide- and Ca^{2+}-independent transfer of phosphate from $[\gamma-^{32}P]ATP$ to histone H1[11]. Alternative assays are based on the induction of meiotic maturation (in the absence of ongoing protein synthesis) in oocytes, or generation of M-phase events such as germinal vesicle (nuclear envelope) breakdown in cell-free extracts[12]. It should be noted that cyclins (A or B) and several other proteins can induce such events without requiring exogenous Cdc2 protein.

Enzyme activators and inhibitors

INHIBITORS	SPECIFICITY	IC_{50} (nM)
Butyrolactone I	Good	680
Staurosporine	Poor	6.9

Pattern of expression

Cdc2 is widely expressed in proliferating tissues and is transcriptionally down-regulated during terminal differentiation or reversible withdrawal from the cell cycle. Transcriptional induction in response to cell cycle re-entry occurs in late G1. In HeLa cells there is a degree of periodicity in Cdc2 expression during continuous proliferation, with Cdc2 transcription, transcript level, *de novo* Cdc2 protein synthesis and degradation all peaking in G2 or M. Cyclin B expression is generally associated with Cdc2 expression, and cyclin B mRNA and protein levels are also periodic during the cell cycle, accumulating during S and G2 and declining rapidly in late mitosis. The Cdc2/cyclin B complex accumulates in the cytoplasm during the S and G2 phases, becoming abruptly relocated to the nuclear compartment at the onset of mitosis.

References

[1] Norbury, C.J. and Nurse, P. (1992) Annu. Rev. Biochem. 61, 441–470 .

[2] Lee, M.G. and Nurse, P. (1987) Nature 327, 31–35.

[3] Pines, J. and Hunter, T. (1989) Cell 58, 833–846.

[4] Hirayama, T. et al. (1991) Gene 105, 159–165.

[5] Krek, W. and Nigg, E.A. (1989) EMBO J. 8, 3071–3078.

[6] Michaelis, C.E. and Weeks, G. (1992) Biochim. Biophys. Acta 1132, 35–42.

[7] Colasanti, J. et al. (1991) Proc. Natl Acad. Sci. USA 88, 3377–3381.

[8] Th'ng, J.P. et al. (1990) Cell 63, 313–24.

[9] Hashimoto, J. et al. (1992) Mol. Gen. Genet. 233, 10–16.

[10] Pickham, K.M. et al. (1992) Mol. Cell. Biol. 12, 3192–3203.

[11] Langan, T.A. et al. (1989) Mol. Cell. Biol. 9, 3860–3868.

[12] Lohka, M. and Maller, J. (1985) J. Cell Biol. 101, 518–523.

J. Wade Harper and Stephen J. Elledge (Baylor College of Medicine, USA)

Cyclin-dependent kinases (Cdks) are members of the **Cdc2** family, and have been implicated in the control of cell cycle transitions in a variety of eukaryotic organisms. Cdk2 is a member of a subfamily that is likely to perform functions distinct from those of **Cdc2**[1]. In particular, Cdk2 has been implicated in control of G1 and S-phase events in the cell cycle, because of the timing of expression and activity of its regulatory subunits, as described below. The human gene was originally identified on the basis of its ability to complement a *Cdc28-4 ts* mutation in *S. cerevisiae*, using a human cDNA expression library[1] and was found to be homologous to the *Xenopus Eg1* gene[2], originally thought to be the *Xenopus cdc2* homologue.

Subunit structure and isoforms

SUBUNIT	AMINO ACIDS	MOL. WT	SDS-PAGE
Cdk2	298	33 892	33 kDa

Cdk2 is catalytically inactive in monomeric form, but is active as a heterodimer in association with its regulatory cyclin subunits[3]. Unlike **Cdc2** which absolutely requires both cyclin association and T161 phosphorylation for activity, Cdk2 can be weakly activated by cyclin A (1.5% maximal activity) in the absence of T160 phosphorylation, and fully activated with it[3]. Cdk2 has been found to associate with cyclin E[4] and cyclin D1[5] during the G1 phase of the cell cycle, and with cyclin A[6,7] during S phase[8]. It has also been detected in complexes with p107-E2F, E1A, and can bind CksHs-1.

Domain structure

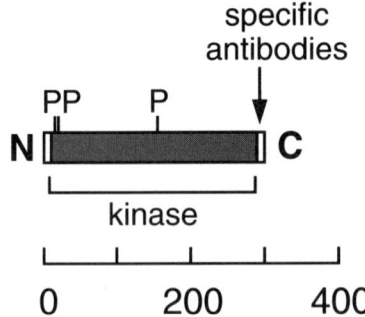

Cdk2 is a known phosphoprotein with phosphorylation demonstrated for amino acids T14, Y15, and T160[9]. T14 and Y15 are in the ATP-binding fold and are thought to negatively regulate kinase activity. Y15 can be phosphorylated *in vitro* by **Wee1**, but it is not known whether this enzyme performs that function *in vivo*. Cdc25 protein phosphatase can dephosphorylate phosphorylated Y15 *in vitro*. Cdc2-activating kinase (CAK, see **Mo15**) can phosphorylate T160 in a cyclin-dependent fashion *in vitro*, but it is not known whether this enzyme performs that function *in vivo*[3]. The

T14 kinase has not been identified. The PSTAIRE motif is present: this region has become a signature for different Cdks. Originally, antipeptide antibodies to PSTAIRE were used for Cdc2 recognition, but then Cdk2 and Cdk3 were discovered to contain the same motif. The PSTAIRE region has been hypothesized to function in cyclin binding, but this has not been completely resolved. Antibodies specific for Cdk2 have been generated using peptides derived from the C-terminal 10 amino acids[6] which are not conserved among related kinases[10].

Database accession numbers

	PIR	SWISSPROT	EMBL/GENBANK	REF
Human			X61622	1

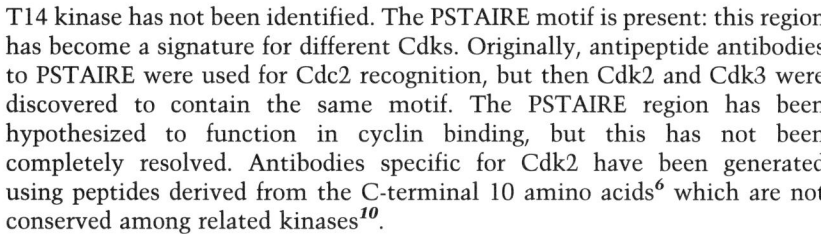

Amino acid sequence of human Cdk2

```
MENFQKVEKI GEGTYGVVYK ARNKLTGEVV ALKKIRLDTE TEGVPSTAIR  50
EISLLKELNH PNIVKLLDVI HTENKLYLVF EFLHQDLKKF MDASALTGIP 100
LPLIKSYLFQ LLQGLAFCHS HRVLHRDLKP QNLLINTEGA IKLADFGLAR 150
AFGVPVRTYT HEVVTLWYRA PEILLGSKYY STAVDIWSLG CIFAEMVTRR 200
ALFPGDSEID QLFRIFRTLG TPDEVVWPGV TSMPDYKPSF PKWARQDFSK 250
VVPPLDEDGR SLLSQMLHYD PNKRISAKAA LAHPFFQDVT KPVPHLRL
```

Sites of interest: T14, Y15, T160, phosphorylation

Homologues in other species

Cdk2 has been sequenced from humans (B-cells and HeLa cells)[1, 7], Xenopus[2] and goldfish[11].

Physiological substrates and specificity determinants

The *in vivo* substrates of Cdk2 are currently unknown. *In vitro*, immune complexes containing Cdk2 and associated cyclins will phosphorylate several protein substrates including histone H1[3, 6–9], the 34 kDa subunit of replication factor A[6], the retinoblastoma protein[12], large T antigen, and p60[c-src] (Dr David Morgan, personal communication). None of the phosphorylation sites in these substrates have been determined. The sequence specificity has also not been investigated in detail, but like **Cdc2**, Cdk2 appears to prefer serine/threonine residues followed by proline (i.e. **XS/TPX**). For example, Cdk2 readily phosphorylates the histone H1 derived peptide AKAKKTPKKAK, but does not appreciably phosphorylate the related peptide in which proline is replaced by glycine (J.W.H., unpublished data).

Assay

Activity is typically measured using histone H1 as substrate[3, 6–9].

Enzyme activators and inhibitors

Formally, cyclins and CAK (**Mo15**) can be considered activators. Staurosporine has been shown to inhibit Cdc2 and other Cdc2-related kinases (which appear to include Cdk2) with an IC_{50} near 5 nM[14]. However this inhibitor is not specific. The 21 kDa Cip1 protein binds directly to Cdk2 and is a potent inhibitor of Cdk2/cyclin E and Cdk2/cyclin A kinase activity: it also inhibits Cdk4/cyclin D1 and, to a lesser extent, Cdc2/cyclin B kinase activity[15]. *CIP1* was also identified as the *WAF1* gene which is positively regulated by p53[16].

Pattern of expression
The localization of Cdk2 through the cell cycle has not been examined directly. However, fractionation studies indicate that the nuclear, but not cytoplasmic, form of cyclin A is associated with Cdk2[13]. Cdk2 mRNA appears to be expressed in all mammalian cell types with the exception of brain, where it is expressed at low to undetectable levels, perhaps due to terminal differentiation[10]. When quiescent T cells are stimulated to re-enter the cell cycle, Cdk2 is expressed in G1, several hours prior to Cdc2 mRNA accumulation[6].

References
1 Elledge, S.J., and Spottswood, M.R. (1991) EMBO J. 10, 2653–2659.
2 Paris, J. et al. (1991) Proc. Natl Acad. Sci. USA. 88, 1039–1043.
3 Connell-Crowley, L. et al. (1993) Mol. Biol. Cell 4, 79–92.
4 Koff, A. et al. (1992) Science 257, 1689–1694.
5 Xiong, Y. et al. (1992) Cell 71, 504–514.
6 Elledge, S.J. et al. (1992) Proc. Natl Acad. Sci. USA 89, 2907–2911.
7 Tsai, L.H. et al. (1991) Nature 353, 174–177.
8 Rosenblatt, J. et al. (1992) Proc. Natl Acad. Sci. USA 89, 2824–2882.
9 Gu, Y et al. (1992) EMBO J. 11, 3995–4005.
10 Meyerson, M. et al. (1992) EMBO J. 11, 2909–2917.
11 Hirai, T. et al. (1992) Devel. Biol. 152, 113–120.
12 Akiyama, T. et al. (1992) Proc. Natl Acad. Sci. USA 89, 7900–7904.
13 Pines, J. and Hunter, T. (1991) J. Cell Biol. 115, 1–17.
14 Gadbois, D.M. et al. (1992) Biochem. Biophys. Res. Commun. 184, 80–85.
15 Harper, J.W. et al. (1993) Cell 75, 805–816.
16 El-Deiry, W.S. et al. (1993) Cell 75, 817–825.

Cdk3

Cyclin-dependent kinase-3 (vertebrates)

Matthew Meyerson, Li-Huei Tsai, Greg H. Enders, Chin-Lee Wu,
Sander van den Heuvel and Ed Harlow
(Massachusetts General Hospital Cancer Center, USA)

cDNA encoding Cdk3 was cloned via homology to the **Cdc2/Cdc28** family. The *CDK3* gene can complement *ts* mutants of *S. cerevisiae cdc28*, suggesting that Cdk3 could be a cell-cycle regulator in vertebrate cells[1]. Dominant negative mutants of *cdk3* arrest mammalian cells in G1[2].

Subunit and domain structure

SUBUNIT	AMINO ACIDS	MOL. WT	SDS-PAGE
Cdk3	305	35045	36 kDa

Like other Cdks, Cdk3 consists essentially of a catalytic domain only. Cyclins are presumed to act as regulatory subunits, but this has not been demonstrated under physiological conditions.

Database accession numbers

	PIR	SWISSPROT	EMBL/GENBANK	REF
Human	S22743		X66357	1

Amino acid sequence of human Cdk3

```
MDMFQKVEKI GEGTYGVVYK AKNRETGQLV ALKKIRLDLE MEGVPSTAIR  50
EISLLKELKH PNIVRLLDVV HNERKLYLVF EFLSQDLKKY MDSTPGSELP 100
LHLIKSYLFQ LLQGVSFCHS HRVIHRDLKP QNLLINELGA IKLADFGLAR 150
AFGVPLRTYT HEVVTLWYRA PEILLGSKFY TTAVDIWSIG CIFAEMVTRK 200
ALFPGDSEID QLFRIFRMLG TPSEDTWPGV TQLPDYKGSF PKWTRKGLEE 250
IVPNLEPEGR DLLMQLLQYD PSQRITAKTA LAHPYFSSPE PSPAARQYVL 300
QRFRH
```

Physiological substrates and specificity determinants

Physiological substrates have not been identified, but transfected Cdk3 can phosphorylate histone H1 (M. Meyerson, unpublished results).

Assay

The enzyme is assayed by transfer of radioactivity from $[\gamma\text{-}^{32}\text{P}]$ATP to histone H1.

Pattern of expression

CDK3 mRNA and Cdk3 protein are expressed in a broad variety of cell types at very low levels.

References
1 Meyerson, M. et al. (1992) EMBO J. 11, 2909–2917.
2 Van der Heuvel, S. and Harlow, E. (1993) Science 262, 2050–2054.

Cdk4

Cyclin-dependent kinase-4 (vertebrates) (PSK-J3)

C.J. Sherr (St Jude Children's Research Hospital, USA)

The cyclin-dependent kinases (Cdks) are 34 kDa catalytic subunits that require regulatory cyclin subunits to manifest their serine/threonine kinase activity. As a group, the Cdks (prototype **Cdc2** (Cdk1)) are believed to regulate key cell cycle transitions. A human Cdk4 cDNA, initially cloned based on homology to nucleotide sequences conserved within known serine/threonine kinases and termed PSK-J3[1], is now recognized to be a major but non-exclusive catalytic partner for mammalian D-type G_1 cyclins[2,3]. The Cdk4 kinase can be activated by cyclins D1, D2 and D3, but not by cyclins A, B1 or E.

Subunit structure and isoforms

SUBUNIT	AMINO ACIDS	MOL. WT	SDS-PAGE
Cdk4	303	33 428	

Domain structure

Like other Cdks, Cdk4 consists essentially of a catalytic domain only. Cyclins are presumed to act as regulatory subunits *in vivo*.

Database accession numbers

	PIR	SWISSPROT	EMBL/GENBANK	REF
Mouse	A40035	P25322	M64403	2
Human	C26368	P11802	M14505	

Amino acid sequence of human Cdk4

```
MATSRYEPVA EIGVGAYGTV YKARDPHSGH FVALKSVRVP NGGGGGGGLP   0
ISTVREVALL RRLEAFEHPN VVRLMDVCAT SRTDREIKVT LVFEHVDQDL 100
RTYLDKAPPP GLPAETIKDL MRQFLRGLDF LHANCIVHRD LKPENILVTS 150
GGTVKLADFG LARIYSYQMA LTPVVVTLWY RAPEVLLQST YATPVDMWSV 200
GCIFAEMFRR KPLFCGNSEA DQLGKIFDLI GLPPEDDWPR DVSLPRGAFP 250
PRGPRPVQSV VPEMEESGAQ LLLEMLTFNP HKRISAFRAL QHSYLHKDEG 300
NPE
```

Homologues in other species

Human Cdk4 has been sequenced from HeLa cells, and mouse Cdk4 from BAC1.2F5 macrophages.

Physiological substrates and specificity determinants

The consensus sequences for phosphorylation by Cdk4 appears to be similar to those of other cyclin-dependent kinases (see **Cdc2**). Physiologic substrates are unknown, but there is a strong preference for the retinoblastoma protein versus histone H1[2,3].

Assay

Sf9 cells coinfected with baculovirus vectors encoding Cdk4 and any of the three known D-type mammalian cyclins can be used as a ready source of enzyme[3]. Because D-type cyclins can bind directly to the product of the retinoblastoma gene (pRb), bacterially produced pRb fusion proteins are particularly good substrates[2, 3].

Pattern of expression

Cdk4 expression in tissues has not been well-documented. The enzyme is expressed in a variety of cultured cell lines, including fibroblasts, T cells, macrophages, and other cells of the myeloid lineage, and may be ubiquitous. Its levels of synthesis oscillate minimally during the cell cycle, being highest during late G1 and early S phase, and the protein is quite stable (half-life 4–6 h). In proliferating cells, Cdk4 forms persistent complexes with regulatory D-type cyclin subunits. In contrast to the catalytic subunit, D-type cyclins in these complexes turn over rapidly (half-life 20 min).

References

[1] Hanks, S.K. (1987) Proc. Natl Acad. Sci. USA 84, 388–392.
[2] Matsushime, H. et al. (1992) Cell 71, 323–334.
[3] Kato, J-Y. et al. (1993) Genes Devel. 7, 331–342.

<table>
<tr><td>Cdk5</td><td>Cyclin-dependent kinase 5 (vertebrates)
(PSSALRE)</td></tr>
</table>

Matthew Meyerson, Li-Huei Tsai, Greg H. Enders, Chin-Lee Wu,
Sander van den Heuvel, Ed Harlow
(Massachusetts General Hospital Cancer Center, USA)

cDNA encoding Cdk5 was cloned via homology to the **Cdc2/Cdc28** family. Although Cdk5 is widely expressed, histone H1 kinase activity has been detected only in brain.

Subunit structure and isoforms

SUBUNIT	AMINO ACIDS	MOL. WT	SDS-PAGE
Cdk5	292	33 304	33 kDa

Domain structure

Like other Cdks, the Cdk5 subunit consists essentially of a catalytic domain only. Putative regulatory subunits include a 35 kDa protein associated with the active kinase in brain[2], and cyclin D1 in fibroblasts and other tissue culture cells[4].

Database accession numbers

	PIR	SWISSPROT	EMBL/GENBANK	REF
Human			X66364, L04658	[1, 4]
Bovine			L04798	[3]
Rat			L02121	[5]

Amino acid sequence of human CDK5

```
MQKYEKLEKI GEGTYGTVFK AKNRETHEIV ALKRVRLDDD DEGVPSSALR  50
EICLLKELKH KNIVRLHDVL HSDKKLTLVF EFCDQDLKKY FDSCNGDLDP 100
EIVKSFLFQL LKGLGFCHSR NVLHRDLKPQ NLLINRNGEL KLADFGLARA 150
FGIPVRCYSA EVVTLWYRPP DVLFGAKLYS TSIDMWSAGC IFAELANAGR 200
PLFPGNDVDD QLKRIFRLLG TPTEEQWPSM TKLPDYKPYP MYPATTSLVN 250
VVPKLNATGR DLLQNLLKCN PVQRISAEEA LQHPYFSDFC PP
```

Homologues in other species
Cdk5 has been sequenced from human (fetal brain and HeLa cells), rat brain, bovine brain, and *C. elegans* (S. van den Heuvel and L.-H. Tsai, unpublished data).

Physiological substrates and specificity determinants

The brain form of Cdk5 can phosphorylate histone H1, microtubule-associated protein 2 and neurofilaments.

Assay
The enzyme is assayed by transfer of radioactivity from $[\gamma\text{-}^{32}P]ATP$ to histone H1.

Pattern of expression

CDK5 is broadly expressed in normal tissue culture cells at high levels[1]. However, active kinase has only been described in brain[2, 3, 6].

References
1. Meyerson, M. et al. (1992) EMBO J. 11, 2909–2917.
2. Lew, J. et al. (1992) J. Biol. Chem. 267, 13383–13390.
3. Lew, J. et al. (1992) J. Biol. Chem. 267, 25922–25926.
4. Xiong, Y. et al. (1992) Cell 71, 505–514.
5. Hellmich, M.R. et al. (1992) Proc. Natl Acad. Sci. USA 89, 10867–10871.
6. Tsai, L.-H. et al. (1993) Development 119, 1029–1040.

Cdk6

Cyclin-dependent kinase-6 (vertebrates)
(PLSTIIRE)

Matthew Meyerson, Li-Huei Tsai, Greg H. Enders, Chin-Lee Wu,
Sander van den Heuvel, Ed Harlow
(Massachusetts General Hospital Cancer Center, USA)

cDNA encoding Cdk6 was cloned via homology to the **Cdc2/Cdc28** family, and originally termed PLSTIRE[1].

Subunit structure and isoforms

SUBUNIT	AMINO ACIDS	MOL. WT	SDS-PAGE
CDK6	326	36938	40 kDa

Domain structure

Like other Cdks, the Cdk6 subunit consists essentially of a catalytic domain only. Cdk6 is physically associated with, and activated by, D-type cyclins, including cyclin D1[2, 3].

Database accession number

	PIR	SWISSPROT	EMBL/GENBANK	REF
Human	S22749		X66365	1

Amino acid sequence of human Cdk6

```
MEKDGLCRAD QQYECVAEIG EGAYGKVFKA RDLKNGGRFV ALKRVRVQTG  50
EEGMPLSTIR EVAVLRHLET FEHPNVVRLF DVCTVSRTDR ETKLTLVFEH 100
VDQDLTTYLD KVPEPGVPTE TIKDMMFQLL RGLDFLHSHR VVHRDLKPQN 150
ILVTSSGQIK LADFGLARIY SFQMALTSVV VTLWYRAPEV LLQSSYATPV 200
DLWSVGCIFA EMFRRKPLFR GSSDVDQLGK ILDVIGLPGE EDWPRDVALP 250
RQAFHSKSAQ PIEKFVTDID ELGKDLLLKC LTFNPAKRIS AYSALSHPYF 300
QDLERCKENL DSHLPPSQNT SELNTA
```

Homologues in other species
A Cdk4/Cdk6 homologue has been isolated from *D. melanogaster* (C. Lehner, personal communication).

Pattern of expression
Cdk6 is expressed in a broad variety of tissues. Among cell lines, Cdk6 expression appears to be elevated in haematopoietic cells (M. Meyerson, unpublished data).

References
1 Meyerson, M. et al. (1992) EMBO J. 11, 2909–2917.
2 Bates, S. et al. (1994) Oncogene 9, 71–79.
3 Meyerson, M. and Harlow, E. (1994) Mol. Cell. Biol. 14, 2077–2086.

Matthew Meyerson, Li-Huei Tsai, Greg H. Enders, Chin-Lee Wu,
Sander van den Heuvel, Ed Harlow
(Massachusetts General Hospital Cancer Center, USA)

cDNAs encoding PCTAIRE-1, -2 and -3 were cloned via homology to the **Cdc2/Cdc28** family. They were named after the amino acid sequence corresponding to the PSTAIRE motif of **Cdc2**.

Subunit structure and isoforms

SUBUNIT	AMINO ACIDS	MOL. WT	SDS-PAGE
PCTAIRE-1	496	55 715	55 kDa
PCTAIRE-2	523	59 636	57 kDa
PCTAIRE-3	451	51 847	

Domain structure

The PCTAIRE kinases have longer N-terminal extensions (120–200 amino acids) than other CDKs (see **Cdc2**).

Database accession numbers

		PIR	SWISSPROT	EMBL/GENBANK	REF
P-1	Human	S22747		X66363	1
	Mouse	S47546		X69025	2
P-2	Human	S22746		X66360	1
P-3	Human			X66362	1
	Mouse			X69026	2

Amino acid sequences of human PCTAIREs

PCTAIRE-1

```
MDRMKKIKRQ LSMTLRGGRG IDKTNGAPEQ IGLDESGGGG GSDPGEAPTR  50
AAPGELRSAR GPLSSAPEIV HEDLKMGSDG ESDQASATSS DEVQSPVRVR 100
MRNHPPRKIS TEDINKRLSL PADIRLPEGY LEKLTLNSPI FDKPLSRRLR 150
RVSLSEIGFG KLETYIKLDK LGEGTYATVY KGKSKLTDNL VALKEIRLEH 200
EEGAPCTAIR EVSLLKDLKH ANIVTLHDII HTEKSLTLVF EYLDKDLKQY 250
LDDCGNIINM HNVKLFLFQL LRGLAYCHRQ KVLHRDLKPQ NLLINERGEL 300
KLADFGLARA KSIPTKTYSN EVVTLWYRPP DILLGSTDYS TQIDMWGVGC 350
IFYEMATGRP LFPGSTVEEQ LHFIFRILGT PTEETWPGIL SNEEFKTYNY 400
PKYRAEALLS HAPRLDSDGA DLLTKLLQFE GRNRISAEDA MKHPFFLSLG 450
ERIHKLPDTT SIFALKEIQL QKEASLRSSS MPDSGRPAFR VVDTEF
```

PCTAIRE-2

```
MKKFKRRLSL TLRGSQTIDE SLSELAEQMT IEENSSKDNE PIVKNGRPPT  50
SHSMHSFLHQ YTGSFKKPPL RRPHSVIGGS LGSFMAMPRN GSRLDIVHEN 100
LKMGSDGESD QASGTSSDEV QSPTGVCLRN RIHRRISMED LNKRLSLPAD 150
IRIPDGYLEK LQINSPPFDQ PMSRRSRRAS LSEIGFGKME TYIKLEKLGE 200
GTYATVYKGR SKLTENLVAL KEIRLEHEEG APCTAIREVS LLKDLKHANI 250
VTLHDIVHTD KSLTLVFEYL DKDLKQYMDD CGNIMSMHNV KLFLYQILRG 300
```

PCTAIRE-2 *continued*

```
LAYCHRRKVL HRDLKPQNLL INEKGELKLA DFGLARAKSV PTKTYSNEVV 350
TLWYRPPDVL LGSSEYSTQI DMWGVGCIFF EMASGRPLFP GSTVEDELHL 400
IFRLLGTPSQ ETWPGISSNE EFKNYNFPKY KPQPLINHAP RLDSEGIELI 450
RKFLQYESKK RVSAEEAMKH VYFRSLGPRI HALPESVSIF SLKEIQLQKD 500
PGFRNSSYPE TGHGKNRRQS MLF
```

Amino acid sequence of mouse PCTAIRE-3

```
MNKMKNFKRR LSLSVPRPET IEESLAEFTE QFNQLHTQTN EDGTDEPEQL  50
SPGMQYQQRQ NQRRFSMEDL NKRLSLPMDI RLPQEFLQKL QLENPGLPKP 100
LTRMSRRASL SDIGFGKLET YVKLDKLGEG TYATVFKGRS KLTENLVALK 150
EIRLEHEEGA PCTAIREVSL LKDLKHANIV TLHDLIHTDR SLTLVFEYLD 200
SDLKQYLDHC GNLMNMHNVK IFMFQLLRGL AYCHHRKILH RDLKPQNLLI 250
NERGELKLAD FGLARAKSVP TKTYSNEVVT LWYRPPDVLL GSTEYSTPID 300
MWGVGCILYE MATGKPLFPG STVKEELHLI FRLLGTPTEE SWPGVTSISE 350
FRAYNFPRYL PQPLLSHAPR LDTEGINLLS SLLLYESKSR MSAEAALNHP 400
YFQSLGDRVH QLHDTASIFS LKEIQLQKDP GYRGLAFQHP GRGKSRRQSI 450
F
```

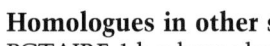

Homologues in other species

PCTAIRE-1 has been cloned from human (fetal brain) and mouse (BAC1.2F5A macrophages). PCTAIRE-2 has been cloned from human (Nalm-6 pre-B lymphoblastic leukaemia cells) and mouse (BAC1.2F5A macrophages). PCTAIRE homologues have been found in *Xenopus laevis* (T. Hunt, personal communication) and *C. elegans* (S. van den Heuvel, unpublished results).

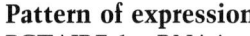

Pattern of expression

PCTAIRE-1 mRNA is readily detected in a broad range of cells and tissues. To date, the highest levels have been identified in post-meiotic sperm (J. Downing, personal communication). PCTAIRE-2 mRNA is expressed in a variety of tissue culture lines; among tissues, the highest levels are in brain and lung. Highest level of expression of PCTAIRE-3 mRNA are in brain, heart, kidney, and intestine. The gene is also expressed in several cell lines.

Assay

PCTAIRE-3 can be assayed by transfer of radioactivity from [γ-^{32}P]ATP to a 12 amino acid peptide from a high molecular weight basic nuclear protein from Northern flounder testis (J. Downing, personal communication).

References

1 Meyerson, M. et al. (1992) EMBO J. 11, 2909–2917.
2 Okuda, T. et al. (1992) Oncogene 7, 2249–2258.

Mo15

Mo15 (*Xenopus laevis*)
(Cdk-activating kinase, CAK)

John Shuttleworth (University of Birmingham, UK)

By sequence comparison the serine/threonine kinase Mo15 is most closely related to the Cdk family of protein kinases[1], with 43% and 49% amino acid sequence identity with **Cdc2** and **Cdk2** respectively, within the catalytic domain[2] (K.M. Pickham, unpublished data). Despite being one of the first vertebrate Cdk-related kinases to be identified, surprising new evidence indicates that Mo15 functions to phosphorylate and activate Cdc2 and Cdk2[3–5].

Subunit structure and isoforms

SUBUNIT	AMINO ACIDS	MOL. WT	SDS-PAGE
Mo15	352	39 700	40 kDa

Active Mo15 exists as a higher molecular weight complex in cell extracts, suggesting that it functions as a catalytic subunit in association with other as yet unidentified proteins[3–5].

Domain structure

Relative to Cdc2, Mo15 has short N- and C-terminal extensions of 14 and 38 amino acids respectively, the significance of which is not yet known. Only 8/15 residues of the highly conserved EGVPSTAIREISLLK motif in Cdc2 are represented in Mo15 (replaced by DGINRTALREIKLLQ), and the protein does not bind to mitotic cyclins or p13*suc1*[1,6]. The sequence also shows a threonine at position 176 (equivalent to T161 in Cdc2) which is probably phosphorylated in the active holoenzyme.

Database accession numbers

	PIR	SWISSPROT	EMBL/GENBANK	REF
Mo15			X53962	[1]

Amino acid sequence of *X. laevis* Mo15

```
MEGIAARGVD VRSRAKQYEK LDFLGEGQFA TVYKARDKNT DRIVAIKKIK  50
LGHRAEANDG INRTALREIK LLQELSHPNI IGLLDAFGHK SNISLVFDFM 100
ETDLEVIIKD TSLVLTPAHI KSYMLMTLQG LEYLHHLWIL HRDLKPNNLL 150
LDENGVLKLA DFGLAKSFGS PNRIYTHQVV TRWYRSPELL FGARMYGVGV 200
DMWAVGCILA ELLLRVPFLP GDSDLDQLTR IFETLGTPTE EQWPGMSSLP 250
DYVAFKSFPG TPLHLIFIAA GDDLLELLQG LFTFNPCARC TASQALRKRY 300
FSNRPAPTPG NLLPRPNCSI EALKEQQNLN LGIKRKRTEG MDQKDIAKKL 350
SF
```

Homologues in other species

Human (E. Levedakou and E.T. Liu, unpublished data) and partial mouse[7] cDNAs encoding proteins with around 80% amino acid sequence identity to Mo15 have been cloned. Also partial amino acid sequence has been obtained from purified preparations of starfish Mo15[4].

Physiological substrates and specificity determinants

The only known substrates for Mo15, T161 on **Cdc2** and T160 on **Cdk2**, have been identified *in vitro*[3–5]. Immunodepletion of cell extracts with antibodies to Mo15 removes virtually all Cdc2/Cdk2 phosphorylating activity suggesting that Mo15 (or a closely related protein) is the major component responsible for Cdc2/Cdk2 activation *in vivo*. Specificity determinants are not yet known, however phosphorylation is not directed by a C-terminal proline residue. Synthetic peptides with sequences which include the T161 of Cdc2 do not serve as substrate (Jean-Claude Cavadore, personal communication). Phosphorylation of Cdc2/Cdk2 T161/160 residues *in vivo* and *in vitro* requires association with cyclin; however bacterially synthesized GST–Cdk2 fusion protein is phosphorylated by Mo15 *in vitro* in the absence of cyclin. Mo15 does not autophosphorylate.

Assay

The activity of the holoenzyme has been assayed by following the phosphorylation and subsequent activation of histone H1 kinase activity in Cdk:cyclin complexes[3, 4].

Pattern of expression

In *X. laevis* oocytes, Mo15 protein and enzyme activity is associated almost exclusively with the nucleus or germinal vesicle[4, 6]. The protein is also present in the nuclei of early and post-blastula embryos[6], but its distibution in other cell types has not yet been studied. Mo15 transcripts are present in *X. laevis* oocytes throughout oogenesis and persist (in a de-adenylated form) through early embryogenesis[1]. Studies on adult tissue are inconclusive; although no transcripts have been detected by Northern blot analysis of several tissues[1], the function of the kinase and the presence of immunoreactive protein in *Xenopus* cell lines (Randy Poon, personal communication) would suggest that all proliferating cells should express Mo15 or a closely related (somatic cell?) protein.

References

1 Shuttleworth, J. et al. (1990) EMBO J. 9, 3233–3240
2 Paris, J. et al. (1991) Proc. Natl Acad. Sci. USA 89, 1039–1043.
3 Poon, R.Y.C. et al. (1993) EMBO J. 12, 3123–3132.
4 Fesquet, D. et al. (1993) EMBO J. 12, 3111–3121.
5 Solomon, M. J. et al. (1993) EMBO J. 12, 3133–3142.
6 Brown, A. et al., manuscript in preparation.
7 Ershler, M.A. et al. (1993) Gene 124, 305–306.

DmCdc2 — Cdc2 protein kinase (*D. melanogaster*)

P.H. O'Farrell (University of California San Francisco, USA)

The *D. melanogaster CDC2* gene encodes a functional and structural homologue of the protein kinase encoded by the *S. pombe cdc2* gene[1,2] **SpCdc2**. Similar kinases are found in all eukaryotic cells[3]. In *D. melanogaster* and a variety of other organisms, activation of this kinase has been shown to drive cell cycle progression from G2 to mitosis[4,5]. Introduction of the *D. melanogaster CDC2* gene into yeasts has demonstrated that it can complement mutations in the *S. pombe cdc2* and *S. cerevisiae CDC28* genes. Mutations in the *D. melanogaster* gene give phenotypes consistent with an arrest of cell cycle progression late in development. A *ts* allele was used to demonstrate that DmCdc2 is also required during early development, but that the maternal supply of DmCdc2 mRNA and/or protein is adequate to support all essential embryonic cell cycles. Immunoprecipitation data demonstrate that the DmCdc2 protein is associated with cyclins (both A and B), and that it has histone H1 kinase activity. Its state of phosphorylation, and histone kinase activity, fluctuate in parallel with later mitoses. By a number of criteria it functions as a mitotic kinase that is essential to progression from G2 to mitosis.

Subunit structure and isoforms

SUBUNIT	AMINO ACIDS	MOL. WT	SDS-PAGE
DmCdc2	297	34 439	34 kDa

Associates with cyclins A and B *in vivo*.

Domain structure

DmCdc2 consists of a minimal kinase domain (see **Cdc2**). Inhibitory phosphorylations occur on T14 and Y15 and reduce electrophoretic mobility on SDS-PAGE, while activating phosphorylation occurs on T161 and increase mobility on SDS-PAGE[4] (note that the degree of mobility shift depends on the conditions of the electrophoresis).

Genetics[5]
Cytological location of gene: 31E.
Mutations:
EMS derived: $cdc2^{B47}$, $cdc2^{D57}$, $cdc2^{E10}$, $cdc2^{E1-9}$, $cdc2^{E1-23}$, $cdc2^{216A}$, and $cdc2^{E1-24}$ (temperature-sensitive).
P element induced: $cdc2^{216P}$.

Database accession numbers

	PIR	FLY BASE	EMBL/GENBANK	REF
DmCdc2	S12006	04106	X57485, X57496	1

Amino acid sequence of *D. melanogaster* Cdc2

```
MEDFEKIEKI GEGTYGVVYK GRNRLTGQIV AMKKIRLESD DEGVPSTAIR  50
EISLLKELKH ENIVCLEDVL MEENRIYLIF EFLSMDLKKY MDSLPVDKHM 100
ESELVRSYLY QITSAILFCH RRRVLHRDLK PQNLLIDKSG LIKVADFGLG 150
RSFGIPVRIY THEIVTLWYR APEVLLGSPR YSCPVDIWSI GCIFAEMATR 200
KPLFQGDSEI DQLFRMFRIL KTPTEDIWPG VTSLPDYKNT FPCWSTNQLT 250
NQLKNLDANG IDLIQKMLIY DPVHRISAKD ILEHPYFNGF QSGLVRN
```

Sites of interest: T14, Y15, T161, phosphorylation

Homologues in other species

DmCdc2 shares ≈60% amino acid sequence identity with other protein kinases in the Cdc2 family that have been shown to be active in a complementation test in *S. pombe*.

Pattern of expression

There is an abundant supply of maternal DmCdc2 mRNA and protein that is adequate to provide function until late larval stages (see phenotype of mutants above). There is zygotic transcription in all parts of the embryo that are actively dividing, and a gradual decay of transcripts in tissues that discontinue division. It is notable that there is no obvious transcription or phenotype seen in the larval tissues that cease division after mitosis 16, but do undergo several rounds of later S phases that lead to the familiar polytene state. In expressing cells the DmCdc2 protein distributes between the nucleus and cytoplasm, and is somewhat concentrated in nuclei.

References

1   Lehner, C.F. and O'Farrell, P.H. (1990) EMBO J. 9, 3573–3581.
2   Jimenez, J. et al. (1990) EMBO J. 9, 3565–3571.
3   Lee, M.G. and Nurse, P. (1987) Nature 327, 31–35.
4   Edgar, B.A. et al. (1994) Genes Devel. 8, 440–452.
5   Stern, B. et al. (1993) Development 117, 219–232.

DmCdc2c

Cdc2 cognate (*D. melanogaster*)

P.H. O'Farrell (University of California San Francisco, USA)

As well as *DmCDC2*, a second gene *(DmCDC2c)* encodes a protein homologous to **SpCdc2** in *D. melanogaster*. Unlike *DmCDC2*, *DmCDC2c* cannot complement mutations in the *S. pombe* gene[1]. A family of Cdc2-related kinases have been found in a number of eukaryotes and have been given the name cyclin-dependent kinases (Cdks). The **Cdk2** protein kinases of human and frog are particularly interesting because of evidence that they are involved in the control of cell cycle progression from G1 to S phase[2]. Unfortunately, it is not yet clear what the relationship of DmCdc2c is to these vertebrate homologues. As yet, there is little biochemical or functional characterization of DmCdc2c.

Subunit structure and isoforms

SUBUNIT	AMINO ACIDS	MOL. WT	SDS-PAGE
DmCdc2	314	35 888	34 kDa

DmCdc2 is presumed to associate with cyclins *in vivo*.

Domain structure

DmCdc2 consists of a minimal kinase domain (see **Cdc2**).

Genetics[1, 3]
Cytological location of gene: 92F.
Mutations not yet defined: the gene is removed by the large deletion *Df(3R)H81*.

Database accession numbers

	PIR	FLY BASE	EMBL/GENBANK	REF
DmCdc2c	S12007	04107	X57486	*1*

Amino acid sequence of *D. melanogaster* Cdc2c

```
MTTILDNFQR AEKIGEGTYG IVYKARSNST GQDVALKKIR LEGETEGVPS  50
TAIREISLLK NLKHPNVVQL FDVVISGNNL YMIFEYLNMD LKKLMDKKKD 100
VFTPQLIKSY MHQILDAVGF CHTNRILHRD LKPQNLLVDT AGKIKLADFG 150
LARAFNVPMR AYTHEVVTLW YRAPEILLGT KFYSTGVDIW SLGCIFSEMI 200
MRRSLFPGDS EIDQLYRIFR TLSTPDETNW PGVTQLPDFK TKFPRWEGTN 250
MPQPITEHEA HELIMSMLCY DPNLRISAKD ALQHAYFRNV QHVDHVALPV 300
DPNAGSASRL TRLV
```

Homologues in other species
DmCdc2 shares about 55% amino acid sequence identity with other Cdc2 kinases.

Pattern of expression

There is an abundant supply of maternal *DmCDC2c* RNA and protein. There is zygotic transcription in all parts of the embryo that are actively dividing, and a gradual decay of transcripts in tissues that discontinue division.

References

1 Lehner, C.F. and O'Farrell, P.H. (1990) EMBO J. 9, 3573–3581.
2 Fang, F. and Newport, J.W. (1991) Cell 66, 731–742.
3 Stern, B. et al. (1993) Development 117, 219–232.

Cdc28

Cyclin-dependent PK (*S. cerevisiae*)

Steven I. Reed (The Scripps Research Institute, USA)

Originally identified based on temperature-sensitive cell division mutations that conferred G1 arrest[1,2], it is now known that this kinase is a central regulatory element for several cell cycle transitions in budding yeast. As the archetype of cyclin-dependent kinases, Cdc28 best exemplifies the paradigm of a single catalytic subunit functionally and enzymologically modified by an array of distinct positive regulatory subunits known as cyclins. In *S. cerevisiae*, G1-specific cyclins are generally called Clns, while S-phase and M-phase specific cyclins are called Clbs.

Subunit structure and isoforms

SUBUNIT	AMINO ACIDS	MOL. WT	SDS-PAGE
Cdc28	298	34 061	34 kDa

Although the exact compositions of the *in vivo* holoenzymes have not been reported, they all contain at least one cyclin of which more than 10 (40–60 kDa) have been identified in *S. cerevisiae*. Active complexes can also contain the 18 kDa protein Cks1. Based on gel filtration chromatography, Cdc28 holoenzymes have native molecular weights of >200 kDa and larger, suggesting oligomerization.

Genetics

Chromosomal location: Right arm of chromosome II.
Mutations: Numerous temperature-sensitive point mutations[2,3].

Domain structure

The Cdc28 subunit consists of a minimal kinase domain. T169 is the site of positive regulatory phosphorylation essential for activity, and Y19 is a target of negative regulatory phosphorylation. The structures of cyclins (positive regulatory subunits) are highly variable but all contain a relatively conserved 150–200 residue "cyclin box", which presumably interacts with the catalytic subunit. For G1 cyclins (Clns), the cyclin box is near the N-terminus and sequences regulating protein stability are C-terminal to the cyclin box. For S-phase and mitotic cyclins (Clbs), the cyclin box is C-terminal but more centred, with sequences controlling turnover located near the N-terminus.

Database accession numbers

	PIR	SWISSPROT	EMBL/GENBANK	REF
Cdc28		P00546	K02648	4

Amino acid sequence of *S. cerevisiae* Cdc28

```
MSGELANYKR LEKVGEGTYG VVYKALDLRP GQGQRVVALK KIRLESEDEG  50
VPSTAIREIS LLKELKDDNI VRLYDIVHSD AHKLYLVFEF LDLDLKRYME 100
GIPKQDPLGA DIVKKFMMQL CKGIAYCHSH RILHRDLKPQ NLLINKDGNL 150
KLGDFGLARA FGVPLRAYTH EIVTLWYRAP EVLLGGKQYS TGVDTWSIGC 200
IFAEMCNRKP IFSGDSEIDQ IFKIFRVLGT PNEAIWPDIV YLPDFKPSFP 250
QWRRKDLSQV VPSLDPRGID LLDKLLAYDP INRISARRAA IHPYFQES
```

Sites of interest: Y19, T169, phosphorylation

Homologues in other species

Cyclin-dependent protein kinases closely related to Cdc28 exist in all eukaryotic organisms. In yeasts, typified by *S. cerevisiae* and *S. pombe* **SpCdc2**, only one isoform exists. However, in vertebrates cyclin-dependent kinases exist as a family of related catalytic subunits. For human, the three most closely related family members, **Cdc2**, **Cdk2** and **Cdk3** have all been shown to be capable of replacing Cdc28 function in *S. cerevisiae*.

Physiological substrates and specificity determinants

Although a detailed analysis of the substrate specificity of the Cdc28 kinase has not been performed, the consensus for closely related kinases that can substitute for Cdc28 (e.g. **Cdc2**) is **S/T-P-X-K/R**. Cdc28 kinase has been shown to be capable of phosphorylating vertebrate histone H1 on sites that agree with this consensus. As with other Cdks, the importance of the downstream basic residues and the degree to which the context of surrounding sequences play a role in substrate determination has not yet been fully explored. Although many biological substrates have been suggested, and a variety of proteins can be phosphorylated by Cdc28 *in vitro*, a convincing case has been made only for the Swi5 transcription factor, where phosphorylation by Cdc28 apparently controls nuclear localization[5]. Although genetic analyses indicate that different cyclins must confer differential substrate specificity on the Cdc28 kinase, this has not been demonstrated biochemically.

Assay

The enzyme is assayed based on transfer of phosphate from $[\gamma\text{-}^{32}P]ATP$ to bovine histone H1. Zn^{2+}, which has been shown to increase the specificity for some substrates, is optional when using histone H1. Usually, reaction products are subjected to SDS-PAGE and the counts incorporated into histone detected by autoradiography or excision followed by direct counting[6].

Enzyme activators and inhibitors

A 40 kDa yeast polypeptide which binds to Cdc28 and can serve as a substrate has been found to have inhibitory properties *in vitro*[7].

Pattern of expression

In *S. cerevisiae*, Cdc28 is expressed invariantly under all physiological conditions analysed. Immunofluorescence microscopy indicates that the protein is dispersed throughout the cytosol and nucleus. However, since the vast majority of the protein is monomeric and inactive, this experiment does not provide information on the location of the active cyclin-bound holoenzymes.

References
1 Hartwell, L.H. et al. (1974) Science 183, 46–51.
2 Reed, S.I. (1980) Genetics 95, 561–577.
3 Lorincz, A.T. and Reed, S.I. (1986) Mol. Cell. Biol. 6, 4099–4103.
4 Lorincz, A.T. and Reed, S.I. (1984) Nature 307, 183–185.
5 Moll, T. et al. (1991) Cell 66, 743–758.
6 Wittenberg, C. and Reed, S.I. (1988) Cell 54, 1061–1072.
7 Mendenhall, M.D. (1993) Science 259, 216–219.

Pho85 | Pho85 protein kinase (*S. cerevisiae*)

Akio Toh-e (University of Tokyo, Japan)

The *PHO85* gene was identified as a negative regulator of the PHO system of *S. cerevisiae*[1,2]. Phosphorylation of unknown substrates by this kinase is necessary to establish repression. This kinase is a member of the Cdc2 family: its amino acid sequence is 51% identical to **Cdc28**, and it contains a PSTAIR sequence, a hallmark of the family. Pho80, another negative regulator of the PHO system, may serve as a regulatory subunit of this kinase.

Subunit structure and isoforms

SUBUNIT	AMINO ACIDS	MOL. WT	SDS-PAGE
Pho85	305	34 881	36 kDa

Genetics

Chromosomal location: left arm of chromosome XVI.
Mutations: Y18F, dominant negative mutation; D88N, negative, leaky; W173Stop, negative.

Domain structure

Pho85 comprises little more than a kinase catalytic domain.

Database accession numbers

	PIR	SWISSPROT	EMBL/GENBANK	REF
Pho85			Y00867	2

Amino acid sequence of *S. cerevisiae* Pho85

```
MSSSSQFKQL EKLGNGTYAT VYKGLNKTTG VYVALKEVKL DSEEGTPSTA  50
IREISLMKEL KHENIVRLYD VIHTENKLTL VFEFMDNDLK KYMDSRTVGN 100
TPRGLELNLV KYFQWQLLQG LAFCHENKIL HRDLKPQNLL INKRGQLKLG 150
DFGLARAFGI PVNTFSSEVV TLWYRAPDVL MGSRTYSTSI DIWSCGCILA 200
EMITGKPLFP GTNDEEQLKL IFDIMGTPNE SLWPSVTKLP KYNPNIQQRP 250
PRDLRQVLQP HTKEPLDGNL MDFLHGLLQL NPDMRLSAKQ ALHHPWFAEY 300
YHHAS
```

Homologues in other species

Homologues of Pho85 have been sequenced from *Kluyveromyces lactis* and *Zygosaccharomyces rouxii*.

Assay

The enzyme is assayed in immunoprecipates by transfer of radioactivity from $[\gamma\text{-}^{32}P]ATP$ to casein[3]. Histone H1, protamine and myelin basic protein are phosphorylated poorly, if at all.

Pattern of expression

A cytosolic distribution was confirmed by indirect immunofluorescence microscopy as well as by subcellular fractionation.

References
1 Uesono, Y. et al. (1987) Nucl. Acids Res. 15, 10299–10309.
2 Toh-e, A. et al. (1988) Mol. Gen. Genet. 214, 162–164.
3 Uesono, Y. et al. (1992) Mol. Gen. Genet. 231, 426–432.
4 Kaffman, A. et al. (1994) Science 183, 1153–1156.
5 Fujino, M. et al. (1994) Phosphate in Microorganisms: Cellular and Molecular Biology. Ed. A. Torriani-Gorini et al. pp. 70–75.
5 Schneider, K.R. et al. (1994) Science 266, 122–126.

Note added in proof

Pho 80 is cyclin and binds Pho 85 to produce an active enzyme[4]. Pho 85 – Pho 80 kinase phosphorylates Pho 4, a transcription factor,[4] and Pho 80[5]. Pho 81, another regulator of the PHO system, is an inhibitor of Pho 85 – Pho 80 kinase[6].

Kin28

Kin28 protein kinase (*S. cerevisiae*)

J.G. Valay, M. Simon and G. Faye (Institut Curie, France)

Kin28 is a putative serine/threonine protein kinase belonging to the Cdc2 family[1]. It is 38% and 37% identical in sequence to **Cdc28** of *S. cerevisiae* and **SpCdc2** of *S. pombe*, respectively[2]. It may have a function in the cell cycle in early G1 phase.

Subunit structure and isoforms

SUBUNIT	AMINO ACIDS	MOL. WT	SDS-PAGE
Kin28	306	35 247	

Kin28 is associated with a cyclin C-like subunit which is 393 amino acids long (mol. wt 45 180)[3].

Genetics

Chromosomal location: IV.

Domain structure

Kin28 consists essentially of a protein kinase catalytic domain with no significant N- or C-terminal extensions. Its C-terminal end is 13 amino acids longer than that of Cdc28. Only 7 out of 16 amino acids of its PSTAIRE motif are identical to that of **Cdc2**.

Database accession numbers

	PIR	SWISSPROT	EMBL/GENBANK	REF
Kin28		P06242	X04423	2

Amino acid sequence of *S. cerevisiae* Kin28

```
MKVNMEYTKE KKVGEGTYAV VYLGCQHSTG RKIAIKEIKT SEFKDGLDMS  50
AIREVKYLQE MQHPNVIELI DIFMAYDNLN LVLEFLPTDL EVVIKDKSIL 100
FTPADIKAWM LMTLRGVYHC HRNFILHRDL KPNNLLFSPD GQIKVADFGL 150
ARAIPAPHEI LTSNVVTRWY RAPELLFGAK HYTSAIDIWS VGVIFAELML 200
RIPYLPGQND VDQMEVTFRA LGTPTDRDWP EVSSFMTYNK LQIYPPPSRD 250
ELRKRFIAAS EYALDFMCGM LTMNPQKRWT AVQCLESDYF KELPPPSDPS 300
SIKIRN
```

References

1 Hanks, S. and Quinn A.M. (1991) Methods Enzymol. 200, 38–81.
2 Simon, M., Séraphin, B. and Faye, G. (1986) EMBO J. 5, 2697–2701.
3 Valay, J. G., Simon, M. and Faye, G. (1993) J. Mol. Biol. 234, 307–310.

SpCdc2 — Cdc2 protein kinase (*S. pombe*)

Chris Norbury (ICRF Molecular Oncology Laboratory, Oxford, UK)

S. pombe Cdc2 protein kinase has two distinct, essential functions in each mitotic cell cycle: the first, in G1 before cell cycle commitment, is biochemically obscure, although conserved in the distantly related yeast *S. cerevisiae* (see **Cdc28**). The second function coincides with the peak of activity of the kinase, determines the timing of entry into mitosis, and is conserved throughout the eukaryotes[1], although substrates for the kinase have yet to be identified in *S. pombe*. Cdc2 function is also required for meiosis. Association with the *cdc13*-encoded cyclin B subunit is essential for generation of the active M-phase form of the kinase, and is correlated with phosphorylation of the Cdc2 subunit at Y15, which blocks kinase activity, and at T167, phosphorylation of which is essential for its activation. Phosphorylation at Y15 is thought to be carried out by the **Wee1** and **Mik1** protein kinases. Final activation at the G2/M boundary results from Cdc25-mediated dephosphorylation of the Cdc2 subunit at Y15. The kinase is inactivated in late M phase through a mechanism probably involving proteolytic degradation of the cyclin B subunit.

Subunit structure

SUBUNIT	AMINO ACIDS	MOL. WT	SDS-PAGE
Cdc2	297	34 336	34 kDa
Cdc13	482	55 596	56 kDa

Cdc2 is a protein kinase subunit barely larger than the canonical catalytic domain. The active form of the kinase at G2/M is thought to consist of a heterodimer with a cyclin B protein encoded by the *cdc13* gene. The p13 product of the *suc1* gene is tightly bound to about 5% of the Cdc2 protein *in vivo*; the biological role of this Cdc2-binding protein, although unclear, may be related to inactivation of the kinase in late M phase.

Genetics
Chromosomal locations:
  cdc2:   II, tightly linked to *his3*
  cdc13:   II, linked to *atb2*

A large number of Cdc2 mutations have been characterized, following random or site-directed mutagenesis. These include heat- *(hs)* or cold- *(cs)* sensitive and temperature-independent *(ti)* loss of function mutations (causing cell cycle arrest in G1 and G2), dominant negative mutations, and partially dominant gain-of-function mutations in which mitosis is initiated prematurely:

Loss of function *(hs)*:	E42K, G43E, P137S, A177T, G183E, P208S, F210L, G212S, L269S, G277C, Y292H
Loss of function *(cs)*:	D90N, D213N, D242N
Loss of function *(ti)*:	K33R, T167V/A/E, E179K/R, D192K/R
Dominant negative:	A48V, E51D/R, D134K/R, D152K/R/N, E215K
Gain of function:	Y15F, C67Y/F, G146D

The Cdc13 C379Y mutation generates a heat-sensitive cyclin, while deletion of the N-terminal 89 amino acids stabilizes the protein and causes mitotic arrest with high Cdc2 kinase activity.

Domain structure

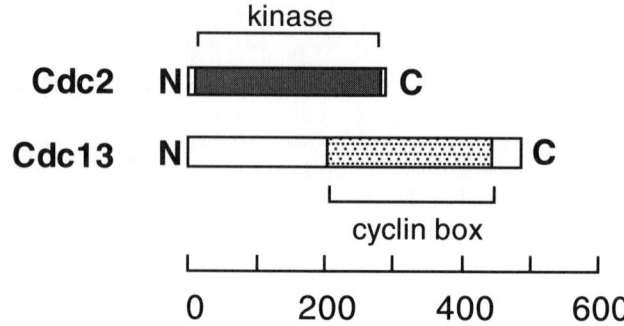

Database accession numbers

	PIR	SWISSPROT	EMBL/GENBANK	REF
Cdc2	A23359	P04551	M12912	2
Cdc13	A34948	P10815	X12557	3

Amino acid sequences of *S. pombe* Cdc2

```
MENYQKVEKI GEGTYGVVYK ARHKLSGRIV AMKKIRLEDE SEGVPSTAIR  50
EISLLKEVND ENNRSNCVRL LDILHAESKL YLVFEFLDMD LKKYMDRISE 100
TGATSLDPRL VQKFTYQLVN GVNFCHSRRI IHRDLKPQNL LIDKEGNLKL 150
ADFGLARSFG VPLRNYTHEI VTLWYRAPEV LLGSRHYSTG VDIWSVGCIF 200
AEMIRRSPLF PGDSEIDEIF KIFQVLGTPN EEVWPGVTLL QDYKSTFPRW 250
KRMDLHKVVP NGEEDAIELL SAMLVYDPAH RISAKRALQQ NYLRDFH
```

Sites of interest: Y15, T167, phosphorylation

Amino acid sequence of *S. pombe* Cdc13 (cyclin B)

```
MTTRRLTRQH LLANTLGNND ENHPSNHIAR AKSSLHSSEN SLVNGKKATV  50
SSTNVPKKRH ALDDVSNFHN KEGVPLASKN TNVRHTTASV STRRALEEKS 100
IIPATDDEPA SKKRRQPSVF NSSVPSLPQH LSTKSHSVST HGVDAFHKDQ 150
ATIPKKLKKD VDERVVSKDI PKLHRDSVES PESQDWDDLD AEDWADPLMV 200
SEYVVDIFEY LNELEIETMP SPTYMDRQKE LAWKMRGILT DWLIEVHSRF 250
RLLPETLFLA VNIIDRFLSL RVCSLNKLQL VGIAALFIAS KYEEVMCPSV 300
QNFVYMADGG YDEEEILQAE RYILRVLEFN LAYPNPMNFL RRISKADFYD 351
IQTRTVAKYL VEIGLLDHKL LPYPPSQQCA AAMYLAREML GRGPWNRNLV 400
HYSGYEEYQL ISVVKKMINY LQKPVQHEAF FKKYASKKFM KASLFVRDWI 450
KKNSIPLGDD ADEDYTFHKQ KRIQHDMKHE EW
```

Homologues in other species

See **Cdc2**.

Assay

Cdc2 activity is usually assayed by the cyclic nucleotide- and Ca^{2+}-independent transfer of phosphate from $[\gamma\text{-}^{32}P]ATP$ to histone H1[4]. In crude *S. pombe* lysates, the *cdc2*-encoded protein kinase is responsible for >90% of such activity.

Pattern of expression

The *cdc2* transcript and protein levels appear to be invariant through the cell cycle, and do not change when cells leave the mitotic cycle in response to nutrient deprivation. Regulation of the kinase is principally at the level of phosphorylation and association with the *cdc13*-encoded cyclin B. Gradual nuclear accumulation of the Cdc2 and Cdc13 proteins in interphase is followed by their rapid disappearance from the nucleus and degradation of the cyclin subunit in late mitosis.

References

[1] Nurse, P. (1990) Nature 344, 503–508.
[2] Hindley, J. and Phear, G. (1984) Gene 31, 129–134.
[3] Hagan, I. et al. (1988) J. Cell Sci. 91, 587–595.
[4] Moreno, S. et al. (1989) Cell 58, 361–372.

Erk1/2

Extracellular signal-regulated kinases 1/2 (vertebrates)
(MAP, MBP, EGF receptor T669 (Ert))

Melanie H. Cobb (University of Texas Southwestern Medical Center, USA)

Many growth factors, mitogens and differentiation-promoting agents, increase the activities of Erk1 and Erk2 via a protein kinase cascade. They have a wide variety of substrates, many of which are important regulatory proteins, leading to the idea that they are pleiotropic in function. They are activated nearly 1000-fold by phosphorylation on a tyrosine and a threonine residue[1, 2]. Both sites must be phosphorylated for high enzymatic activity. These two phosphorylations appear to be catalysed by a single type of protein kinase known as **Mek** or MAP kinase kinase[3, 4]. Erk1 and Erk2 autophosphorylate to low stoichiometry on the tyrosine residue, but the regulatory significance of this is not known.

Other names

Mitogen-activated protein (MAP) kinases, p44/p42 MAP kinases, microtubule-associated protein-2 kinases, myelin basic protein (MBP) kinases, EGF receptor T669 (Ert) kinases.

Subunit structure and isoforms

SUBUNIT	AMINO ACIDS	MOL. WT	SDS-PAGE
Erk1 (rat)	379	43 156	43 (44) kDa
Erk2 (rat)	358	41 281	41 (42) kDa

The major isoforms, Erk1 and Erk2 (p44 and p42 MAP kinase), both exist as monomers[5–8]. Figures in parentheses are the apparent molecular weights of the doubly phosphorylated forms, which migrate slightly more slowly on SDS-PAGE.

Genetics

Site-directed mutants:
Erk1: K71R (inactive); K71R and K72R (inactive), T202A (inactive), Y204F (inactive), T202A and Y204F (inactive).
Erk2: K52R (5–10% active), T183A (inactive), T183E (10–20% active), Y185F (inactive), T183A and Y185F (inactive), S39D (active), Y191F (active).

Domain structure

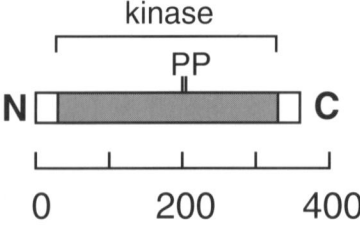

Erk1/2 consist of a kinase domain with short N- and C-terminal extensions. The crystal structure of Erk2 has been solved[9].

Database accession numbers

		PIR	SWISSPROT	EMBL/GENBANK	REF
Erk1	Rat			X65198	5
	Mouse			X14249	17
	Human			M84490, Z11696	8, 16
Erk2	Rat			M64300	6
	Mouse			X58712	7
	Human			M84489, Z11694	8, 16

Amino acid sequence of human Erk1

```
MAAAAAQGGG GGEPRRTEGV GPGVPGEVEM VKGQPFDVGP RYTQLQYIGE  50
GAYGMVSSAY DHVRKTRVAI KKISPFEHQT YCQRTLREIQ ILLRFRHENV 100
IGIRDILRAS TLEAMRDVYI VQDLMETDLY KLLKSQQLSN DHICYFLYQI 150
LRGLKYIHSA NVLHRDLKPS NLLSNTTCDL KICDFGLARI ADPEHDHTGF 200
LTEYVATRWY RAPEIMLNSK GYTKSIDIWS VGCILAEMLS NRPIFPGKHY 250
LDQLNHILGI LGSPSQEDLN CIINMKARNY LQSLPSKTKV AWAKLFPKSD 300
SKALDLLDRM LTFNPNKRIT VEEALAHPYL EQYYDPTDEP VAEEPFTFAM 350
ELDDLPKERL KELIFQETAR FQPGVLEAP
```

Sites of interest: T202, Y204, regulatory phosphorylation

Amino acid sequence of rat Erk2

```
MAAAAAAGPE MVRGQVFDVG PRYTNLSYIG EGAYGMVCSA YDNLNKVRVA  50
IKKISPFEHQ TYCQRTLREI KILLRFRHEN IIGINDIIRA PTIEQMKDVY 100
IVQDLMETDL YKLLKTQHLS NDHICYFLYQ ILRGLKYIHS ANVLHRDLKP 150
SNLLLNTTCD LKICDFGLAR VADPDHDHTG FLTEYVATRW YRAPEIMLNS 200
KGYTKSIDIW SVGCILAEML SNRPIFPGKH YLDQLNHILG ILGSPSQEDL 250
NCIINLKARN YLLSLPHKNK VPWNRLFPNA DSKALDLLDK MLTFNPHKRI 300
EVEQALAHPY LEOYYDPSDE PIAEAPFKFD MELDDLPKEK LKELIFEETA 350
FRQPGYRS
```

Sites of interest: T183, Y185, regulatory phosphorylation

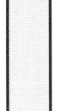

Homologues in other species

Erk1 has been sequenced from rat (brain), mouse (3T3-L1 cells), human (T cells, HeLa), and Chinese hamster (fibroblasts). Erk2 has been sequenced from rat (brain), mouse (3T3 cells), and human (T cells, HeLa). Erk homologues have been identified in *X. laevis*, *D. melanogaster*, plants and *S. cerevisiae*.

Physiological substrates and specificity determinants

The consensus sequence is **S/TP**. P is also common at the P-2 residue. Substrates whose properties are altered following phosphorylation by Erk1/2 include: **Rsk**, tau, c-Jun, c-Myc, Tal1, phospholipase A2, EGF receptor, ternary complex factor (c-Elk). Evidence indicating that these proteins are substrates of Erks *in vivo* is thus far limited. A selection of sequences around sites phosphorylated by the kinases are listed below[10]:

Myelin basic protein:	VTPRTPPPS[11]
Epidermal growth factor receptor:	VEPLTPSGE[12]
Tyrosine hydroxylase:	EAIMSPRFK[13]
c-Myc:	TPPLSPSRR[12]
c-Elk:	IAPRSPAKL[14]
Phospholipase A2:	SYPLSPLSD[15]

Assay
The enzymes may be assayed as a transfer of radioactivity from $[\gamma\text{-}^{32}P]ATP$ to myelin basic protein (MBP)[18], microtubule-associated protein-2, or peptide substrates such as from MBP (RRNIVTPRTPPPSQCKGR), the EGF receptor (RRELVEPLTPSGEAPNQALLR) or tyrosine hydroxylase (EAIMSPRFK).

Pattern of expression
Ubiquitously distributed in tissues and cell lines with largest amounts in the nervous system. Developmentally regulated. Erk1 and Erk2 differ in subcellular distribution, although both are found in cytoplasm, in nuclei, and are associated with membranes and the cytoskeleton.

References
1 Robbins, D.J. et al. (1993) J. Biol. Chem. 268, 5097–5106.
2 Payne, D.M. et al. (1991) EMBO J. 10, 885–892.
3 Seger, R. et al. (1992) J. Biol. Chem. 267, 25628–25631.
4 Crews, C.M. et al. (1992) Proc. Natl Acad. Sci. USA 89, 8205–8209.
5 Boulton, T.G. et al. (1990) Science 249, 64–67.
6 Boulton, T.G. et al. (1991) Cell 65, 663–675.
7 Her, J.-H. et al. (1991) Nucl. Acids Res. 19, 3743.
8 Gonzalez, F.A. et al. (1992) FEBS Lett. 304, 170–178.
9 Zhang, F. et al. (1994) Nature 367, 704–711.
10 Davis, R. (1991) J. Biol. Chem. 268, 14553, 14559.
11 Erickson, A.K. et al. (1990) J. Biol. Chem. 265, 19728–19735.
12 Alvarez, E. et al. (1991) J. Biol. Chem. 266, 15277–15285.
13 Haycock, J.W. et al. (1992) Proc. Natl Acad. Sci. USA 89, 2365–2369.
14 Gille, H. et al. (1992) Nature 358, 414–416.
15 Seth, A. et al. (1992) J. Biol. Chem. 267, 24796–24804.
16 Owaki, H. et al. (1992) Biochem. Biophys. Res. Commun. 182, 1416–1422.
17 Tanner et al. (1993) Biochim. Biophys. Acta 182, 1416–1422.
18 Boulton, T.G. et al. (1991) Biochemistry 30, 278–286.

Erk3 Extracellular signal-regulated kinase 3 (vertebrates)

Melanie H. Cobb (University of Texas Southwestern Medical Center, USA)

A cDNA encoding Erk3 was isolated as an Erk1 homologue. Erk3 is 50% identical to Erk1 in the catalytic domain. The protein has been detected with antibodies but its function is unknown. The recombinant enzyme is phosphorylated on serine residues by MAP kinase kinase (**Mek**). Thus, Erk3 may be regulated by the same wide variety of stimuli as Erk1 and Erk2. Unlike these enzymes it does not appear to be phosphorylated on tyrosine.

Subunit structure and isoforms

SUBUNIT	AMINO ACIDS	MOL. WT	SDS-PAGE
Erk3	543	62 652	62 kDa

Domain structure

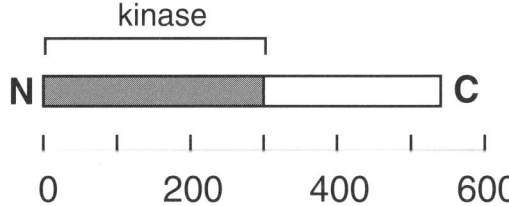

The catalytic domain is followed by a C-terminal region of nearly 200 amino acids. Deletion of this region does not eliminate autophosphorylation (M. Cheng, unpublished data). Neither the tyrosine nor the threonine which are phosphorylated in Erk1/2 are conserved in Erk3.

Database accession numbers

		PIR	SWISSPROT	EMBL/GENBANK	REF
Erk3	Rat			M64301	1
	Human			X59727	2

Amino acid sequence of rat Erk3

```
MAEKFESLMN IHGFDLGSRY MDLKPLGCGG NGLVFSAVDN DCDKRVAIKK  50
IVLTDPQSVK HALREIKIIR RLDHDNIVKV FEILGPSGSQ LTDDVGSLTE 100
LNSVYIVQEY METDLANVLE QGPLLEEHAR LFMYQLLRGL KYIHSANVLH 150
RDLKPANLFI NTEDLVLKIG DFGLARIMDP HYSHKGHLSE GLVTKWYRSP 200
RLLLSPNNYT KAIDMWAAGC IFAEMLTGKT LFAGAHELEQ MQLILESIPV 250
VHEEDRQELL SVIPVYIRND MTEPHKPLTQ LLPGISREAL DFLEQILTFS 300
PMDRLTAEEA LSHPYMSIYS FPTDEPISSH PFHIEDEVDD ILLMDETHSH 350
IYNWERYHDC QFSEHDWPIH NNFDIDEVQL DPRALSDVTD EEEVQVDPRK 400
YLDGDREKYL EDPAFDTSYS AEPCWQYPDH HENKYCDLEC SHTCNYKTRS 450
PSYLDNLVWR ESEVNHYYEP KLIIDLSNWK EQSKDKSDKR GKSKCERNGL 500
VKRRLRLRKR PSSWLRGRGA KALTLMPSSQ APFSSVPSVS LLT
```

Homologues in other species

A protein similar to rat Erk3 has been sequenced from human HeLa cells[2]. The kinase domain is 72% identical to that of rat Erk3, but the C-terminal domain is only 27% identical.

Physiological substrates and specificity determinants

Erk3 has a substrate specificity distinct from its relatives Erk1 and Erk2, in that it does not phosphorylate either myelin basic protein or microtubule-associated protein-2. No physiological substrates have yet been identified.

Pattern of expression

Ubiquitously distributed in tissues and cell lines with largest amounts in the nervous system. Developmentally regulated. The majority is found in nuclei.

References

1. Boulton, T.G. et al. (1991) Cell 65, 663–675.
2. Gonzalez, F.A. et al. (1992) FEBS Lett. 304, 170–178.

DmErkA — Extracellular signal-regulated kinase (*D. melanogaster*)

William Biggs and S. Lawrence Zipursky (UCLA School of Medicine, USA)

The Erks, or MAP kinases, are activated in response to a dizzying array of extracellular ligands which act through a variety of cell surface receptor tyrosine kinases (RTKs)[1]. The *D. melanogaster ERK-A* gene is a closely related member of the MAP kinase family[2]. DmErkA is encoded at the *rolled* locus and is a genetically required member of several *D. melanogaster* RTK signal transduction pathways.

Subunit structure and isoforms

SUBUNIT	AMINO ACIDS	MOL. WT	SDS-PAGE
DmErkA	376	43 160	43 kDa

By analogy with vertebrate MAP kinases, probably a monomer. There is evidence of an additional closely related MAP kinase in *D. melanogaster* (W.H. Biggs and S.L. Zipursky, unpublished observation).

Genetics

The *ERK-A* gene was localized cytologically on salivary gland polytene chromosomes to the centromeric heterochromatin of the second chromosome. The *rolled (rl)* locus had been previously mapped to this same region[3]. Molecular and genetic analysis has demonstrated that the *rl* locus encodes DmErkA[4]. In addition, the gain-of-function mutation *sevenmaker (rl^Sev)* is found to encode a G→A change in codon 334, encoding a D→N substitution[5]. In separate genetic screens it was found that *rl* lies downstream of the **Sev** *(sevenless)*, **Torso** and **DmEGFR** RTKs, Ras and **DmRaf**[4,5] (G.M. Rubin, personal communication). Together these data demonstrate the position of, as well as a requirement for, DmErkA within at least two developmentally important RTK signal transduction pathways. The DmErk gene is rather large for *D. melanogaster*, spanning ~120 kb of genomic sequence. In addition, regions of the gene appear to have undergone several rounds of duplication generating a number of pseudogenes (W.H. Biggs *et al.*, unpublished data).

Domain structure

DmErkA comprises of a catalytic domain, with short N- and C-terminal extensions[6].

Database accession numbers

	PIR	SWISSPROT	EMBL/GENBANK	REF
DmErk			M95124	2

Amino acid sequence of *D. melanogaster* ErkA

```
MEEFNSSGSV VNGTGSTEVP QSNAEVIRGQ IFEVGPRYIK LAYIGEGAYG   50
MVVSADDTLT NQRVAIKKIS PFEHQTYCQR TLREITILTR FKHENIIDIR  100
DILRVDSIDQ MRDVYIVQCL METDLYKLLK TQRLSNDHIC YFLYQILRGL  150
KYIHSANVLH RDLKPSNLLL NKTCDLKICD FGLARIADPE HDHTGFLTEY  200
VATRWYRAPE IMLNSKGYTK SIDIWSVGCI LAEMLSNRPI FPGKHYLDQL  250
NHILGVLGSP SRDDLECIIN EKARNYLESL PFKPNVPWAK LFPNADALAL  300
DLLGKMLTFN PHKRIPVEEA LAHPYLEQYY DPGDEPVAEV PFRINMENDD  350
ISRDALKSLI FEETLKFKER QPDNAP
```

Sites of interest: T198, Y200, phosphorylation; R133–K177, deleted in rl^{4H6}; R353–P376, deleted in rl^{64}; D334, changed to N in rl^{Sem}

Homologues in other species

The deduced amino acid sequence of DmErkA is 79% and 80% identical to rat Erk1 and Erk2 (**Erk1/2**), respectively. DmErkA demonstrates a similar degree of identity, ~79%, to *Xenopus laevis* XP42 and **Mpk1**. The identity of DmErkA to the *S. cerevisiae* kinases **Fus3** and **Kss1** is slightly lower (47%).

Physiological substrates and specificity determinants

Genetic analysis has indicated that there are several proteins required in similar developmental decisions to DmErkA, but which are likely to be downstream from it. These proteins include the transcription factors Sina and Yan, which positively and negatively regulate R7 development, respectively. Biochemical experiments will be necessary to evaluate whether these proteins are targets of DmErkA.

Assay

The kinase activity of DmErk has not been directly assessed. Using the *D. melanogaster* S2 embryonic cell line, we have shown that DmErkA is rapidly (<5 min) phosphorylated on tyrosine in response to treatment of the cell line with bovine insulin[2].

Enzyme activators and inhibitors

The activation of MAP kinases has been demonstrated to require the phosphorylation of both threonine and tyrosine residues (in DmErkA, T198 and Y200). This phosphorylation has been shown to be carried out by members of the Ste7 family of protein kinases[7]. The *D. melanogaster Dsor1* locus has been shown to encode a close homologue of the Ste7 family of kinases[8]. It has been demonstrated that Dsor1, as with DmErkA, is a member of both the **Sev** and **DmEGFR** signal transduction pathways (Y. Nishida, personal communication).

Pattern of expression

Western blot analysis indicated that DmErkA is expressed in a variety of tissues throughout development, including embryonic, larval, eye imaginal disc, adult head, and adult body tissues[2]. Immunohistochemical staining of the eye primordia, the eye imaginal disc, revealed that DmErkA was expressed in all cells of the disc.

References
1 Cobb, M.H. et al. (1991) Cell Regul. 2, 965–978.
2 Biggs, W.H. and Zipursky, S.L. (1992) Proc. Natl Acad. Sci. USA 89, 6295–6299.
3 Hilliker, A.J. (1976) Genetics 83, 765–782.
4 Biggs, W.H. et al. (1994) EMBO J. 13, 1628–1635.
5 Brunner, D. et al. (1994) Cell 76, 875–888.
6 Hanks, S.K. et al. (1988) Science 241, 72–82.
7 Ahn, N. et al. (1992) Curr. Opin. Cell. Biol. 4, 992–999.
8 Tsuda, L. et al. (1993) Cell 72, 407–414.

Kss1
Kinase suppressor of sst2-1 (*S. cerevisiae*)
(Erk/MAP kinase homologue)

Jeanette Gowen Cook, Doreen Ma and Jeremy Thorner
(University of California, Berkeley, USA)

The *KSS1* gene was originally isolated on the basis of its ability, when over-expressed, to promote the resumption of growth in certain *MAT*a haploid cells (*sst2-1* mutants) that had been arrested in the G1 phase of the cell cycle by prior exposure to the mating pheromone, α-factor[1]. Kss1 was the first member identified of the family now known as the MAP kinases or Erks[2]. Kss1 is required in haploid cells for efficient signal transduction in response to mating pheromone and is partially redundant in function with a homologue, **Fus3**[3] (for a review of the role of protein kinases in the mating pheromone response pathway, see ref. 4).

Subunit structure and isoforms

SUBUNIT	AMINO ACIDS	MOL. WT	SDS-PAGE
Kss1	368	42 697	43 kDa

The native active form(s) of Kss1 has not been purified or characterized biochemically.

Genetics

Chromosomal location: right arm, VII (9 centimorgans from centromere)[1].
Site-directed mutations[5]:

Y24F:	inactive for mating, but still promotes recovery when over-produced
K42R:	inactive for mating, but still promotes recovery when over-produced
K43R:	normal function preserved
T183A:	inactive for mating, but still promotes recovery when over-produced
Y185F:	inactive for mating, but still promotes recovery when over-produced
T183A ⎫ Y185F ⎭	(double mutant) inactive for mating, but still promotes recovery when overproduced
Δ321–368:	inactive for mating, does not promote recovery, excluded from the nucleus

Domain structure

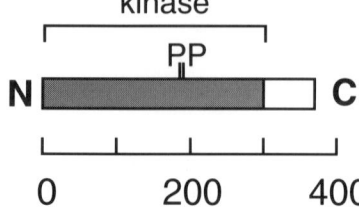

Kss1 consists of a kinase domain followed by an acidic (net charge −12) C-terminal extension. Kss1 is phosphorylated *in vivo* on both T183 and Y185[5, 6], as well as on serine residue(s) at unknown location(s)[5].

Database accession numbers

	PIR	SWISSPROT	EMBL/GENBANK	REF
Kss1		P14681	M26398	1

Amino acid sequence of *S. cerevisiae* Kss1

```
MARTITFDIP SQYKLVDLIG EGAYGTVCSA IHKPSGIKVA IKKIQPFSKK  50
LFVTRTIREI KLLRYFHEHE NIISILDKVR PVSIDKLNAV YLVEELMETD 100
LQKVINNQNS GFSTLSDDHV QYFTYQILRA LKSIHSAQVI HRDIKPSNLL 150
LNSNCDLKVC DFGLARCLAS SSDSRETLVG FMTEYVATRW YRAPEIMLTF 200
QEYTTAMDIW SCGCILAEMV SGKPLFPGRD YHHQLWLILE VLGTPSFEDF 250
NQIKSKRAKE YIANLPMRPP LPWETVWSKT DLNPDMIDLL DKMLQFNPDK 300
RISAAEALRH PYLAMYHDPS DEPEYPPLNL DDEFWKLDNK IMRPEEEEEV 350
PIEMLKDMLY DELMKTME.
```

Sites of interest: T183, Y185, phosphorylation

Homologues in other species

Members of the MAP kinase family have been isolated from *S. pombe* (**Spk1**)[7], sea star (*Pisaster ochraceus*), *Xenopus laevis, D. melanogaster* (**DmErkA**), and several mammalian species (**Erk1/2, Erk3**). In every organism examined to date, MAP kinases become activated in response to a wide variety of extracellular stimuli, depending on the cell type, by phosphorylation at positions analogous to T183 and Y185 in Kss1[8].

Physiological substrates and specificity determinants

No substrates for Kss1 (either *in vivo* or *in vitro*) have been identified.

Enzyme activators and inhibitors

Kss1 becomes rapidly phosphorylated at T183 and Y185 in cells exposed to mating pheromone[5, 6]. At least three other protein kinases (Ste20, **Ste11** and **Ste7**) appear to act upstream of Kss1[5]. **Ste7** is homologous to MAP kinase kinases (see **Mek**), which have been identified in several other systems[9].

Pattern of expression

KSS1 is expressed in all three yeast cell types (*MAT*a haploids, *MAT*α haploids and *MAT*a/*MAT*α diploids) and is not a pheromone-responsive gene[1]. Kss1 is localized primarily in the nucleus, even in the absence of exposure of the cells to pheromone, and is found almost exclusively in the particulate fraction of cell lysates[5].

References
1 **Courchesne, W.E., Kunisawa, R. and Thorner, J. (1989) Cell 58, 1107–1119.**
2 Boulton et al. (1990) Science 249, 64–67.

[3] Elion, E.A. et al. (1991) Cold Spring Harbor Symp. Quant. Biol. 56, 41–49.

[4] Sprague, G.F., Jr. and Thorner, J. (1992) In The Molecular and Cellular Biology of the Yeast Saccharomyces: Gene Expression, Vol. II, Cold Spring Harbor Laboratory Press, pp. 657–744.

[5] Ma, D. et al. (1995), submitted.

[6] Gartner, A. et al. (1992) Genes Devel. 6, 1280–1292.

[7] Toda, T. et al. (1991) Genes Devel. 5, 60–73.

[8] Pelech, S.L. and Sanghera, J.S. (1992) Trends Biochem. Sci. 17, 233–238.

[9] Crews, C.M. and Erikson, R.L. (1992) Proc. Natl Acad. Sci. USA 89, 8205–8209.

Fus3

Fus3 protein kinase (*S. cerevisiae*)

Elaine A. Elion and Gerald R. Fink (Harvard Medical School and Whitehead Institute for Biomedical Research, USA)

Fus3 is an **Erk1/2** (MAP kinase) homologue[1] that acts in a receptor/G protein-coupled signal transduction cascade to promote mating in response to growth factors[3]. Fus3 promotes mating through at least two functions, one that activates a transcription factor of mating-specific genes, and one that promotes G1 arrest by inhibition of G1 cyclins[3, 4]. Fus3 is functionally redundant with a second MAP kinase homologue, **Kss1**[2], for a step in the signal transduction cascade that activates transcription, but appears to inhibit the G1 cyclins through a unique function[4]. Fus3 has multiple *in vivo* substrates including components of the signalling cascade and the negative growth regulator Far1[5, 6].

Subunit structure and isoforms

SUBUNIT	AMINO ACIDS	MOL. WT	SDS-PAGE
Fus3	353	40 476	

Fus3 is a monomer.

Domain structure

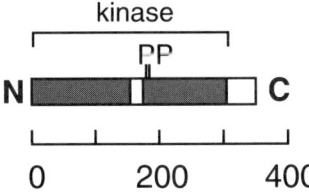

Fus3 exists in active phosphorylated and inactive dephosphorylated forms[7]. The active form is phosphorylated on T180 and Y182 by **Ste7**[6] which is structurally related to MAP kinase kinase (**Mek**)[8]. Fus3 appears to autophosphorylate on Y182[5]. Just N-terminal to T180 is a unique insert of 18 residues (unshaded box). The catalytic domain is ≈50% identical to **Erk1/2** and 30% identical to **SpCdc2**.

Database accession numbers

	PIR	SWISSPROT	EMBL/GENBANK	REF
Fus3	S28548	P16892	M31132	4

Amino acid sequence of *S. cerevisiae* Fus3

```
MPKRIVYNIS SDFQLKSLLG EGAYGVVCSA THKPTGEIVA IKKIEPFDKP  50
LFALRTLREI KILKHFKHEN IITIFNIQRP DSFENFNEVY IIQELMQTDL 100
HRVISTQMLS DDHIQYFIYQ TLRAVKVLHG SNVIHRDLKP SNLLINSNCD 150
LKVCDFGLAR IIDESAADNS EPTGQQSGMT EYVATRWYRA PEVMLTSAKY 200
SRAMDVWSCG CILAELFLRR PIFPGRDYRH QLLLIFGIIG TPHSDNDLRC 250
IESPRAREYI KSLPMYPAAP LEKMFPRVNP KGIDLLQRML VFDPAKRITA 300
KEALEHPYLQ TYHDPNDEPE GEPIPPSFFE FDHHKEALTT KDLKKLIWNE 350
IFS
```

Sites of interest: T180, Y182, phosphorylation

Homologues in other species

Protein kinases (MAP kinases) highly related to Fus3 have been sequenced from many organisms including rat, mouse (**Erk1/2**) *S. cerevisiae* (**Fus3**), *S. pombe* (**Spk1**), and *Xenopus laevis*[9].

Physiological substrates and specificity determinants

Substrates include Far1[5, 6], and **Ste7**[6]. Recognition sequences not yet determined.

Assay

Assayed by immune complex kinase assays using epitope-tagged derivatives of Fus3[5, 6]. The enzyme is assayed as a pheromone-dependent activity that transfers [γ-^{32}P]ATP to various substrates. Exogenous substrates include Far1 fusion protein[5], N-terminal fragment of Far1[6], **Ste7**[6], myelin basic protein[5] and casein[5]. Fus3 also phosphorylates a group of physiologically relevant coprecipitated substrates[5].

Enzyme activators and inhibitors

Activated by phosphorylation by **Ste7**[6].

Pattern of expression

Transcription of Fus3 is haploid-specific and increased several-fold by pheromone induction[2]. The level of Fus3 protein parallels mRNA[5, 7]. The protein is largely cytosolic and soluble.

References

[1] Boulton, T.G. et al. (1991) Science 249, 476–489.
[2] Courchesne, W.E. et al. (1989) Cell 58, 1107–1119.
[3] Elion, E. A. et al. (1991) Proc. Natl Acad. Sci. USA 88, 9392–9396.
[4] Elion, E.A. et al. (1990) Cell 60, 649–664.
[5] Elion, E.A. et al. (1993) Mol. Biol. Cell 4, 495–510.
[6] Errede, B. et al. (1993) Nature 362, 261–264.
[7] Gartner, A. et al. (1992) Genes Devel. 6, 1280–1292.
[8] Crews, C.M. et al. (1992) Science 358, 478.
[9] Cobb, M.H. et al. (1991) Cell Regul. 2, 965–978.

Mpk1

MAP kinase 1 (*S. cerevisiae*)
(Slt2)

David E. Levin (John Hopkins University, USA)

Mpk1 mediates signalling by the protein kinase C isozyme encoded by PKC1 (ScPKC)[1]. Genetic analysis of MPK1 revealed a role in cell wall construction. Deletion mutants display a temperature-dependent cell lysis defect that is suppressed by osmotic stabilizing agents[2, 3]. The MPK1 gene was isolated as a dosage-dependent suppressor of a deletion mutant of BCK1, which encodes a protein kinase whose mutational activation suppresses the cell lysis defect associated with loss of PKC1 function[4]. Mpk1 is proposed, based on genetic epistasis experiments, to function at the end of a protein kinase cascade in which the PKC1 gene product ScPKC activates Bck1, which activates a redundant pair of MAP kinase kinases Mkk1/2[5], which activate Mpk1. Mpk1 kinase activity is stimulated by heat shock and reduced osmolarity.

Subunit structure and isoforms

SUBUNIT	AMINO ACIDS	MOL. WT	SDS-PAGE
Mpk1	484	55,636	67 kDa

Genetics

Chromosomal location: VIII, tightly linked to PUT2[2]
Mutations: Y192F, T190A (inactive)

Domain structure

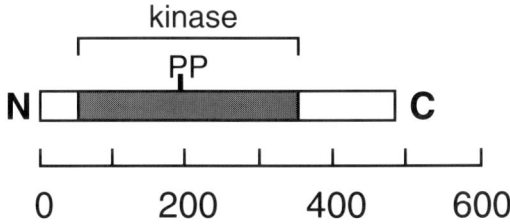

Database accession numbers

	PIR	SWISSPROT	EMBL/GENBANK	REF
Mpk1			X59262	2, 3

Amino acid sequence of Mpk1

```
MADKIERHTF KVFNQDFSVD KRFQLIKEIG HGAYGIVCSA RFAEAAEDTT  50
VAIKKLTNVF SKTLLCKRSL RELKLLRHFR GHKNITCLYD MDIVFYPDGS 100
INGLYLYEEL MECDMHQIIK SGQPLTDAHY QSFTYQILCG LKYIHSADVL 150
HRDLKPGNLL VNADCQLKIC DFGLARGYSE NPVENSQFLT EYVATRWYRA 200
PEIMLSYQGY TKAIDVWSAG CILAEFLGGK PIFKGKDYVN QLNQILQVLG 250
TPPDETLRRI GSKNVQDYIH QLGFIPKVPF VNLYPNANSQ ALDLLEQMLA 300
FDPQKRITVD EALEHPYLSI WHDPADEPVC SEKFEFSFES VNDMEDLKQM 350
VIQEVQDFRL FVRQPLLEEQ RQLQLQQQQQ QQQQQQQQQQ QPSDVDNGNA 400
AASEENYPKQ MATSNSVAPQ QESFGIHSQN LPRHDADFPP RPQESMMEMR 450
PATGNTADIP PQNDNGSLLD LEKELEFGLD RKYF
```

Sites of interest: putative sites phosphorylated by Mkk1/2

References

[1] Errede, B. and Levin, D.E. (1993) Curr. Opinion Cell Biol. 5, 254–260.
[2] Lee, K.S. et al. (1993) Mol. Cell. Biol. 13, 3067–3075.
[3] Torres, L. et al. (1991) Mol. Microbiology 5, 2845–2854.
[4] Lee, K.S. and Levin, D.E. (1992) Mol. Cell. Biol. 12, 172–182.
[5] Irie, K. et al. (1993) Mol. Cell. Biol. 13, 3076–3083.

 **Spk1** MAP kinase homologue (*S. pombe*)

Toda (Imperial Cancer Research Fund, UK)

The *spk1⁺* gene encodes a protein kinase highly homologous to vertebrate MAP kinases (**Erk1/2**), and works as a key mediator in the mating pheromone signal transduction pathway in the fission yeast, *S. pombe*. Deletion of *spk1⁺* results in defects in both conjugation and sporulation. The Spk1 protein becomes phosphorylated after receipt of the mating signal. As with mammalian MAP kinases, Spk1 is part of a protein kinase cascade; the upstream kinases (equivalent to MAP kinase kinase and MAP kinase kinase kinase) are **Byr1**/Ste1 and **Byr2**/Ste8 respectively.

Subunit structure and isoforms

SUBUNIT	AMINO ACIDS	MOL. WT	SDS-PAGE
Spk1	372	42 003	45 kDa

Mammalian MAP kinases are monomers, but the native structure of Spk1 is not known.

Domain structure

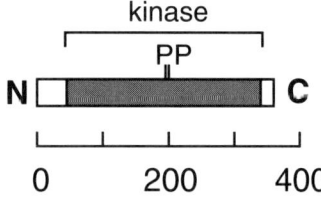

The Spk1 gene product consists of a protein kinase domain with short N- and C-terminal extensions. The various members of the MAP kinase family show amino acid sequence identity of ≈50% to each other, and are also homologous (≈40% identity) to **Cdc2**. T199 and Y201 are conserved in almost all MAP kinases and probably become phosphorylated upon activation of the Spk1 kinase.

Database accession numbers

	PIR	SWISSPROT	EMBL/GENBANK	REF
Spk1		P27638	X57334	*1*

 ## Amino acid sequence of *S. pombe* Spk1

```
MASATSTPTI ADGNSNKESV ATSRSPHTHD LNFELPEEYE MINLIGQGAY  50
GVVCAALHKP SGLKVAVKKI HPFNHPVFCL RTLREIKLLR HFRHENIISI 100
LDILPPPSYQ ELEDVYIVQE LMETDLYRVI RSQPLSDDHC QYFTYQILRA 150
LKAMHSAGVV HRDLKPSNLL LNANCDLKVA DFGLARSTTA QGGNPGFMTE 200
YVATRWYRAP EIMLSFREYS KAIDLWSTGC ILAEMLSARP LFPGKDYHSQ 250
ITLILNILGT PTMDDFSRIK SARARKYIKS LPFTPKVSFK ALFPQASPDA 300
IDLLEKLLTF NPDKRITAEE ALKHPYVAAY HDASDEPTAS PMPPNLVDLY 350
CNKEDLEIPV LKALIFREVN FR
```

Sites of interest: T199, Y201, phosphorylation (by analogy with MAP kinase)

Homologues in other species

MAP kinase homologues have been sequenced from human (**Erk1/2**), hamster, rat, *Xenopus laevis* and *S. cerevisiae* (**Kss1, Fus3**). It has been shown that rat MAP kinase (Erk2) can partially suppress *spk1* disruptants. Expression of *spk1⁺* can complement the mating defect of budding yeast *fus3/kss1* double mutants, and also the cell-cycle arrest defect of *fus3* single mutants.

Physiological substrates and specificity determinants

The consensus sequence for phosphorylation by MAP kinases is (P)XS/TP. Physiological substrates of Spk1 have not been identified in *S. pombe*.

Assay

The activity of Spk1 was detected in immunocomplexes precipitated with anti-Spk1 antibodies from whole cell extracts of *S. pombe*, using a standard protein kinase assay with [γ-^{32}P]ATP and histone H1 as substrates. Casein is a poor substrate for the Spk1 kinase, while myelin basic protein has not been tested.

Pattern of expression

Immunofluorescence using anti-Spk1 antibodies has shown that Spk1 is a nuclear protein. Cellular fractionation using Ficoll gradients has also confirmed this result. The level of the Spk1 mRNA is highly stimulated upon nitrogen starvation. Consistent with this notion, the 5′-flanking region of the *spk1⁺* gene contains a thymidine-rich *cis*-element which exists in the upstream regions of a number of *S. pombe* genes induced by nitrogen starvation. The expression induced by nitrogen starvation is diminished in the *ste11* mutant background. *ste11⁺* encodes a transcription factor which recognizes and binds the thymidine-rich *cis*-element. The Spk1 mRNA and protein is also induced by addition of staurosporine, a kinase inhibitor.

References
[1] Toda, T. et al. (1991) Genes Devel. 5, 60–73.
[2] Neiman, A. M. et al. (1993). Mol. Biol. Cell 4, 107–120.
[3] Nielsen, O. (1993) Trends Cell Biol. 3, 60–65.
[4] Gotoh, Y. et al. (1993) Mol. Cell. Biol. (13, 6427–6434).

GSK3	Glycogen synthase kinase-3 (vertebrates) (Factor F$_A$, ATP-citrate lyase kinase)

James Woodgett (Ontario Cancer Institute, Canada)

In mammals, glycogen synthase kinase-3 comprises two highly related proteins (GSK3α and GSK3β) that phosphorylate a wide variety of target proteins. Since the kinases appear to be constitutively active in resting cells, and phosphorylation of several of the characterized substrates is inhibitory, these protein kinase may repress the function of proteins that are activated by dephosphorylation[1]. GSK3 is identical to Factor A (F$_A$), a protein required for activation of the MgATP-dependent protein phosphatase (PP1$_I$). The latter is a complex between the protein phosphatase-1 catalytic subunit and inhibitor-2, and activation occurs via phosphorylation of the latter[1] (see below). GSK3α is identical to the multi-functional ATP-citrate lyase kinase[2]. GSK3β is identical to tau protein kinase-1 (Tpk1)[3].

Subunit structure and isoforms

SUBUNIT	AMINO ACIDS	MOL. WT	SDS-PAGE
α	482	50 995	51 kDa
β	419	46 712	47 kDa

Both α and β isoforms are monomeric and the products of distinct genes[4].

Domain structure

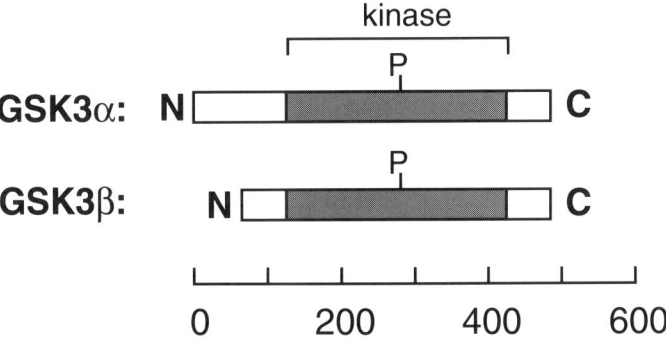

While the catalytic domains are 93% identical between the two forms, GSK3α has an additional 40 residues, N-terminal to the kinase domain, which is glycine- and proline-rich.

Database accession numbers

	PIR	SWISSPROT	EMBL/GENBANK	REF
GSK3α			X53427	4
GSK3β			X53428	4

Amino acid sequences of rat GSK3

GSK3α

```
SGGGPSGGGP GGSGRARTSS FAEPGGGGGG GGGGPGGSAS GPGGTGGGKA  50
SVGAMGGGVG ASSSGGGPSG SGGGGSGGPG AGTSFPPPGV KLGRDSGKVT 100
TVVATLGQGP ERSQEVAYTD IKVIGNGSFG VVYQARLAET RELVAIKKVL 150
QDKRFKNREL QIMRKLDHCN IVRLRYFFYS SGEKKDELYL NLVLEYVPET 200
VYRVARHFTK AKLIIPIIYV KVYMYQLFRS LAYIHSQGVC HRDIKPQNLL 250
VDPDTAVLKL CDFGSAKQLV RGEPNVSYIC SRYYRAPELI FGATDYTSSI 300
DVWSAGCVLA ELLLGQPIFP GDSGVDQLVE IIKVLGTPTR EQIREMNPNY 350
TEFKFPQIKA HPWTKVFKSR TPPEAIALCS SLLEYTPSSR LSPLEACAHS 400
FFDELRSLGT QLPNNRPLPP LFNFSPGELS IQPSLNAILI PPHLRSPSGP 450
ATLTSSSQAL TETQTGQDWQ APDATPTLTN SS
```

Sites of interest: Y278, phosphorylation

GSK3β

```
SGRPRTTSFA ESCKPVQQPS AFGSMKVSRD KDGSKVTTVV ATPGQGPDRP  50
QEVSYTDTKV IGNGSFGVVY QAKLCDSGEL VAIKKVLQDK RFKNRELQIM 100
RKLDHCNIVR LRYFFYSSGE KKDEVYLNLV LDYVPETVYR VARHYSRAKQ 150
TLPVIYVKLY MYQLFRSLAY IHSFGICHRD IKPQNLLLDP DTAVLKLCDF 200
GSAKQLVRGE PNVSYICSRY YRAPELIFGA TDYTSSIDMW SAGCVLAELL 250
LGQPIFPGDS GVDQLVEIIK VLGTPTREQI REMNPNYTEF KFPQIKAHPW 300
TKVFRPRTPP EAIALCSRLL EYTPTARLTP LEACAHSFFD ELRDPNVKLP 350
NGRDTPALFN FTTQELSSNP PLATILIPPH ARIQAAASPP ANATAASDTN 400
AGDRGQTNNA ASASASNST
```

Sites of interest: Y215, phosphorylation

Homologues in other species

Both forms of GSK3 have been sequenced from human cells (HepG2 cells, E. Nikolakaki and J.R.W., unpublished) and rat (brain cDNA[4]). A GSK3β homologue has been sequenced from chicken (M. Corchoran and E. Nikolakaki, unpublished). The cognate genes exhibit >96% identity between human and chicken. GSK3 is functionally homologous to the *shaggy/zeste-white*[5,6] (**Sgg**)homeotic gene product in *D. melanogaster*. **Mck1** from *S. cerevisiae* displays about 35% identity and lies on the same branch as the GSK3 family upon phylogenetic analysis. In addition, a closer homologue (75% identity) has been isolated from this organism (Genbank L12761)[7]. Homologues have also been cloned from *D. discoideum*, *S. pombe* (S. Plyte, K.L. Gould, J. Williams and J.R.W., in preparation) and *Arabidopsis thaliana*[8]. Genomic zoo blots and sequence analysis indicate expression of a β-like form of GSK3 in all eukaryotes examined[1].

Physiological substrates and specificity determinants

The substrate recognition parameters for phosphorylation by GSK3 are complicated and substrate-dependent. For some target proteins, prior phosphorylation (double underlined) by a distinct protein kinase, 4 residues C-terminal to the GSK3 site (bold) is necessary[1,9]. In the case of glycogen synthase, the priming protein kinase is casein kinase-2 (**CK2**), whereas cAMP-dependent protein kinase (**PKA**) provides this function in the G subunit of protein phosphatase-1. Some substrates (e.g. c-Jun, c-Myc, L-Myc) exhibit no priming requirement. Most sites are embedded in proline-rich sequences:

Glycogen synthase:	RPASVPP**S**PSLS**R**HS**S**PHQ<u>S</u>EDEE
ATP-citrate lyase:	STS**T**PAP**S**RTA<u>S</u>F**S**ESR
G subunit:	PGF**S**PQP**S**RRG<u>S</u>E**S**SEEVY
Inhibitor-2:	EPS**T**PYHSMIGDDDDAY
c-Jun:	EPQ**T**VPEMPGE**T**PPL**S**PIDME**S**QER
c-Myc:	LLP**T**PPL**S**PSRRSG
L-Myc:	LVP**S**PPT**S**PPWGL
tau:	VVR**T**PPKSPSSAK

Assay
GSK3 can be assayed by following incorporation of ^{32}P from [γ-^{32}P]ATP into glycogen synthase, myelin basic protein, or synthetic peptides containing phosphorylation sites, provided priming sites are pre-phosphorylated.

Enzyme activators and inhibitors
Although apparently constitutively active in cells, all forms of GSK3 require tyrosine phosphorylation for activity, in a position analogous to that seen for the MAP kinases[10]. In addition, GSK3β is phosphorylated and inactivated by certain isoforms of protein kinase C[11], p70 S6 kinase and Rsk-1[12].

Pattern of expression
GSK3 is predominantly cytoplasmic, although significant amounts are associated with membranes and the nucleus, as determined by confocal microscopy, and assay of subcellular fractions. Both forms are widely distributed both spatially and developmentally, although independent expression of each form is observed in certain cultured cell lines.

References
1 Plyte, S. et al. (1992) Biochim. Biophys. Acta 1114, 147–162.
2 Hughes, K. et al. (1992) Biochem. J. 288, 309–314.
3 Ishiguro et al. (1993) FEBS Letts. 325, 167–172.
4 Woodgett, J. R. (1990) EMBO J. 9, 2431–2438.
5 Seigfried, E. et al. (1992) Cell 71, 1167–1179.
6 Ruel, L. et al. (1993) Nature 362, 557–560.
7 Bianchi, M. et al. (1993) Gene 134, 51–56.
8 Bianchi, M. et al. (1995) Mol. Gen. Genet. (in press).
9 Fiol, C. J. et al. (1990) J. Biol. Chem. 265, 6061–6065.
10 Hughes, K. et al. (1993) EMBO J. 12, 803–808.
11 Goode, N. et al. (1992) J. Biol. Chem. 267, 16878–16882.
12 Sutherland, C. and Cohen, P. (1993) Biochem. J. 296, 15–19.

Sgg	**Shaggy protein kinase (*D. melanogaster*)** (Zeste-white 3)

Marc Bourouis (LGME-CNRS, Strasbourg, France)

SGG has multiple functions defined by the effects of mutant flies. The gene is also required for intrasegmental (epidermal) cell fate specification[1,2,3], as a segment polarity gene (maternal effect) and during the determination of the neural cells in the embryo[1] (maternal effect). The growth and differentiation of larval tissues (of neural and non-neural cell types), and of imaginal tissues are affected[1] (zygotic effect). Mutants cause a hyperplasia of the adult sensory structures (a neurogenic-like property) as a result of the failure to transduce an intercellular, neural inhibitory signal[4] (zygotic effect).

Subunit structure and isoforms

SUBUNIT	AMINO ACIDS	MOL. WT	SDS-PAGE
Sgg10/Zw3A	514	53 800	58 000
SggY/Zw3C	496	52 100	56 000
Sgg39	575	58 700	68 000
SggX			90 000
Sgg46/Zw3B	1067	114 400	140 000

Five forms of Sgg protein kinases are known, resulting from alternative splicing. All forms contain the same catalytic domain. One form, Sgg10, resembles the vertebrate **GSK3** isoforms. This protein may exist as a monomer by analogy to GSK3β.

Genetics

Chromosomal location: 1.
Meiotic position: 1.32 map units.
Cytological position: 3B1.
Many mutant alleles of the gene are known. They all behave as zygotic lethals and define a single complementation group. True null mutations were defined by the absence of detectable Sgg proteins[5].

Domain structure

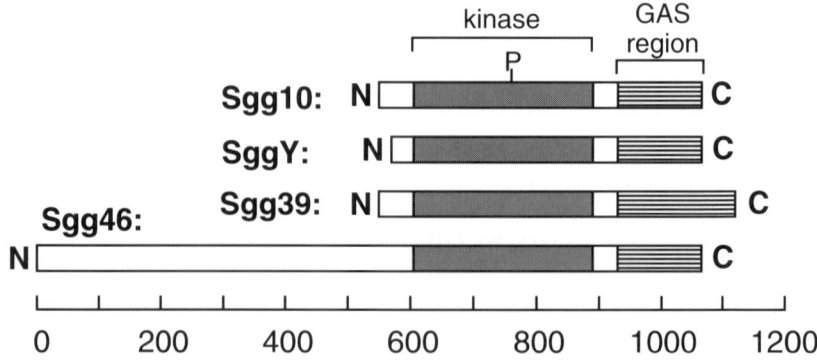

The different Sgg proteins comprise a common kinase domain with distinct N- or C-terminal regions. The GAS region contains 75% of G, A and S residues. Similarities between the Sgg proteins and mammalian **GSK3** stop abruptly at the end of the hinge segment found between the two domains. Activity of Sgg10 and **GSK3** requires the phosphorylation of a tyrosine residue[6], which could potentially regulate activity of all forms.

Database accession numbers

	PIR	SWISSPROT	EMBL/GENBANK	REF
Sgg10	S10931	P18431	X53332	1, 5
SggY				3
Sgg39			X70863	5
Sgg46			X70864	5

Amino acid sequences of *D. melanogaster* Sgg/Zw3

Sgg10/Zw3A
```
MSGRPRTSSF AEGNKQSPSL VLGGVKTCSR DGSKITTVVA TPGQGTDRVQ  50
EVSYTDTKVI GNGSFGVVFQ AKLCDTGELV AIKKVLQDRR FKNRELQIMR 100
KLEHCNIVKL LYFFYSSGEK RDEVFLNLVL EYIPETVYKV ARQYAKTKQT 150
IPINFIRLYM YQLFRSLAYI HSLGICHRDI KPQNLLLDPE TAVLKLCDFG 200
SAKQLLHGEP NVSYICSRYY RAPELIFGAI NYTTKIDVWS AGCVLAELLL 250
GQPIFPGDSG VDQLVEVIKV LGTPTREQIR EMNPNYTEFK FPQIKSHPWQ 300
KVFRIRTPTE AINLVSLLLE YTPSARITPL KACAHPFFDE LRMEGNHTLP 350
NGRDMPPLFN FTEHELSIQP SLVPQLLPKH LQNASGPGGN RPSAGGAASI 400
AASGSTSVSS TGSGASVEGS AQPQSQGTAA AAGSGSGGAT AGTGGASAGG 450
PGSGNNSSSG GASGAPSAVA AGGANAAVAG GAGGGGAGA ATAAATATGA 500
IGATNAGGAN VTDS
```

Site of interest: Y214, phosphorylation

SggY/Zw3C
```
MLINRGSLLE SRDGSKITTV VATPGQGTDR VQEVSYTDTK VIGNGSFGVV  50
FQAKLCDTGE LVAIKKVLQD RRFKNRELQI MRKLEHCNIV KLLYFFYSSG 100
EKRDEVFLNL VLEYIPETVY KVARQYAKTK QTIPINFIRL YMYQLFRSLA 150
YIHSLGICHR DIKPQNLLLD PETAVLKLCD FGSAKQLLHG EPNVSYICSR 200
YYRAPELIFG AINYTTKIDV WSAGCVLAEL LLGQPIFPGD SGVDQLVEVI 250
KVLGTPTREQ IREMNPNYTE FKFPQIKSHP WQKVFRIRTP TEAINLVSLL 300
LEYTPSARIT PLKACAHPFF DELRMEGNHT LPNGRDMPPL FNFTEHELSI 350
QPSLVPQLLP KHLQNASGPG GNRPSAGGAA SIAASGSTSV SSTGSGASVE 400
GSAQPQSQGT AAAAGSGSGG ATAGTGGASA GGPGSGNNSS SGGASGAPSA 450
VAAGGANAAV AGGAGGGGGA GAATAAATAT GAIGATNAGG ANVTDS
```

Sgg39
```
MSGRPRTSSF AEGNKQSPSL VLGGVKTCSR DGSKITTVVA TPGQGTDRVQ  50
EVSYTDTKVI GNGSFGVVFQ AKLCDTGELV AIKKVLQDRR FKNRELQIMR 100
KLEHCNIVKL LYFFYSSGEK RDEVFLNLVL EYIPETVYKV ARQYAKTKQT 150
IPINFIRLYM YQLFRSLAYI HSLGICHRDI KPQNLLLDPE TAVLKLCDFG 200
SAKQLLHGEP NVSYICSRYY RAPELIFGAI NYTTKIDVWS AGCVLAELLL 250
```

Sgg39   continued

```
GQPIFPGDSG VDQLVEVIKV LGTPTREQIR EMNPNYTEFK FPQIKSHPWQ 300
KVFRIRTPTE AINLVSLLLE YTPSARITPL KACAHPFFDE LRMEGNHTLP 350
NGRDMPPLFN FTEHELSIQP SLVPQLLPKH LQNASGPGGN RPSAGGAASI 400
AASGSTSVSS TGSGASVEGS AQPQSQGTAA AAGSGSGGAT AGTGGASAGG 450
PGSGNNSSSG GASGAPSAVA AGGANAAVAG GAGGGGGAGA ATAAATATGA 500
IGATNAGGAN VTGSQSNSAL NSSGSGGSGN GEAAGSGSGS GSGSGGGNGG 550
DNDAGDSGAI ASGGGAAETE AAASG
```

Sgg46(ZW3B)

```
MATTTTTQRA GAAPALNLLP ASNNNINNTL INNNNNNNNT SNSNNNNNNV   50
ISQPIKIPLT ERFSSQTSTG SADSGVIVSS ASQQQLQLPP PRSSSGSLSL  100
PQAPPGGKWR QKQQRQQLLL SQDSGIENGV TTRPSKAKDN QGAGKASHNA  150
TSSKESGAQS NSSSESLGSN CSEAQEQQRV RASSALELSS VDTPVIVGGV  200
VSGGNSILRS RIKYKSTNST GTQGFDVEDR IDEVDICDDD DVDCDDRGSE  250
IEEEEEEEED DGVNVDDDVE EADNQSDNQS GIIINLKSQT EQEEEVDEVD  300
AKPKNRLLPP DQAELTVAAA MARRRDAKSL ATDGHIYFPL LKISEDPHID  350
SKLINRKDGL QDTMYYLDEF GSPKLREKFA RKQKQLLAKQ QKQLMKRERR  400
SEEQRKKRNT TVASNLAASG AVVDDTKDDY KQQPHCDTSS RSKNNSVPNP  450
PSSHLHQNHN HLVVDVQEDV DDVNVVATSD VDSGVVKMRR HSHDNHYDRI  500
PRSNAATITT RPQIDQQSSH HQNTEDVEQG AEPQIDGEAD LDADADADSD  550
GSGENVKTAK LARTQSCVSW TKVVQKFKNI LGRDGSKITT VVATPGQGTD  600
RVQEVSYTDT KVIGNGSFGV VFQAKLCDTG ELVAIKKVLQ DRRFKNRELQ  650
IMRKLEHCNI VKLLYFFYSS GEKRDEVFLN LVLEYIPETV YKVARQYAKT  700
KQTIPINFIR LYMYQLFRSL AYIHSLGICH RDIKPQNLLL DPETAVLKLC  750
DFGSAKQLLH GEPNVSYICS RYYRAPELIF GAINYTTKID VWSAGCVLAE  800
LLLGQPIFPG DSGVDQLVEV IKVLGTPTRE QIREMNPNYT EFKFPQIKSH  850
PWQKVFRIRT PTEAINLVSL LLEYTPSARI TPLKACAHPF FDELRMEGNH  900
TLPNGRDMPP LFNFTEHELS IQPSLVPQLL PKHLQNASGP GGNRPSAGGA  950
ASIAASGSTS VSSTGSGASV EGSAQPQSQG TAAAGSGSG GATAGTGGAS 1000
AGGPGSGNNS SSGGASGAPS AVAAGGANAA VAGGAGGGGG AGAATAAATA 1050
TGAIGATNAG GANVTDS
```

Homologues in other species

The Sgg proteins have an unusually high degree of similarity to vertebrate **GSK3** (both α and β isoforms). Similarities extend over the catalytic domains which are 80–85% identical, and on both sides of it. GSK3β represents a true functional homologue of Sgg10, as assayed by complementation[4]. GSK3 homologues have been found in a number of other vertebrate and invertebrate species including yeast (see **GSK3**). The **Mck1** protein kinase from *S. cerevisiae* has lower, ≈40%, identity to Sgg and GSK3. Mck1 may represent a distant member of the Sgg/GSK3 family.

Physiological substrates and specificity determinants

Enzymatic properties of Sgg10 are likely to be similar or identical to those of the mammalian homologue (**GSK3**). The substrates of GSK3 are in general nodal points of regulation in signalling cascades (i.e. regulated enzymes or transcription factors). Sgg39, but not Sgg46, displays redundant functions to Sgg10[5]. The unique N-terminal domain of Sgg46 may regulate the catalytic function on this form.

Assay
Identical to the assay for **GSK3**.

Pattern of expression
No tissue restriction of Sgg isoforms has been found at the level of RNA expression. However developmental regulation occurs[1, 5]. The Sgg proteins appear to be expressed in most, if not all, embryonic and larval tissues. The splicing variants Sgg10 and Sgg39 exhibit some tissue restriction of expression[5]. A PEST region found in Sgg46 may account for its low aboundance. Subcellular location includes cytosol, nuclei and membranes with variations between tissues.

References
1 Bourouis, M. et al. (1989) Nature 341, 442–444.
2 Perrimon, N. and Smouse, D. (1989) Devel. Biol. 135, 287–305.
3 Seigfried, E. et al. (1992) Cell 71, 1167–1179.
4 Ruel, L. et al. (1993) Nature 362, 557–560.
5 Ruel, L. et al. (1993) EMBO J. 12, 1657–1669.
6 Hughes, K. et al. (1993) EMBO J. 12, 803–808.

Mck1

Meiosis and centromere regulatory PK (*S. cerevisiae*) (Ypk1)

Lenore Neigeborn (Rutgers University, USA)

The *MCK1* locus was identified, independently, by two functional dosage-dependent assays: (1) a screen for positive regulators of the sporulation pathway[1] and (2) a screen for suppressors of a centromere mutation[2]. In addition, the protein (then called Ypk1) was uncovered in a general screen for phosphotyrosyl proteins in *S. cerevisiae in vivo*[3]. Genetic and molecular biological analyses revealed that Mck1 plays at least three independent roles in cellular regulation: as a positive, transcriptional regulator of the meiotic activator Ime1; as a stimulator of ascus maturation; and as a promoter of centromere activity. Mck1 is a dual-specificity protein kinase[3] (L. Neigeborn, unpublished results).

Subunit structure and isoforms

SUBUNIT	AMINO ACIDS	MOL. WT	SDS-PAGE
Mck1	375	43 136	

Mck1 is a monomer.

Domain structure

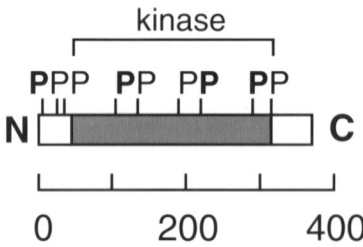

Mck1 has several potential phosphorylation sites for **PKC** (P) and **CK2** (**P**), but the physiological relevance of these are not known.

Database accession numbers

	PIR	SWISSPROT	EMBL/GENBANK	REF
Mck1	A39623	P21965	X55054	1
	A39622		M55984	2, 3

Amino acid sequence of *S. cerevisiae* Mck1

```
MSTEEQNGVP LQRGSEFIAD DVTSNKSNNT RRMLVKEYRK IGRGAFGTVV  50
QAYLTQDKKN WLGPFAIKKV PAHTEYKSRE LQILRIADHP NIVKLQYFFT 100
HLSPQDNKVY QHLAMECLPE TLQIEINRYV TNKLEMPLKH IRLYTYQIAR 150
GMLYLHGLGV CHRDIKPSNV LVDPETGVLK ICDFGSAKKL EHNQPSISYI 200
CSRFYRAPEL IIGCTQYTTQ IDIWGLGCVM GEMLIGKAIF QGQEPLLQLR 250
EIAKLLGPPD KRFIFFSNPA YDGPLFSKPL FSGSSQQRFE KYFGHSGPDG 300
IDLLMKILVY EPQQRLSPRR ILAHQFFNEL RNDDTFLPRG FTEPIKLPNL 350
FDFNDFELQI LGEFADKIKP TKVAE
```

Homologues in other species

Mck1 is most closely related to the β isoform of rat **GSK3**[4], which has been shown to be the functional homologue of shaggy/zeste-white 3 (**Sgg**) of *D. melanogaster*[5, 6]. The biological relevance of this homology is currently under investigation.

Physiological substrates and specificity determinants

In vitro studies reveal serine, threonine and tyrosine autophosphorylation, but the target residues are unknown (L. Neigeborn, unpublished). The protein has been shown to be phosphorylated on both serine and tyrosine *in vivo*; these sites are also uncharacterized[3]. Currently, the only known substrates *in vitro* (other than autophosphorylation) are poly(Glu80–Tyr20)[3] (L. Neigeborn, unpublished) and PKA-phosphorylated CREB2 peptide[7].

Assay

Autophosphorylation is monitored by incorporation of radioactivity from $[\gamma\text{-}^{32}\text{P}]\text{ATP}$ (L. Neigeborn, unpublished). Tyrosine kinase activity is monitored by the incorporation of radioactivity from $[\gamma\text{-}^{32}\text{P}]\text{ATP}$ into the synthetic substrate poly(Glu80–Tyr20)[3].

Pattern of expression

Preliminary indirect immunofluorescence analysis utilizing anti-β-galactosidase antibodies directed against a bifunctional Mck1–lacZ fusion protein revealed uniform distribution throughout the cell. The distribution of antigen was identical in vegetative cells and cells subjected to conditions favouring sporulation (L. Neigeborn, unpublished). Both the *MCK1* mRNA and Mck1 protein are expressed constitutively at low levels in all three yeast cell types and under a variety of nutritional and growth conditions, including those prompting entry into the sporulation pathway[1] (L. Neigeborn, unpublished).

References

1. Neigeborn, L. and Mitchell, A.P. (1991) Genes Devel. 5, 533–548.
2. Shero, J.H. and Hieter, P. (1991) Genes. Devel. 5, 549–560.
3. Dailey, D. et al. (1990) Mol. Cell. Biol. 10, 6244–6256.
4. Woodgett, J.R. (1990) EMBO J. 9, 2431–2438.
5. Siegfried, E. et al. (1990) Nature 345, 825–829.
6. Siegfried, E. et al. (1993) Cell 71, 1167–1179.
7. Puziss, J.W. et al. (1994) Mol. Cell. Biol. 14, 831–839.

CK2 Casein kinase II (vertebrates)

Herman Meisner and Michael P. Czech
(University of Massachusetts Medical Center, USA)

CK2 is a protein kinase that phosphorylates serine or threonine residues that are N-terminal to acidic amino acids, using either ATP or GTP as phosphoryl donors. CK2 phosphorylates many proteins *in vitro*. The enzyme appears to be constitutively active *in vivo*, although extracellular signals may further activate the enzyme 30–200%. The function of a limited number of substrates, generally nuclear transcription factors, has been shown to be changed by phosphorylation. The kinase may be involved in regulation of cell proliferation.

Subunit structure and isoforms

SUBUNIT	AMINO ACIDS	MOL. WT	SDS-PAGE
α	391	45 160	42 kDa
α'	350	41 400	40 kDa
β	215	24 925	25 kDa

The holoenzyme (130 kDa) is a heterotetramer consisting of two α (catalytic) and two β (regulatory) subunits in an $\alpha_2\beta_2$ configuration. An isoform of the catalytic subunit (α') may be complexed as either $\alpha'_2\beta_2$ or $\alpha'\alpha\beta_2$, although the relative tissue distribution of possible heterotetramers is not known.

Domain structure

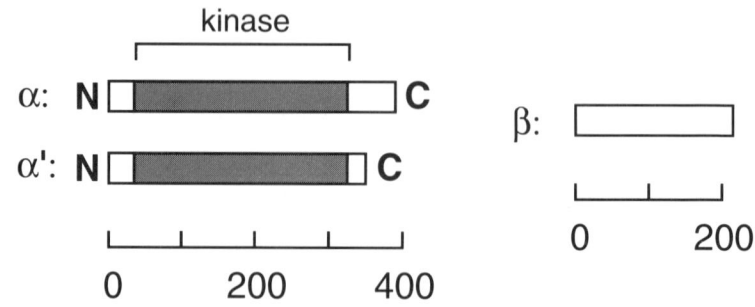

The β subunit is phosphorylated, probably on S3, either by an as yet uncharacterized protein kinase, or by autophosphorylation.

Database accession numbers

		PIR	SWISSPROT	EMBL/GENBANK	REF
α	Human			J02853	1
α'	Human			M55268, J02924	2
β	Human			M30448	3

Amino acid sequences of human CK2

α subunit

```
MSGPVPSRAR VYTDVNTHRP REYWDYESHV VEWGNQDDYQ LVRKLGRGKY  50
SEVFEAINIT NNEKVVVKIL KPVKKKKIKR EIKILENLRG GPNIITLADI 100
VKDPVSRTPA LVFEHVNNTD FKQLYQTLTD YDIRFYMYEI LKALDYCHSM 150
GIMHRDVKPH NVMIDHEHRK LRLIDWGLAE FYHPGQEYNV RVASRYFKGP 200
ELLVDYQMYD YSLDMWSLGC MLASMIFRKE PFFHGHDNYD QLVRIAKVLG 250
TEDLYDYIDK YNIELDPRFN DILGRHSRKR WERFVHSENQ HLVSPEALDF 300
LDKLLRYDHQ SRLTAREAME HPYFYTVVKD QARMGSSSMP GGSTPVSSAN 350
MMSGISSVPT PSPLGPLAGS PVIAAANPLG MPVPAAAGAQ Q
```

α′ subunit

```
MPGPAAGSRA RVYAEVNSLR SREYWDYEAH VPSWGNQDDY QLVRKLGRGK  50
YSEVFEAINI TNNERVVVKI LKPVKKKKIK REVKILENLR GGTNIIKLID 100
TVKDPVSKTP ALVFEYINNT DFKQLYQILT DFDIRFYMYE LLKALDYCHS 150
KGIMHRDVKP HNVMIDHQQK KLRLIDWGLA EFYHPAQEYN VRVASRYFKG 200
PELLVDYQMY DYSLDMWSLG CMLASMIFRR EPFFHGQDNY DQLVRIAKVL 250
GTEELYGYLK KYHIDLDPHF NDILGQHSRK RWENFIHSEN RHLVSPEALD 300
LLDKLLRYDH QQRLTAKEAM EHPYFYPVVK EQSQPCADNA VLSSGLTAAR 350
```

β subunit

```
MSSSEEVSWI SWFCGLRGNE FFCEVDEDYI QDKFNLTGLN EQVPHYRQAL  50
DMILDLEPDE ELEDNPNQSD LIEQAAEMLY GLIHARYILT NRGIAQMLEK 100
YQQGDFGYCP RVYCENQPML PIGLSDIPGE AMVKLYCPKC MDVYTPKSSR 150
HHHTDGAYFG TGFPHMLFMV HPEYRPKRPA NQFVPRLYGF KIHAMAYQLQ 200
LQAASNFKSP VKTIR
```

Sites of interest: S3, putative phosphorylation

Homologues in other species

The full coding region of the α subunit of human (HepG2 cells) and nearly complete coding region of rat (adipose tissue) has been sequenced. The coding region of the β subunit from human (HepG2 and HeLa cells) and mouse (brain) has been sequenced. The deduced amino acid sequence is about 99% conserved between mammalian species. Homologues have also been sequenced from *D. melanogaster* (**DmCK2**), *C. elegans*, *Theileria parva* (a protozoan parasite), *D. discoideum*, *S. cerevisiae* (**ScCK2**) and maize (**ZmCK2**).

Physiological substrates and specificity determinants[4–6]

The general consensus sequence phosphorylated by CK2 is SXX<u>E</u>/<u>D</u>, although SX<u>E</u>/<u>D</u>, S<u>D</u>, and variations of these sequences are also phosphorylated. Threonine can also be utilized, but the affinity is less. The following represents a partial list of sites that have been identified by both *in vitro* and *in vivo* phosphorylation. A corresponding change in substrate function has been demonstrated in most cases:

PU.1 :	$\left\{\begin{array}{l}\text{SGDE}\\\text{SGDE}\end{array}\right.$
DARPP-32:	SEEE
c-ErbA:	SEPE
hsp90a:	SDD
c-myb:	SSDD/ED/E
SRF:	SGEE
B23:	SEDE

Assay

CK2 is assayed[7] by transfer of radioactivity from $[\gamma\text{-}^{32}\text{P}]$ATP or GTP to the synthetic peptide RRREESEEE or RRREEETEEE.

Enzyme activators and inhibitors

Enzyme activity is increased by polycations and decreased by polyanions.

ACTIVATORS	K_a	INHIBITORS	K_i
Polyamines	280 μM	Heparin	1.4 nM
		poly(Glu4:Tyr1)	10 nM
		3,5,6-Dichloro-1-(β-D ribofuranosyl)-benzimidazole	23 μM
		2,3-phosphoglyceric acid	1 mM

Pattern of expression

Although expressed in all tissues, higher concentrations are found in brain, spleen, embryonic tissue and transformed cells. CK2 is found in cytosol, nuclei and nucleolus.

References

[1] Meisner, H. et al. (1989) Biochemistry 28, 4072–4076.
[2] Lozeman, F.J. et al. (1990) Biochemistry 29, 8436–8447.
[3] Heller-Harrison, R.A. et al. (1989) Biochemistry 28, 9053–9058.
[4] Pinna, L. (1990) Biochim. Biophys. Acta 1054, 267–284.
[5] Tuazon, P. and Traugh, J. (1991) Adv. Second Messenger and Phosphoprotein Res. 23, 123–162.
[6] Meisner, H., and Czech, M.P. (1991) Curr. Opin. Cell Biol. 3, 474–483.
[7] Sommercorn, J. and Krebs, E. (1987) J. Biol. Chem. 262, 3839–3843.

Claiborne V.C. Glover (University of Georgia, USA)

Casein kinase II (CK2) is a Ser/Thr protein kinase of broad substrate specificity which is widely distributed among eukaryotic organisms[1]. Both the structure and the function of the enzyme are highly conserved. The physiological role of CK2 is poorly understood, but the enzyme is essential for viability in *S. cerevisiae*[2]. DmCK2 behaves similarly to its vertebrate counterpart *in vitro*[3] and is able to substitute for the yeast enzyme *in vivo*[2].

Subunit structure and isoforms

SUBUNIT	AMINO ACIDS	MOL. WT	SDS-PAGE
α	336	39 833	37 kDa
β	215	24 700	28 kDa
Stellate	172	19 507	

DmCK2 consists of a catalytic α subunit and a regulatory β subunit[3, 4]. An α′ isoform similar to that found in vertebrates or *S. cerevisiae* has not been identified. The product of the Dm *Stellate* locus is 38% identical to DmCK2β[5] and may or may not represent a β subunit isoform. The native form of the enzyme is an $\alpha_2\beta_2$ tetramer. Although expressed DmCK2 α subunit exists as a catalytically active monomer *in vitro*[6], free catalytic subunit has not been observed *in vivo*. The native tetrameric form polymerizes into linear filaments under physiological conditions *in vitro*[7].

Genetics
Cytological location:

CK2α:	80A
CK2β:	10E1-2
Stellate:	12F1-2

Domain structure

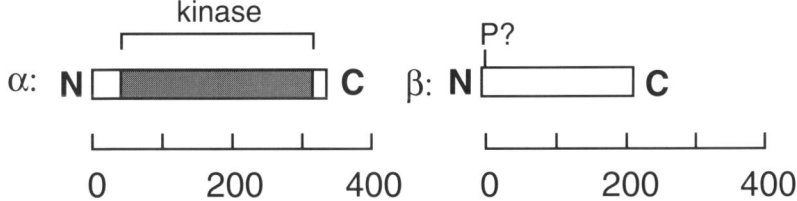

The α subunit consists of a kinase domain flanked by short N- and C-terminal extensions. The N-terminal extension contains two motifs, ARVY and EYWDYE, which are highly conserved among CK2 catalytic subunits. The β subunit contains an N-terminal CK2 recognition site, SSSEE, which probably represents the autophosphorylation site. A variant of this site is present in Stellate. Both β and Stellate contain a possible metal binding motif, CPXXC–22 aa–CPXC. Whether either protein binds a metal ion is unknown. The β subunit but not Stellate contains an acidic region which could conceivably serve as a pseudosubstrate site and/or as a site for interaction with polybasic activators.

Database accession numbers

	PIR	SWISSPROT	EMBL/GENBANK	REF
DmCK2α	A26688	P08181	M16534	4
DmCK2β		P08182	M16535	4
Stellate	S08120	P15021	X15899	5

Amino acid sequences of *D. melanogaster* CK2

α subunit

```
MTLPSAARVY TDVNAHKPDE YWDYENYVVD WGNQDDYQLV RKLGRGKYSE  50
VFEAINITTT EKCVVKILKP VKKKKIKREI KILENLRGGT NIITLLAVVK 100
DPVSRTPALI FEHVNNTDFK QLYQTLTDYE IRYYLFELLK ALDYCHSMGI 150
MHRDVKPHNV MIDHENRKLR LIDWGLAEFY HPGQEYNVRV ASRYFKGPEL 200
LVDYQMYDYS LDMWSLGCML ASMIFRKEPF FHGHDNYDQL VRIAKVLGTE 250
ELYAYLDKYN IDLDPRFHDI LQRHSRKRWE RFVHSDNQHL VSPEALDFLD 300
KLLRYDHVDR LTAREAMAHP YFLPIVNGQM NPNNQQ
```

β subunit

```
MSSSEEVSWV TWFCGLRGNE FFCEVDEDYI QDKFNLTGLN EQVPNYRQAL  50
DMILDLEPED ELEDNPLQSD MTEQAAEMLY GLIHARYILT NRGIAQMIEK 100
YQTGDFGHCP RVYCESQPML PLGLSDIPGE AMVKTYCPKC IDVYTPKSSR 150
HHHTDGAYFG TGFPHMLFMV HPEYRPKRPT NQFVPRLYGF KIHSLAYQIQ 200
LQAAANFKMP LRAKN
```

Sites of interest: S2–E6, putative autophosphorylation site; C109–C114, C137–C140, putative metal-binding motif

Stellate gene product

```
MSSSQNNNSS WIDWFLGIKG NQFLCRVPTD YVQDTFNQMG LEYFSEILDV  50
ILKPVIDSSS GLLYGDEKKW YGMIHARYIR SERGLIAMHR KYLRGDFGSC 100
PNISCDRQNT LPVGLSAVWG KSTVKIHCPR CKSNFHPKSD TQLDGAMFGP 150
SFPDIFFSLL PNLTSPLDDP RT
```

Sites of interest: C100–C105, C128–C131, putative metal-binding motif

Homologues in other species

Casein kinase II appears likely to exist in all eukaryotic organisms. Complete sequences of the catalytic subunit (α and/or α') are available for human (**CK2**), bovine, chicken, *X. laevis*, *C. elegans*, *D. discoideum*, *S. cerevisiae* (**ScCK2**), *T. parva*, and maize (**ZmCK2**). Complete sequences of the β subunit are available for human, mouse, chicken and *C. elegans* CK2.

Physiological substrates and specificity determinants

The substrate specificity of DmCK2 *in vitro* is essentially indistinguishable from that of bovine CK2. The consensus sequence for phosphorylation is thus presumably the same, i.e. a serine or threonine followed by an acidic cluster (underlined). Well characterized substrates in *D. melanogaster* include topoisomerase II and chromosomal protein D1 (a homologue of mammalian HMG I)[8]. Sites of phosphorylation on these proteins have not been mapped, but both contain several canonical sites:

Dm topoisomerase II: E S̱DDDDIEIDEDDDDD
                        F S̱DEEED
                        A S̱GDEVDEFD
Dm chromosomal protein D1:   D S̱ENEDDQDEDDD
                              E S̱NGDGE
                              N S̱DGENDAND

Assay
DmCK2 is routinely assayed using $[\gamma\text{-}^{32}\text{P}]$ATP as phosphoryl donor and partially hydrolysed, partially dephosphorylated casein as substrate[3]. Heparin-sensitive casein kinase activity can be used as a measure of CK2 activity in extracts, but it is important to realize that the heparin-sensitivity of CK2 can be overcome by basic compounds such as histones. The DmCK2 holoenzyme can be assayed specifically with the synthetic peptide RRREEETEEE, but free DmCk2 α exhibits a very low V_{max} with this substrate[9]. A solid phase assay for DmCK2 has also been devised[10].

Enzyme activators and inhibitors
Like its vertebrate counterpart, DmCK2 is inhibited by polyanions such as heparin and activated by large polycations such as protamine and polylysine. The enzyme is also moderately stimulated by polyamines, particularly spermine. Heparin appears to act directly on the Dm α subunit, but the activation exerted by polylysine and protamine is largely mediated via the β subunit. The regulatory role of the β subunit is complex and substrate-dependent. It stimulates the activity of the α subunit against most peptide and protein substrates tested, but it strongly inhibits phosphorylation of calmodulin[9].

Pattern of expression
Neither the subcellular localization nor the tissue- and developmental stage-specific expression of DmCK2 has been described. Interestingly, the Stellate protein is specifically expressed in testis[5].

References
[1] Pinna, L.A. (1990) Biochim. Biophys. Acta 1054, 267–284.
[2] Padmanabha, R. et al. (1990) Mol. Cell. Biol. 10, 4089–4099.
[3] Glover, C.V.C. et al. (1983) J. Biol. Chem. 258, 3258–3265.
[4] Saxena, A. et al. (1987) Mol. Cell. Biol. 7, 3409–3417.
[5] Livak, K.J. (1989) Genetics 124, 303–316.
[6] Bidwai, A.P. et al. (1992) J. Biol. Chem. 267, 18790–18796.
[7] Glover, C.V.C. (1986) J. Biol. Chem. 261, 14349–14354.
[8] Ashley, C.T. et al. (1989) J. Biol. Chem. 264, 8394–8401.
[9] Bidwai, A.P et al. (1993) Arch. Biochem. Biophys. 300, 265–270.
[10] Glover, C.V.C. and Allis, C. D. (1991) Meth. Enzymol. 200, 85–90.

Claiborne V.C. Glover (University of Georgia, USA)

Casein kinase II (CK2) is a Ser/Thr protein kinase of broad substrate specificity which is widely distributed among eukaryotic organisms[1]. The structure and properties of CK2 are highly conserved, but the physiological role of the enzyme is poorly understood. ScCK2 has been characterized both biochemically and genetically[2-4]. CK2 is essential for viability in *S. cerevisiae*, and the catalytic subunit of *D. melanogaster* CK2 (**DmCK2**) can functionally substitute for the yeast catalytic subunits *in vivo*[4].

Subunit structure and isoforms

SUBUNIT	AMINO ACIDS	MOL. WT	SDS-PAGE
α (Cka1)	372	44 673	42 kDa
α′ (Cka2)	339	39 408	35 kDa
β (Ckb1)			41 kDa
β′ (Ckb2)			32 kDa

ScCK2 is composed of two catalytic subunits, α and α′, and two regulatory subunits, β and β′[2]. The α and α′ subunits are encoded by the *CKA1* and *CKA2* genes and the β and β′ subunits by the *CKB1* and *CKB2* genes, respectively[3, 4] (Bidwai, Reed and Glover, unpublished). Disruption of either *CKA1* or *CKA2* yields no overt phenotype, but simultaneous disruption of both is lethal[4]. The molecular mass of native ScCK2 is 150 000 Da, consistent with a tetramer consisting of two catalytic and two regulatory subunits. Whether every tetramer has an αα′ββ′ composition is not known. Free catalytic subunit has not been observed *in vivo*.

Genetics

Chromosomal location:
    CKA1:   IX
    CKA2:   XV

Domain structure

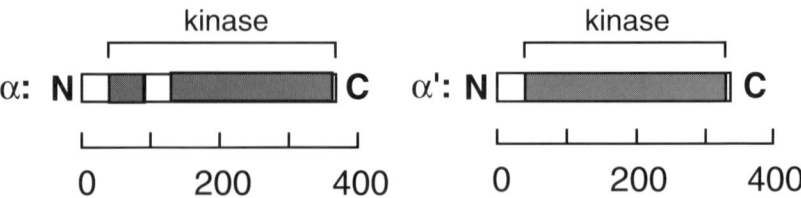

Both catalytic subunits consist of a kinase domain flanked by short N- and C-terminal extensions. The N-terminal extension contains two motifs, ARVY and EYWDYE, which are highly conserved among CK2 catalytic subunits. The ScCK2 α subunit contains a 38 amino acid insertion between subdomains III and IV of the catalytic domain. An insertion at this position is not found in other CK2 catalytic subunits sequenced to date.

Database accession numbers

	PIR	*SWISSPROT*	*EMBL/GENBANK*	*REF*
ScCK2α	A31564	P15790	M22473	3
ScCK2α′	S11192	P19454	M33759	4

Amino acid sequences of *S. cerevisiae* CK2

α subunit

```
MKCRVWSEAR VYTNINKQRT EEYWDYENTV IDWSTNTKDY EIENKVGRGK  50
YSEVFQGVKL DSKVKIVIKM LKPVKKKKIK REIKILTDLS NEKVPPTTLP 100
FQKDQYYTNQ KEDVLKFIRP YIFDQPHNGH ANIIHLFDII KDPISKTPAL 150
VFEYVDNVDF RILYPKLTDL EIRFYMFELL KALDYCHSMG IMHRDVKPHN 200
VMIDHKNKKL RLIDWGLAEF YHVNMEYNVR VASRFFKGPE LLVDYRMYDY 250
SLDLWSFGTM LASMIFKREP FFHGTSNTDQ LVKIVKVLGT SDFEKYLLKY 300
EITLPREFYD MDQYIRKPWH RFINDGNKHL SGNDEIIDLI DNLLRYDHQE 350
RLTAKEAMGH PWFAPIREQI EK
```

Sites of interest: N91–N128, insert in α subunit

α′ subunit

```
MPLPPSTLNQ KSNRVYSVAR VYKNACEERP QEYWDYEQGV TIDWGKISNY  50
EIINKIGRGK YSEVFSGRCI VNNQKCVIKV LKPVKMKKIY RELKILTNLT 100
GGPNVVGLYD IVQDADSKIP ALIFEEIKNV DFRTLYPTFK LPDIQYYFTQ 150
LLIALDYCHS MGIMHRDVKP QNVMIDPTER KLRLIDWGLA EFYHPGVDYN 200
VRVASRYHKG PELLVNLNQY DYSLDLWSVG CMLAAIVFKK EPFFKGSSNP 250
DQLVKIATVL GTKELLGYLG KYGLHLPSEY DNIMRDFTKK SWTHLLLTSET 300
KLAVPEVVDL IDNLLRYDHQ ERLTAKEAMD HKFFKTKFE
```

Homologues in other species

See **CK2** and **DmCK2** entries.

Physiological substrates and specificity determinants

Given that ScCK2 can be functionally replaced by DmCK2 *in vivo*, the consensus sequence for phosphorylation is presumably identical to that of higher eukaryotic CK2, i.e. a serine or threonine followed by an acidic cluster (underlined). The best characterized substrate in *S. cerevisiae* is topoisomerase II. Sites of *in vivo* phosphorylation on Sc topoisomerase II correspond well with sites modified by ScCK2 *in vitro*, and phosphorylation of Sc topoisomerase II *in vivo* is blocked at the non-permissive temperature in a strain temperature-sensitive for CK2 activity[5]. Sites exhibiting a good match to the consensus sequence include the following:

Sc topoisomerase II:
```
VSFNEED
DSFIEDDEEE
ESDLEILD
```

Assay

ScCK2 is routinely assayed using [γ-^{32}P]ATP as phosphoryl donor and partially hydrolysed, partially dephosphorylated casein as substrate[2]. Heparin-sensitive casein kinase activity can be taken as a measure of CK2 activity in extracts, but it is important to realize that the heparin-sensitivity of CK2 can be overcome by basic compounds such as histones. ScCK2 holoenzyme can also be assayed specifically with the synthetic peptide RRREEETEEE.

Enzyme activators and inhibitors

Like its vertebrate counterpart, ScCK2 is moderately stimulated by polyamines, particularly spermine, and strongly activated by large polycations such as protamine and polylysine. The enzyme is inhibited by a variety of polyanions, particularly heparin.

References

1 Pinna, L.A. (1990) Biochim. Biophys. Acta 1054, 267–284.
2 Padmanabha, R. and Glover, C.V.C. (1987) J. Biol. Chem. 262, 1829–1835.
3 Chen-Wu, J.L.-P. et al. (1988) Mol. Cell. Biol. 8, 4981–4990.
4 Padmanabha, R. et al. (1990) Mol. Cell. Biol. 10, 4089–4099.
5 Cardenas, M.E. et al. (1992) EMBO J. 11, 1785–1796.

ZmCK2	Casein kinase II (*Zea mays*) (mCK-2α)

Olaf-Georg Issinger (Universität des Saarlandes, Germany)

Cloning of a casein kinase II from a maize cDNA library (*mCK-2α*) and subsequent expression led to the isolation of a protein which shows 75% amino acid sequence identity with the human CK2α[1]. The expressed recombinant protein shares most biochemical properties so far investigated with the recombinant human counterpart[2]. From biochemical data the cloned *mCK-2α* resembles CK2A, a casein kinase from maize seedlings whose sequence has not yet been elucidated[3]. Interestingly, as yet no β subunit has been detected in maize, although the recombinant mCK-2α can reconstitute by self-assembly *in vitro* with human CK2β, to form a functional hybrid holoenzyme[2].

Subunit structure and isoforms

SUBUNIT	AMINO ACIDS	MOL. WT	SDS-PAGE
mCK-2α	332	39 228	36 kDa

Domain structure

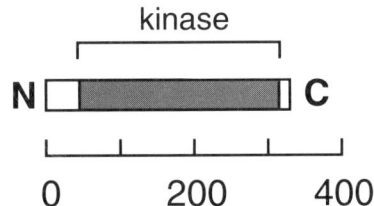

Database accession numbers

	PIR	SWISSPROT	EMBL/GENBANK	REF
mCK-2α	S16387	P28523	X61387	*1*
	S19726			

Amino acid sequence of *Z. mays* CK2

```
MSKARVYADV NVLRPKEYWD YEALTVQWGE QDDYEVVRKV GRGKYSEVFE  50
GINVNNNEKC IIKILKPVKK KKIKREIKIL QNLCGGPNIV KLLDIVRDQH 100
SKTPSLIFEY VNNTDFKVLY PTLTDYDIRY YIYELLKALD YCHSQGIMHR 150
DVKPHNVMID HELRKLRLID WGLAEFYHPG KEYNVRVASR YFKGPELLVD 200
LQDYDYSLDM WSLGCMFAGM IFRKEPFFYG HDNHDQLVKI AKVLGTDGLN 250
VYLNKYRIEL DPQLEALVGR HSRKPWLKFM NADNQHLVSP EAIDFLDKLL 300
YYDHQERLTA LEAMTHPYFQ QVRAAENSRT RA
```

Homologues in other species

CK2α has also been cloned and sequenced from *Arabidopsis thaliana*[4]. The deduced amino acid sequences of the two *A. thaliana* clones show ≈90% amino acid identity with the recombinant maize enzyme. Taking into account that divergence of monocots and dicots occurred 200 million years ago and that the *Cruciferae* (formerly *Brassicaceae*) are a relatively young family (appearance 5–10 million years ago) the observed differences, which are found mostly towards the C-terminus, are not surprising.

Physiological substrates and specificity determinants

The consensus sequence for phosphorylation by ZmCK2 is **S/TXX<u>D</u>/<u>E</u>**, by analogy to mammalian CK2[5]. So far only one plant substrate (RAB-17) has been shown to become phosphorylated by CK2 *in vivo* and *in vitro*. RAB-17 is rapidly induced in young embryos by abcisic acid[6]. The phosphorylation sites reside in the central region (residues 56–89)[7]. The potential seryl phosphates are shown in bold face:

RAB-17 phosphopeptide: HKTGGILHRSGSSSS**SSS**<u>EDD</u>GM

Assay

The enzyme is usually assayed as a transfer of radioactivity from [γ-^{32}P]ATP or [γ-^{32}P]GTP to dephosphorylated casein or phosvitin. In crude extracts the use of casein as a substrate is ambiguous because the assay will also detect casein kinase I. Therefore, it is preferable to use the peptide RRRDDDSDDD[8]. The assay is performed as described previously[2] in the absence of monovalent salt. Monovalent salt concentrations as they are used for the determination of CK2 holoenzyme (200 mM) inhibit CK2α, both from man and maize[9].

Enzyme inhibitors

	SPECIFICITY	IC_{50}
Heparin	Poor	28 ng/ml

Pattern of expression

The distribution of ZmCK2 has not been determined, but its substrate, RAB-17, is found in embryos of *Z. mays* during embryogenesis and during maturation; and in leaves under conditions of water deficit[10].

References

1 Dobrowolska, G. et al. (1991) Biochim. Biophys. Acta 1129, 139–140.
2 Boldyreff, B. et al. (1993) Biochim. Biophys. Acta 1173, 32–38.
3 Dobrowolska, G. et al. (1987) Biochim. Biophys. Acta 931, 188–195.
4 Miziguchi, T. et al. (1993) Plant Mol. Biol. 21, 279–289.
5 Pinna, L.A. (1990) Biochim. Biophys. Acta 1054, 267–284.
6 Davies, M.J. and Mansfield, T.A. (1983) In: Abcisic Acid, Addicott, F.T., ed., Praeger, New York, pp. 237–268.

[7] Plana, M. et al. (1991) J. Biol. Chem. 266, 22510–22514.
[8] Kuenzel, E. et al. (1987) J. Biol. Chem. 262, 9136–9140.
[9] Grankowski, N. et al. (1991) Eur. J. Biochem. 198, 25–30.
[10] Plana, M. et al. (1989) Plant Mol. Biol. 13, 385–394.

Cdc28-like kinase (vertebrates)
(Serine/threonine/tyrosine kinase, Sty)

Karen Colwill, Vladimir Lhotak and Tony Pawson
(Mount Sinai Hospital, Canada)

Clk was originally isolated in an anti-P-Tyr antibody expression screen for cDNAs encoding tyrosine kinases[1,2]. Although the function of Clk is unknown, its widespread expression in mouse tissues implies a function basic to many cells. A *D. melanogaster* homologue of Clk, Doa, is an essential gene product since *D. melanogaster* homozygous for loss-of-function *doa* mutations die during embryogenesis[3]. When expressed in bacteria, Clk auto-phosphorylates on serine, threonine and tyrosine, suggesting that it is a dual specificity kinase.

Subunit structure and isoforms

SUBUNIT	AMINO ACIDS	MOL. WT	SDS-PAGE
Clk	483	57 093	

Alternate splicing results in two transcripts: one encodes a full-length protein, the other encodes a truncated polypeptide lacking the full kinase domain, suggesting that there may be two different isoforms (John Bell, personal communication).

Domain structure

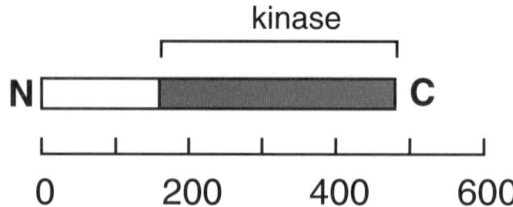

The N-terminal region is likely to play a regulatory role. There is a putative nuclear localization sequence (KRKKR) at residues 29–33. The sites of auto-phosphorylation have not been identified.

Database accession numbers

	PIR	SWISSPROT	EMBL/GENBANK	REF
Clk/Sty	S13364		X57186	1,2
	A39676		M38381	

Amino acid sequence of mouse Clk

```
MRHSKRTYCP DWDERDWDYG TWRSSSSHKR KKRSHSSARE QKRCRYDHSK  50
TTDSYYLESR SINEKAYHSR RYVDEYRNDY MGYEPGHPYG EPGSRYQMHS 100
SKSSGRSGRS SYKSKHRSRH HTSQHHSHGK SHRRKRSRSV EDDEEGHLIC 150
QSGDVLSARY EIVDTLGEGA FGKVVECIDH KVGGRRVAVK IVKNVDRYCE 200
AAQSEIQVLE HLNTTDPHST FRCVQMLEWF EHRGHICIVF ELLGLSTYDF 250
IKENSFLPFR MDHIRKMAYQ ICKSVNFLHS NKLTHTDLKP ENILFVKSDY 300
```

Amino acid sequence of mouse Clk *continued*

```
TEAYNPKMKR DERTIVNPDI KVVDFGSATY DDEHHSTLVS TRHYRAPEVI 350
LALGWSQPCD VWSIGCILIE YYLGFTVFPT HDSREHLAMM ERILGPLPKH 400
MIQKTRKRRY FHHDRLDWDE HSSAGRYVSR RCKPLKEFML SQDAEHELLF 450
DLIGKMLEYD PAKRITLKEA LKHPFFYPLK KHT
```

Homologues in other species
Clk/Sty has been sequenced from human T cells[4]. Another family member, PskG1, has also been cloned from humans[5]. Other family members have been identified in *D. melanogaster* (Doa[3]) *C. elegans* (AFC1, AFC2, AFC3)[6] and *S. cerevisiae* (Kns1[7]).

Assay
After bacterial expression, Clk has been assayed via autophosphorylation, and by phosphorylation of the exogenous substrate poly(Glu,Tyr).

Pattern of expression
The location of Clk has been shown to be nuclear after expression in monkey *cos* cells and rat fibroblasts (John Bell, personal communication). By Northern analysis, Clk is expressed in all mammalian cells and tissues tested. There are three different transcripts: the 3.2 and 6.6 kb transcripts are predominant in normal mouse tissues, whereas in malignant lymphoid and erythroid cell lines the 1.8 kb transcript is more highly expressed.

References
1   Ben-David, Y. et al. (1991) EMBO J. 10, 317–325.
2   Howell, B.W. et al. (1991) Mol. Cell. Biol. 11, 568–572
3   Rabinow, L. and Birchler, J.A. (1989) EMBO J., 8, 879–889.
4   Johnson, K.W. and Smith, K.A. (1991) J. Biol. Chem. 266, 3402–3407.
5   Hanks, S.K. and Quinn, A.M. (1991) Methods Enzymol. 200, 38–81.
6   Bender, J. and Fint, G.R. (1994) Proc. Natl Acad. Sci. USA 91, 12105–12109.
7   Padmanabha, R. et al. (1991) Mol. Gen. Genet. 229, 1–9.

 **Yak1** Yak1 protein kinase (*S. cerevisiae*)

S. Garrett (Duke University Medical Center, USA)

In the yeast *S. cerevisiae*, cyclic AMP-dependent protein kinase activity is essential for growth and progression through G_1. Although the molecular details of this requirement are not understood, it can be alleviated by the loss of another protein kinase, Yak1. Thus, a major role for cyclic AMP-dependent protein kinase is to antagonize the growth-restricting activity of a second protein kinase, Yak1.

Subunit structure and isoforms

SUBUNIT	AMINO ACIDS	MOL. WT	SDS-PAGE
Yak1	807	92 000	

Domain structure

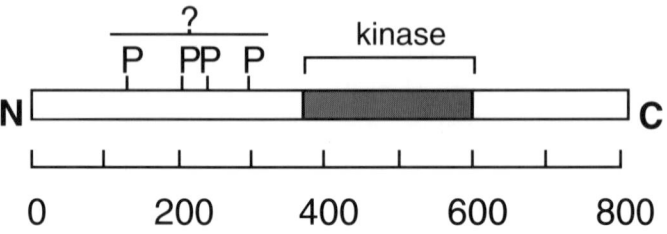

The Yak1 protein consists of a central kinase domain (375–600) with N-terminal and C-terminal domains of unknown function[1]. The N-terminal domain contains four consensus phosphorylation sites for cAMP-dependent protein kinase, and several long stretches of contiguous glutamines, but the relevance of either to Yak1 activity or function are unknown[2].

Database accession numbers

	PIR	SWISSPROT	EMBL/GENBANK	REF
Yak1			X16056	[1]

Amino acid sequence of *S. cerevisiae* Yak1

```
MNSSNNNDSS SSNSNMNNSL SPTLVTHSDA SMGSGRASPD NSHMGRGIWN  50
PSYVNQGSQR SPQQQHQNHH QQQQQQQQQQ QQNSQFCFVN PWNEEKVTNS 100
QQNLVYPPQY DDLNSNESLD AYRRRKSSLV VPPARAPAPN PFQYDSYPAY 150
TSSNTSLAGN SSGQYPSGYQ QQQQQVYQQG AIHPSQFGSR FVPSLYDRQD 200
FQRRQSLAAT NYSSNFSSLN SNTNQGTNSI PVMSPYRRLS AYPPSTSPPL 250
QPPFKQLRRD EVQGQKLSIP QMQLCNSKND LQPVLNATPK FRRASLNSKT 300
ISPLVSVTKS LITTYSLCSP EFTYQTSKNP KRVLTKPSEG KCNNGFDNIN 350
SDYILYVNDV LGVEQNRKYL VLDILGQGTF GQVVKCQNLL TKEILAVKVV 400
KSRTEYLTQS ITEAKILELL NQKIDPTNKH HFLRMYDSFV HKNHLCLVFE 450
LLSNNLYELL KQNKFHGLSI QLIRTFTTQI LDSLCVLKES KLIHCDLKPE 500
```

Amino acid sequence of *S. cerevisiae* Yak1   *continued*

```
NILLCAPDKP ELKIIDFGSS CEEARTVYTY IQSRFYRAPE IILGIPYSTS 550
IDMWSLGCIV AELFLGIPIF PGASEYNQLT RIIDTLGYPP SWMIDMGKNS 600
GKFMKKLAPE ESSSSTQKHR MKTIEEFCRE YNIVEKPSKQ YFKWRKLPDI 650
IRNYRYPKSI QNSQELIDQE MQNRECLIHF LGGVLNLNPL ERWTPQQAML 700
HPFITKQEFT GEWFPPGSSL PGPSEKHDDA KGQQSEYGSA NDSSNNAGHN 750
YVYNPSSATG GADSVDIGAI SKRKENTSGD ISNNFAVTHS VQEGPTSAFN 800
KLHIVEE
```

Sites of interest: S127, S206, S240, S295, putative phosphorylation

Assay

The enzyme is assayed as a transfer of radioactivity from $[\gamma\text{-}^{32}P]ATP$ to a casein hydrolysate. The enzyme will not phosphorylate histone H1 or enolase, but does autophosphorylate[2].

Pattern of expression

By immunofluorescence, the enzyme appears to be found only in the cytoplasm (although it can be visualized only on overproduction).

References

[1]   Garrett, S. and Broach, J.R. (1989. Genes Devel. 3, 1336–1348.
[2]   Garrett, S. et al. (1991) Mol. Cell. Biol. 11, 4045–4052.

Mak
Male germ cell-associated PK (vertebrates)

Masabumi Shibuya (University of Tokyo, Japan)

The *MAK* gene is specifically expressed in testicular germ cells at and after meiosis. Mak protein kinase has significant homology with the Cdc2/Cdc28 protein kinase family including MAP kinase (**Erk1/2**)[1]. Mak exists as a complex with cellular proteins[2]. Based on these observations, Mak is suggested to have an important role in the process of meiosis in the testis.

Subunit structure and isoforms

SUBUNIT	AMINO ACIDS	MOL. WT	SDS-PAGE
p66	622	69 925	66 kDa
p60	581	65 419	60 kDa

Mak exists as two forms, with p60 being due to a deletion of 41 amino acids from the C-terminal region.

Domain structure

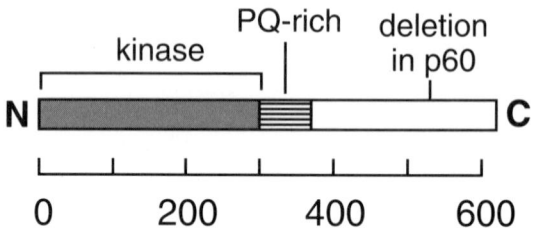

The kinase domain has significant similarity with the **Cdc2** family, **Sme1** and MAP kinase (**Erk1/2**) (35–40% identity). The PQ-rich region is 50% P/Q. The C-terminal region shows no significant similarity with other known proteins.

Database accession numbers

	PIR	SWISSPROT	EMBL/GENBANK	REF
Mak			M35862, M34569	*1*

Amino acid sequence of rat p66 Mak

```
MNRYTTMRQL GDGTYGSVLM GKSNESGELV AIKRMKRKFY SWDECMNLRE  50
VKSLKKLNHA NVIKLKEVIR ENDHLYFIFE YMKENLYQLM KDRNKLFPES 100
VIRNIMYQIL QGLAFIHKHG FFHRDMKPEN LLCMGPELVK IADFGLAREL 150
RSQPPYTDYV STRWYRAPEV LLRSSVYSSP IDVWAVGSIM AELYTFRPLF 200
PGTSEVDEIF KICQVLGTPK KSDWPEGYQL ASSMNFRFPQ CIPINLKTLI 250
PNASSEAIQL MTEMLNWDPK KRPTASQALK HPYFQVGQVL GPSAHHLDAK 300
QTLHKQLQPP EPKPSSSERD PKPLPNILDQ PAGQPQPKQG HQPLQAIQPP 350
QNTVVQPPPK QQGHHKQPQT MFPSIVKTIP TNPVSTVGHK GARRRWGQTV 400
FKSGDSCDNI EDCDLGASHS KKPSMDAFKE KKKKESPFRF PEAGLPVSNH 450
LKGENRNLHA SLKSDTNLST ASTAKQYYLK QSRYLPGVNP KNVSLVAGGK 500
DINSHSWNNQ LFPKSLGSMG ADLAFKRSNA AGNLGSYSAY SQTGCVPSFL 550
KKEVGSAGQR IHLAPLGASA ADYTWSTKTG RGQFSGRTYN PTAKNLNIVN 600
RTQPVPSVHG RTDWVAKYGG HR
```

Sites of interest: A531–A571, deleted in p60

Physiological substrates and specificity determinants

A 210 kDa cellular protein is tightly associated with the Mak kinase, and in immunocomplex assays using anti-Mak antibody, the 210 kDa protein is efficiently phosphorylated on serine and threonine residues[2]. The 210 kDa protein is also phosphorylated on serine and threonine *in vivo*. The Mak kinase complex elutes during gel filtration at an apparent molecular mass of 500–700 kDa, with no activity corresponding to the monomer at 60–70 kDa.

Assay

Mak may be assayed in immune complexes or after SDS-PAGE by phosphorylation of p210 or myelin basic protein[2]. A reaction mixture containing 10 mM Mn^{2+} ion gives a higher activity than 10 mM Mg^{2+}.

Pattern of expression

MAK mRNA consists of two forms, 3.8 and 2.6 kb in length. Neither form is detectable before puberty in rats, suggesting that the *MAK* gene is not involved in somatic cell division of testicular germ cells. The 3.8 kb mRNA is expressed a little earlier than the 2.6 kb type, and the expression of 2.6 kb mRNA continues even after meiotic division (at the stage of round spermatids). Testicular germ cell fractionation[1] and *in situ* hybridization studies[3] indicate that the *MAK* transcript is most abundant around the late pachytene stage of spermatogenesis. The expression of the Mak proteins (p66 and p60) and their kinase activities correlate with the 3.8 kb type mRNA[2]. Mak kinases are hardly detectable after meiosis, although some amount of the 2.6 kb mRNA is still present in round spermatids. Among different spermatogenesis-deficient mouse strains, the *qk/qk* strain expresses both the *MAK* transcripts, the T16H and T37H strains express only the 3.8 kb transcript, whereas the *Sxr* strain do not express any *MAK* RNA[1].

References
[1] Matsushime, H. et al. (1990) Mol. Cell. Biol. **10**, 2261–2268.
[2] Jinno, A. et al. (1993) Mol. Cell. Biol. 13, 4146–4156.
[3] Koji, T. et al. (1992) Cell Biochem. Funct. 10, 273–279.

Ched

Cholinesterase-related cell division controller (vertebrates)

(Information assembled by the Editors)

Ched DNA was cloned by chance when a human glioblastoma cDNA library was screened with butylcholinesterase-specific oligodeoxynucleotide probes[1]. A complete cDNA was found to encode a homologue of **Cdc2** and other members of the Cdc2 subfamily. Expression of Ched antisense oligonucleotides in differentiating murine bone marrow cells reduced the proportion of late, polynuclear, mature megakaryocytes, suggesting a possible role for Ched in haematopoiesis.

Subunit structure and isoforms

SUBUNIT	AMINO ACIDS	MOL. WT	SDS-PAGE
Ched	418	48 180	

Domain structure

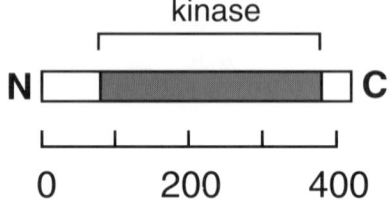

Ched has somewhat longer N- and C-terminal extensions than other members of the Cdc2 subfamily.

Database accession numbers

	PIR	SWISSPROT	EMBL/GENBANK	REF
Human			M80629	[1]

Amino acid sequence of human Ched

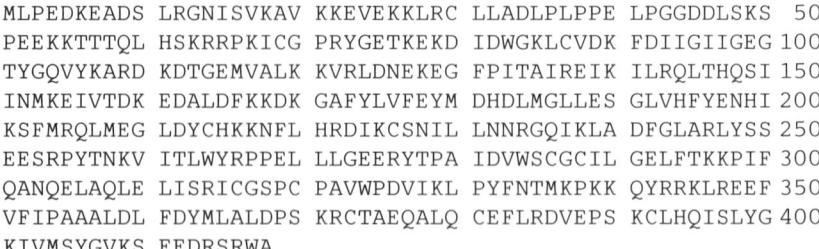

```
MLPEDKEADS LRGNISVKAV KKEVEKKLRC LLADLPLPPE LPGGDDLSKS  50
PEEKKTTTQL HSKRRPKICG PRYGETKEKD IDWGKLCVDK FDIIGIIGEG 100
TYGQVYKARD KDTGEMVALK KVRLDNEKEG FPITAIREIK ILRQLTHQSI 150
INMKEIVTDK EDALDFKKDK GAFYLVFEYM DHDLMGLLES GLVHFYENHI 200
KSFMRQLMEG LDYCHKKNFL HRDIKCSNIL LNNRGQIKLA DFGLARLYSS 250
EESRPYTNKV ITLWYRPPEL LLGEERYTPA IDVWSCGCIL GELFTKKPIF 300
QANQELAQLE LISRICGSPC PAVWPDVIKL PYFNTMKPKK QYRRKLREEF 350
VFIPAAALDL FDYMLALDPS KRCTAEQALQ CEFLRDVEPS KCLHQISLYG 400
KIVMSYGVKS EEDRSRWA
```

Homologues in other species

Within the kinase domain, Ched is 34–42% identical to **Cdc2**, **SpCdc2**, and **Cdc28**.

Pattern of expression
Ched mRNA was detected on Northern blots in human fetal brain, liver and muscle, and neuroblastoma and glioblastoma tumours.

Reference
[1] Lapidot-Lifson, Y. et al. (1992) Proc. Natl Acad. Sci. USA 89, 579–583.

PITSLRE	PITSLRE protein kinase (vertebrates) (p58[Gta], cdc2L1)

Vincent J. Kidd (St Jude Children's Research Hospital, Memphis, USA)

PITSLRE (p58[Gta]) was isolated as an associated component of mammalian β-1,4-galactosyltransferase preparations[1]. The predicted structure of this protein kinase indicated that it was most closely related to the cell division control protein kinase **Cdc2**. Minimal overexpression of p58[Gta] led to a delay in the CHO cell cycle at late telophase[1]. Additionally, transient over-expression of the cDNA in COS cells resulted in an increase in β-1,4-galacto-syltransferase enzyme activity[2]. At this time, however, the exact function of this protein kinase has not been established.

Subunit structure and isoforms

SUBUNIT	AMINO ACIDS	MOL. WT	SDS-PAGE
β1	439	49 000	58 kDa
β2-1	786	86 000	106 kDa
β2-2	775	85 000	106 kDa
α1	471	51 000	65 kDa
α2-1	777	85 000	106 kDa
α2-2	775	85 000	106 kDa
α2-3	766	84 000	105 kDa
α2-4	562	62 000	81 kDa
γ1			75 kDa
γ2			50 kDa

At present, ten different PITSLRE isoforms have been identified by cross-reactivity with p58 antibodies and by molecular cloning from human cell lines[3]. Five of these isoforms arise by alternative splicing from one gene, while the others appear to arise from highly duplicated sequences.

Genetics
Chromosome locations:
  Human: 1p36 (all isoforms)[3, 4]
  Mouse: 4q[5]

Domain structure

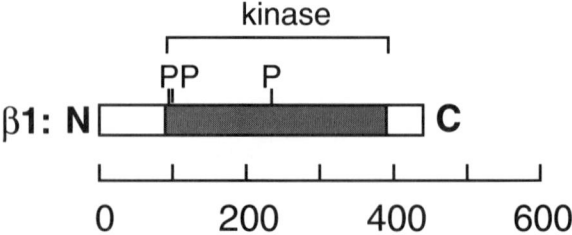

The central kinase domain of PITSLREβ1 is 42% identical to **Cdc2**, and contains residues equivalent to T14, Y15 and T161 which are involved in the regulation of Cdc2. Isoforms differ in the extent and nature of the N-terminal domains. The α and β isoforms also have four distinct amino acid differences in the C-terminal domain.

Database accession numbers

		PIR	SWISSPROT	EMBL/GENBANK	REF
p58[Gta]	Human	A38282	P21127	M37712	1, 6
β1	Human			U04819	3
β2-1	Human			U07704	3
β2-2	Human			U07705	3
α1	Human			U04815	3
α2-1	Human			U04824	3
α2-2	Human			U04816	3
α2-3	Human			U04817	3
α2-4	Human			U04818	3
α	Mouse		P24788	M58633	7

Amino acid sequences of human PITSLRE

β1

```
MSEDEERENE NHLLVVPESR FDRDSGESEE AEEEVGEGTP QSSALTEGDY  50
VPDSPALSPI ELKQELPKYL PALQGCRSVE EFQCLNRIEE GTYGVVYRAK 100
DKKTDEIVAL KRLKMEKEKE GFPITSLREI NTILKAQHPN IVTVREIVVG 150
SNMDKIYIVM NYVEHDLKSL METMKQPFLP GEVKTLMIQL LRGVKHLHDN 200
WILHRDLKTS NLLLSHAGIL KVGDFGLARE YGSPLKAYTP VVVTQWYRAP 250
ELLLGAKEYS TAVDMWSVGC IFGELLTQKP LFPGNSEIDQ INKVFKELGT 300
PSEKIWPGYS ELPAVKKMTF SEHPYNNLRK RFGALLSDQG FDLMNKFLTY 350
FPGRRISAED GLKHEYFRET PLPIDPSMFP TWPAKSEQQR VKRGTSPRPP 400
EGGLGYSQLG DDDLKETGFH LTTTNQGASA AGPGFSLKF
```

Sites of interest: T92, Y93, T239, putative phosphorylation

β2-1

```
MGDELKSWKV KTLDEILQEK KRRKEQEEKA EIKRLKNSDD RDSKRDSLEE  50
GELRDHTMEI TIRNSPYRRE DSMEDRGEED DSLAIKPPQQ MSRKEKTHHR 100
KDEKRAEKWR HRSHSAEGGK HARVKEREHE RRKRHREEQD KARREWERQK 150
RREMAREHSR RERGNDGVLL FRDRLEQLER KRERERKMRE QQKEQREQKE 200
RERRAEERRK EREARREVSA HHRTMREDYS DKVKASHWSR SPPRPPRERF 250
ELGDGRKPVK EEKMEERDLL SDLQDISDSE RKTSSAESSS AESGSGSEEE 300
EEEEEEEEEE GSTSEESEEE EEEEEEETGS NSEEASEQSA EEVSEEEMSE 350
DEERENENHF LVVPESRFDR DSGESEEAEE EVGEGTPQSS ALTEGDYVPD 400
SPALSPIELK QELPKYLPAL QGCRSVEEFQ CLNRIEEGTY GVVYRAKDKK 450
TDEIVALKRL KMEKEKEGFP ITSLREINTI LKAQHPNIVT VREIVVGSNM 500
DKIYIVMNYV EHDLKSLMET MKQPFLPGEV KTLMIQLLRG VKHLHDNWIL 550
HRDLKTSNLL LSHAGILKVG DFGLAREYGS PLKAYTPVVV TLWYRAPELL 600
LGAKEYSTAV DMWSVGCIFG ELLTQKPLFP GNSEIDQINK VFKELGTPSE 650
```

β2-1 *continued*

```
KIWPGYSELP AVKKMTFSEH PYNNLRKRFG ALLSDQGFDL MNKFLTYFPG 700
RRISAEDGLK HEYFRETPLP IDPSMFPTWP AKSEQQRVKR GTSPRPPEGG 750
LGYSQLGDDD LKETGFHLTT TNQGASAAGP GFSLKF
```

Sites of interest: K20–K24, R131–R136, putative nuclear localization signals;
T439, Y440, T586, putative phosphorylation

β2-2

```
MGDELKSWKV KTLDEILQEK KRRKEQEEKA EIKRLKNSDD RDSKRDSLEE  50
GELRDHTMEI TIRNSPYRRE DSMEDRGEED DSLAIKPPQQ MSRKEKTHHR 100
KDEKRAEKKH ARVKEREHER RKRHREEQDK ARREWERQKR REMAREHSRR 150
ERGNDGVCLF RDRLEQLERK RERERKMREQ QKEQREQKER ERRAEERRKE 200
REARREVSAH HRTMREDYSD KVKASHWSRS PPRPPRERFE LGDGRKPVKE 250
EKMEERDLLS DLQDISDSER KTSSAESSSA ESGSGSEEEE EEEEEEEEG 300
STSEESEEEE EEEEEETGSN SEEASEQSAE EVSEEEMSED EERENENHFL 350
VVPESRFDRD SGESEEAEEE VGEGTPQSSA LTEGDYVPDS PALSPIELKQ 400
ELPKYLPALQ GCRSVEEFQC LNRIEEGTYG VVYRAKDKKT DEIVALKRLK 450
MEKEKEGFPI TSLREINTIL KAQHPNIVTV REIVVGSNMD KIYIVMNYVE 500
HDLKSLMETM KQPFLPGEVK TLMIQLLRGV KHLHDNWILH RDLKTSNLLL 550
SHAGILKVGD FGLAREYGSP LKAYTPVVVT LWYRAPELLL GAKEYSTAVD 600
MWSVGCIFGE LLTQKPLFPG NSEIDQINKV FKELGTPSEK IWPGYSELPA 650
VKKMTFSEHP YNNLRKRFGA LLSDQGFDLM NKFLTYFPGR RISAEDGLKH 700
EYFRETPLPI DPSMFPTWPA KSEQQRVKRG TSPRPPEGGL GYSQLGDDDL 750
KETGFHLTTT NQGASAAGPG FSLKF
```

Sites of interest: K20–K24,   R120–R125,   putative   nuclear   localization
sequences; T428, Y429, T575, putative phosphorylation

α1

```
METGSNSEEA SEQSAEEVSE EEMSEDEERE NENHFLVVPE SRFDRDSGES  50
EEAEEEVGEG TPQSSALTEG DYVPDSPALS PIELKQELPK YLPAPQGCRS 100
VEEFQCLNRI EEGTYGVVYR AKDKKTDEIV ALKRLKMEKE KEGFPITSLR 150
EINTILKAQH PNIVTVREIV VGSNMDKIYI VMNYVEHDLK SLMETMKQPF 200
LPGEVKTLMI QLLRGVKHLH DNWILHRDLK TSNLLLSHAG ILKVGDFGLA 250
REYGSPLKAY TPVVVTLWYR APELLLGAKE YSTAVDMWSV GCIFGELLTQ 300
KPLFPGKSEI DQINKVFKDL GTPSEKIWPG YSELPAVKKM TGSEHPYNNL 350
RKRFGALLSD QGFDLMNKFL TYFPGRRISA EDGLKHEYFR ETPLPIDPSM 400
FPTWPAKSEQ QRVKRGTSPR PPEGGLGYSQ LGDDDLKETG FHLTTTNQGA 450
SAAGPGFSLK F
```

Sites of interest: S7, S14, T114, Y115, T261, putative phosphorylation

α2-1

```
MGDELKSWKV KTLDEILQEK KRRKEQEEKA EIKRLKNSDD RDSKRDSLEE  50
GELRDHTMEI TIRNSPYRRE DSMEDRGEED DSLAIKPPQQ MSRKEKTHHR 100
KDEKRAEKRR HRSHSAEGGK HARVKKKERE HERRKRHREE QDKARREWER 150
QKRREMAREH SRRERDRLEQ LERKRERERK MREQQKEQRE QKERERRAEE 200
RRKEREARRE VSAHHRTMRE DYSDKVKASH WSRSPPRPPR ERFELGDGRK 250
PVKEEKMEER DLLSDLQDIS DSERKTSSAE SSSAESGSGS EEEEEEEEEE 300
```

α2-1 *continued*

```
EEEGSTSEES EEEEEEEEEE TGSNSEEASE QSAEEVSEEE MSEDEERENE 350
NHFLVVPESR FDRDSGESEE AEEEVGEGTP QSSALTEGDY VPDSPALSPI 400
ELKQELPKYL PALQGCRSVE EFQCLNRIEE GTYGVVYRAK DKKTDEIVAL 450
KRLKMEKEKE GFPITSLREI NTILKAQHPN IVTVREIVVG SNMDKIYIVM 500
NYVEHDLKSL METMKQPFLP GEVKTLMIQL LRGVKHLHDN WILHRDLKTS 550
NLLLSHAGIL KVGDFGLARE YGSPLKAYTP VVVTLWYRAP ELLLGAKEYS 600
TAVDMWSVGC IFGELLTQKP LFPGKSEIDQ INKVFKDLGT PSEKIQPGYS 650
ELPAVKKMTG SEHPYNNLRK RFGALLSDQG FDLMNKFLTY FPGRRISAED 700
GLKHEYFRET PLPIDPSMFP TWPAKSEQQR VKRGTSPRPP EGGLGYSQLG 750
DDDLKETGFH LTTTNQGASA AGPGFSLKF
```

Sites of interest: K20–K24, R131–R136, putative nuclear localization signals;
T432, Y433, T579, putative phosphorylation

α2-2

```
MGDELKSWKV KTLDEILQEK KRRKEQEEKA EIKRLKNSDD RDSKRDSLEE  50
GELRDHTMEI TIRNSPYRRE DSMEDRGEED DSLAIKPPQQ MSRKEKTHHR 100
KDEKRAEKKH ARVKKKEREH ERRKRHREEQ DKARREWERQ KRREMAREHS 150
RRERGNDGVC LFRDRLEQLE RKRERERKMR EQQKEQREQK ERERRAEERR 200
KEREARREVS AHHRTMREDY SDKVKASHWS RSPPRPPRER FELGDGRKPV 250
KEEKMEERDL LSDLQDISDS ERKTSSAESS SAESGSGSEE EEEEEEEEEE 300
EGSTSEESEE EEEEEEEETG SNSEEASEQS AEEVSEEEMS EDEERENENH 350
FLVVPESRFD RDSGESEEAE EEVGEGTPQS SALTEGDYVP DSPALSPIEL 400
KQELPKYLPA LQGCRSVEEF QCLNRIEEGT YGVVYRAKDK KTDEIVALKR 450
LKMEKEKEGF PITSLREINT ILKAQHPNIV TVREIVVGSN MDKIYIVMNY 500
VEHDLKSLME TMKQPFLPGE VKTLMIQLLR GVKHLHDNWI LHRDLKTSNL 550
LLSHAGILKV GDFGLAREYG SPLKAYTPVV VTLWYRAPEL LLGAKEYSTA 600
VDMWSVGCIF GELLTQKPLF PGKSEIDQIN KVFKDLGTPS EKIQPGYSEL 650
PAVKKMTGSE HPYNNLRKRF GALLSDQGFD LMNKFLTYFP GRRISAEDGL 700
KHEYFRETPL PIDPSMFPTW PAKSEQQRVK RGTSPRPPEG GLGYSQLGDD 750
DLKETGFHLT TTNQGASAAG PGFSLKF
```

Sites of interest: K20–K24, R120–R125, putative nuclear localization
sequences; T430, Y431, T577, putative phosphorylation

α2-3

```
MGDELKSWKV KTLDEILQEK KRRKEQEEKA EIKRLKNSDD RDSKRDSLEE  50
GELRDHTMEI TIRNSPYRRE DSMEDRGEED DSLAIKPPQQ MSRKEKTHHR 100
KDEKRAEKKH ARVKKKEREH ERRKRHREEQ DKARREWERQ KRREMAREHS 150
RRERDRLEQL ERKRERERKM REQQKEQREQ KERERRAEER RKEREARREV 200
SAHHRTMRED YSDKVKASHW SRSPPRPPRE RFELGDGRKP VKEEKMEERD 250
LLSDLQDISD SERKTSSAES SSAESGSGSE EEEEEEEEEE EEGSTSEESE 300
EEEEEEEEET GSNSEEASEQ SAEEVSEEEM SEDEERENEN HFLVVPESRF 350
DRDSGESEEA EEEVGEGTPQ SSALTEGDYV PDSPALSPIE LKQELPKYLP 400
ALQGCRSVEE FQCLNRIEEG TYGVVYRAKD KKTDEIVALK RLKMEKEKEG 450
FPITSLREIN TILKAQHPNI VTVREIVVGS NMDKIYIVMN YVEHDLKSLM 500
ETMKQPFLPG EVKTLMIQLL RGVKHLHDNW ILHRDLKTSN LLLSHAGILK 550
VGDFGLAREY GSPLKAYTPV VVTLWYRAPE LLLGAKEYST AVDMWSVGCI 600
```

α2-3 *continued*

```
FGELLTQKPL FPGKSEIDQI NKVFKDLGTP SEKIQPGYSE LPAVKKMTGS 650
EHPYNNLRKR FGALLSDQGF DLMNKFLTYF PGRRISAEDG LKHEYFRETP 700
LPIDPSMFPT WPAKSEQQRV KRGTSPRPPE GGLGYSQLGD DDLKETGFHL 750
TTTNQGASAA GPGFSLKF
```

Sites of interest: K20–K24, R120–R125, putative nuclear localization sequences; T421, Y422, T568, putative phosphorylation

α2-4

```
MREDYSDKVK ASHWSRSPPR PPRERFELGD GRKPVKEEKM EERDLLSDLQ  50
DISDSERKTS SAESSSAESG SGSEEEEEEE EEEEEEGSTS EESEEEEEEE 100
EEETGSNSEE ASEQSAEEVS EEEMSEDEER ENENHFLVVP ESRFDRDSGE 150
SEEAEEEVGE GTPQSSALTE GDYVPDSPAL SPIELKQELP KYLPALQGCR 200
SVEEFQCLNR IEEGTYGVVY RAKDKKTDEI VALKRLKMEK EKEGFPITSL 250
REINTILKAQ HPNIVTVREI VVGSNMDKIY IVMNYVEHDL KSLMETMKQP 300
FLPGEVKTLM IQLLRGVKHL HDNWILHRDL KTSNLLLSHA GILKVGDFGL 350
AREYGSPLKA YTPVVVTLWY RAPELLLGAK EYSTAVDMWS VGCIFGELLT 400
QKPLFPGKSE IDQINKVFKD LGTPSEKIQP GYSELPAVKK MTGSEHPYNN 450
LRKRFGALLS DQGFDLMNKF LTYFPGRRIS AEDGLKHEYF RETPLPIDPS 500
MFPTWPAKSE QQRVKRGTSP RPPEGGLGYS QLGDDDLKET GFHLTTTNQG 550
ASAAGPGFSL KF
```

Sites of interest: T215, Y216, T362 putative phosphorylation

Homologues in other species

PITSLRE has been sequenced from human fetal liver, human HeLa cells, human genomic DNA (placental and lymphoid), mouse (1881 cells, lymphoid), and chicken (DT40, lymphoid).

Physiological substrates and specificity determinants

The consensus sequence for phosphorylation by PITSLRE is not known, but histone H1 can serve as an inefficient substrate[1]. A somewhat better substrate *in vitro* is the dodecapeptide, SPMRSRSPSRSK, which is also phosphorylated by Cdc2-like kinases. A physiological substrate for this kinase has not been determined.

Assay

The enzyme has been assayed using $[\gamma\text{-}^{32}P]ATP$ and histone H1[1].

Pattern of expression

Northern blot analysis indicates expression of multiple PITSLRE mRNAs (1.9, 3.5, 3.7 kb) in all human tissues and cell lines as well as all murine tissues[1,3,6]. The presence of multiple mRNAs is consistent with the various PITSLRE isoforms isolated thus far[3]. Differences in mRNA sizes in mouse tissues may indicate preferential expression of mRNA species[6]. Differential expression of PITSLRE isoforms has been observed[3].

References

1 Bunnell, B.A. et al. (1990) Proc. Natl Acad. Sci. USA 87, 7467–7471.
2 Bunnell, B.A. et al. (1990) Biochem. Biophys. Res. Commun. 171, 196–203.
3 Xiang, J. et al. (1994) J. Biol. Chem., 269, 15786–15794.
4 Eipers, P.E. et al. (1991) Genomics 11, 621–629.
5 Mock, B. et al. (1994) Mammalian Genome 5, 191–192.
6 Eipers, P.E. et al. (1992) Genomics 13, 613–621.
7 Kidd, V.J. et al. (1991) Cell Growth Diff. 2, 85–93.

Matthew Meyerson, Li-Huei Tsai, Greg H. Enders, Chin-Lee Wu,
Sander van den Heuvel and Ed Harlow
(Massachusetts General Hospital Cancer Center, USA)

cDNA encoding KKIALRE was cloned via homology to the **Cdc2/Cdc28** family. It was named after the amino acid sequence corresponding to the PSTAIRE motif of **Cdc2**.

Subunit structure and isoforms

SUBUNIT	AMINO ACIDS	MOL. WT	SDS-PAGE
KKIALRE	358	41 834	42 kDa

Domain structure

Like other Cdks, KKIALRE consists essentially of a catalytic domain only. Cyclins are presumed to act as regulatory subunits, but this has not been demonstrated under physiological conditions.

Database accession numbers

	PIR	SWISSPROT	EMBL/GENBANK	REF
Human	S22744, S22745		X66358, X66359, X66361	1

Amino acid sequence of human KKIALRE

```
MMEKYEKIGK IGEGSYGVVF KCRNRDTGQI VAIKKFLESE DDPVIKKIAL  50
REIRMLKQLK HPNLVNLLEV FRRKRRLHLV FEYCDHTVLH ELDRYQRGVP 100
EHLVKSITWQ TLQAVNFCHK HNCIHRDVKP ENILITKHSV IKLCDFGFAR 150
LLTGPSDYYT DYVATRWYRS PELLVGDTQY GPPVDVWAIG CVFAELLSGV 200
PLWPGKSDVD QLYLIRKTLG DLIPRHQQVF STNQYFSGVK IPDPEDMEPL 250
ELKFPNISYP ALGLLKGCLH MDPTERLTCE QLLHHPYFEN IREIEDLAKE 300
HDKPTRKTLR KSRKHHCFTE TSKLQYLPQL TGSSILPALD NKKYYCDTKK 350
LNYRFPNI
```

Pattern of expression

KKIALRE is expressed in HeLa and NGP cells. In tissues, highest levels are in brain, lung, and kidney.

Reference
1 Meyerson, M. et al. (1992) EMBO J. 11, 2909–2917.

Sme1

PK essential for meiosis (*S. cerevisiae*)
(Ime2)

I. Yamashita (Hiroshima University, Japan)

The Sme1 kinase is required for the initiation of meiosis and sporulation in the yeast *S. cerevisiae*[1]. Forced expression of the kinase can bypass the genetic (mating-type) and environmental restrictions on the entry into meiosis[1,2]. The kinase activates transcription of downstream meiosis-specific genes[3-5]. The *SME1* gene encoding the kinase is identical to *IME2*[1,2].

Subunit structure and isoforms

SUBUNIT	AMINO ACIDS	MOL. WT	SDS-PAGE
Sme1	645	73 592	74 and 85 kDa

The 85 kDa form may be the result of post-translational modification.

Genetics
Chromosomal loction: X, tightly linked to *URA2*[2].

Domain structure

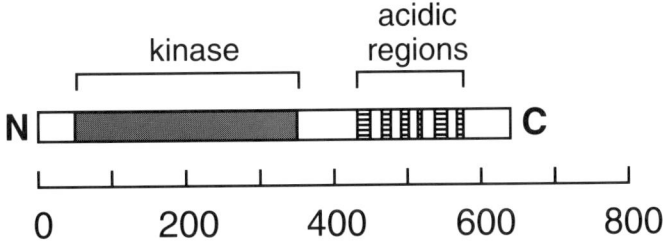

The C-terminal domain contains six highly acidic subregions and has a negative role in kinase activity.

Database accession numbers

	PIR	SWISSPROT	EMBL/GENBANK	REF
Sme1			X53262	*1*

Amino acid sequence of *S. cerevisiae* Sme1

```
MVEKRSRQSS SSGSEFSVPP DVDNPPLSIP LKTLSDRYQL IEKLGAGSFG  50
CVTLAKAQFP LSNILGKQHD IRGTLMDQPK NGHQNYITKT QGVVAIKTMM 100
TKLHTLQDYT RVREIKFILA IPANDHLIQI FEVFIDSENY QLHIVMECME 150
QNLYQMMKHR RRRVFSIPSL KSILSQILAG LKHIHEHNFF HRDLKPENIL 200
ITPSTQYFEK EYMNQIGYQD NYVIKLADFG LARHVENKNP YTAYVSTRWY 250
RSPEILLRSG YYSKPLDIWA FGCVAVEVTV FRALFPGANE IDQIWKILEV 300
LGTPIKRSDF VNTNHITAPP PGGFWDDASN LVHKLNLKLP YVEGSSLDHL 350
LSSSQLSDLS EVVKKCLRWD PNERATAQEL CEMPFFENTV ASQVDARGNV 400
```

Amino acid sequence of *S. cerevisiae* Sme1 *continued*

```
TNTEQALIFA GINPVATNTK PIYFNSSTKL PAETESNDID ISNNDHDSHA 450
MCSPTLNQEK LTLVEFLNEF VEEDNDDHSI PDVGTDSTIS DSIDETELSK 500
EIRNNLALCQ LPDEEVLDHS LSNIRQLTND IEIINKDEAD NMEQLFFDLE 550
IPEKDEFQRK QPFNEHADID EDIVLPYVNN SNYTHTDRSH HRGDNVLGDA 600
SLGDSFNSMP DFTPRNFLIP TLKKSREKFE PHLSNSNQHF GNVTF
```

Sites of interest: E433–D447, E465–D477, D491–E501, D513–D518, D537–E556, E565–D572, acidic regions

Assay[6]
The enzyme is assayed as the transfer of radioactivity from $[\gamma\text{-}^{32}\text{P}]\text{ATP}$ to histone H1, κ-casein or myelin basic protein. Autophosphorylation can also be detected.

Pattern of expression
A Sme1–lacZ fusion protein is located in the nucleus, suggesting that Sme1 is a nuclear protein[6].

References

1 Yoshida, M. et al. (1990) Mol. Gen. Genet. 221, 176–186.
2 Smith, H.E. and Mitchell, A.P. (1989) Mol. Cell. Biol. 9, 2142–2152.
3 Kihara, K. et al. (1991) Mol. Gen. Genet. 226, 383–392.
4 Mitchell, A.P. et al. (1990) Mol. Cell. Biol. 10, 2104–2110.
5 Kawaguchi, et al. (1992) Biosci. Biotech. Biochem. 56, 289–297.
6 Kominami, K. et al. (1993) Biosci. Biotech. Biochem. 57, 1731–1735.

Sgv1

Suppressor of Gpa1^{val50} (*S. cerevisiae*)

Kenji Irie and Kunihiro Matsumoto (Nagoya University, Japan)

The *GPA1* gene of *S. cerevisiae* encodes a Gα subunit that plays a positive role in the transduction of signal stimulating recovery from pheromone-induced cell cycle arrest[1]. The *GPA1^{val50}* mutation (G50V) causes hyperadaptation to pheromone[1]. An *sgv1* mutation suppresses the hyperadaptive response caused by *GPA1^{val50}* and also confers cold- and temperature-sensitive growth[2]. The *SGV1* gene encodes a protein kinase homologous to **Cdc28/SpCdc2**: Sgv1 is 42% identical to Cdc28[2]. An activated mutation of G1 cyclin, *CLN3-2*, partially suppresses the growth defect of the *sgv1* mutation[2].

Subunit structure and isoforms

SUBUNIT	AMINO ACIDS	MOL. WT	SDS-PAGE
Sgv1	657	74 238	

Genetics
Chromosomal location: XVI.

Domain structure

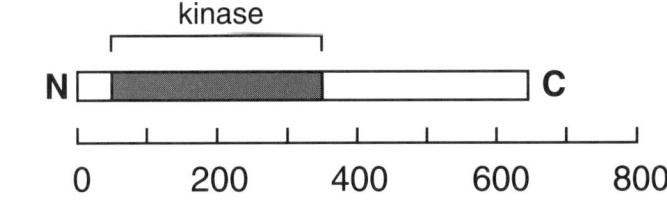

Database accession numbers

	PIR	SWISSPROT	EMBL/GENBANK	REF
Sgv1			D90317	2

Amino acid sequence of *S. cerevisiae* Sgv1

```
MSDNGSPAVL PKTEFNKYKI GKVKSTPAIQ RDAKTNLTYI KLRKRSSEKV  50
YGCTVFQNHY REDEKLGQGT FGEVYKGIHL ETQRQVAMKK IIVSVEKDLF 100
PITAQREITI LKRLNHKNII KLIEMVYDHS PDITNAASSN LHKSFYMILP 150
YMVADLSGVL HNPRINLEMC DIKNMMLQIL EGLNYIHCAK FMHRDIKTAN 200
ILIDHNGVLK LADFGLARLY YGCPPNLKYP GGAGSGAKYT SVVVTRWYRA 250
PELVLGDKQY TTAVDIWGVG CVFAEFFEKK PILQGKTDID QGHVIFKLLG 300
TPTEEDWAVA RYLPGAELTT TNYKPTLRER FGKYLSETGL DFLGQLLALD 350
PYKRLTAMSA KHHPWFKEDP LPSEKITLPT EESHEADIKR YKEEMHQSLS 400
QRVPTAPRGH IVEKGESPVV KNLGAIPRGP KKDDASFLPP SKNVLAKPPP 450
SKIRELHQNP RPYHVNSGYA KTAIPPPAAP AGVNRYGPNN SSRNNRFSGN 500
STAPNNSRNP VNRFHPETNV SSKYNKVPLP LGPQSRYQGN SNESRYKNSP 550
NDSRYHNPRY VNKPETNFNR QPQKYSRQES NAPINKNYNP SNGSRNMAGD 600
HHQGSRPSHP QFPISPSQGQ HQLTSKPIEK KNGSFKDERA KPDESKEFQN 650
SDIADLY
```

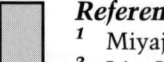

References

[1] Miyajima, I. et al. (1989) Mol. Cell. Biol. 9, 2289–2297.
[2] Irie, K. et al. (1991) Cell 65, 785–795.

Snk   Serum-inducible kinase (vertebrates)

Daniel L. Simmons (Brigham Young University, USA)

Serum-inducible kinase is rapidly induced by mitogens in immortalized and non-immortalized fibroblasts[1]. Induction is due to increased *SNK* gene transcription and occurs within minutes following the stimulus, suggesting that Snk functions in cell cycle regulation. Snk is a member of a subfamily of putative cell division-associated kinases which includes Cdc5/Msd2, Polo, and Polo-like kinases.

Subunit structure

SUBUNIT	AMINO ACIDS	MOL. WT	SDS-PAGE
Snk	682	77811	76–80 kD

Genetics

Chromosomal location (mouse): 13, linked to Ctla-3 and Gp130.
(Nancy Jenkins, Frederick Cancer Research & Development Center, personal communication).

Domain structure

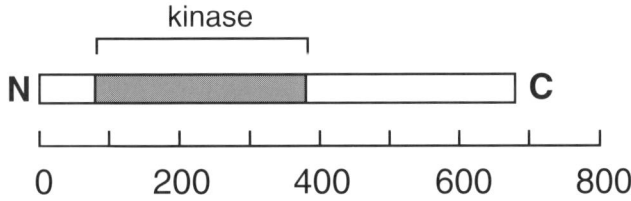

Database accession numbers

		PIR	SWISSPROT	EMBL/GENBANK	REF
Snk	Mouse			M96163	1

Amino acid sequence of mouse Snk

```
MELLRTITYQ PAAGTKMCEQ ALGKACGGDS KKKRPQQPSE DGQPQAQVTP  50
AAPHHHHHHS HSGPEISRII VDPTTGKRYC RGKVLGKGGF AKCYEMTDLT 100
NNKVYAAKII PHSRVAKPHQ REKIDKEIEL HRLLHHKHVV QFYHYFEDKE 150
NIYILLEYCS RRSMAHILKA RKVLTEPEVR YYLRQIVSGL KYLHEQEILH 200
RDLKLGNFFI NEAMELKVGD FGLAARLEPL EHRRRTICGT PNYLSPEVLN 250
KQGHGCESDI WALGCVMYTM LLGRPPFETT NLKETYRCIR EARYTMPSSL 300
LAPAKHLIAS MLSKNPEDRP SLDDIIRHDF FLQGFTPDRL SSSCCHTVPD 350
FHLSSPAKNF FKKAAAALFG GKKDKARYND THNKVSKEDE DIYKLRHDLK 400
KVSITQQPSK HRADEEPQPP PTTVARSGTS AVENKQQIGD AIRMIVRGTL 450
GSCSSSSECL EDSTMGSVAD TVARVLRGCL ENMPEADCIP KEQLSTSFQW 500
VTKWVDYSNK YGFGYQLSDH TVGVLFNNGA HMSLLPDKKT VHYYAELGQC 550
SVFPATDAPE QFISQVTVLK YFSHYMEENL MDGGDLPSVT DIRRPRLYLL 600
QWLKSDKALM MLFNDGTFQV NFYHDHTKII ICNQSEEYLL TYINEDRIST 650
TFRLTTLLMS GCSLELKNRM EYALNMLLQR CN
```

Homologues in other species

Murine Snk is related to the Polo kinase of *D. melanogaster* and to the probable murine homologue of Polo, Plk (Polo-like kinase)[2, 3]. Polo regulates imaginal disc formation during embryogenesis. The function of Plk is unknown, but its expression is restricted to fetal and neonatal tissues in the developing mouse and to haematopoietic tissues, ovaries, and testes in the adult. Snk also shares high sequence similarity with Cdc5/Msd2 kinase of *S. cerevisiae*. Cdc5/Msd2 regulates segregation of specific yeast chromosomes during meiosis[4]. A serum/glucocorticoid-inducible kinase, Sgk, has been reported. Despite its similarity in being serum inducible, Sgk shows less sequence similarity to Snk than the above kinases, and does not appear to be in the same kinase subfamily[5].

Assay

Snk expressed in *E. coli* phosphorylates myelin basic protein and casein (Mei-Ann Liu, Harvard University, unpublished data).

Pattern of expression

SNK mRNA is transcriptionally elevated >10-fold within 1 h of stimulation of immortalized fibroblasts with serum or phorbol ester. Mitogen treatment in the presence of cycloheximide causes superinduction of *SNK* mRNA. Dexamethasone inhibits the induction of *SNK* by serum as does v-*Ki-ras* in the DT cell line. The v-*src* oncogene and/or serum induce *SNK* in chicken embryo fibroblasts. *SNK* mRNA has been detected at low levels in mouse brain, lung and heart.

References

[1] Simmons, D.L. et al. (1992) Mol. Cell. Biol. 12, 4164–4169.
[2] Llamazares, S. et al. (1991) Genes Devel. 5, 2153–2165.
[3] Clay, F.J. et al. (1993) Proc. Natl Acad. Sci. USA 90, 4882–4886.
[4] Sharon, G. and Simchen, G. (1990) Genetics 125, 475–485.
[5] Webster, M.K. et al. (1993) Mol. Cell. Biol. 13, 2031–2040.

Polo

Polo protein kinase (*D. melanogaster*)

Salud Llamazares and David M. Glover (University of Dundee, UK)

The *polo* gene product is essential for mitosis and meiosis[1]. *polo*[1] was identified as a female sterile allele. Embryos derived from homozygous mothers show defective mitosis. Homozygous animals survive to adulthood, but show mitotic defects in larval brain, and non-disjunction in male meiosis[1]. *polo*[2] is a larval lethal that results from a P-element insertion[2].

Subunit structure and isoforms

SUBUNIT	AMINO ACIDS	MOL. WT	SDS-PAGE
Polo	576	66 947	

Genetics

Chromosomal location: 3, 77A.

Domain structure

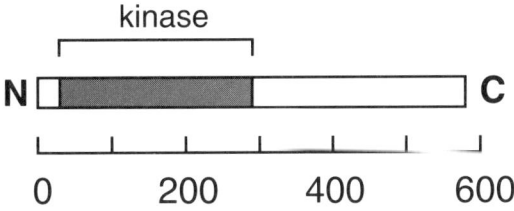

Database accession numbers

	PIR	SWISSPROT	EMBL/GENBANK	REF
Polo			X63361	2

Amino acid sequence of *D. melanogaster* Polo

```
MAAKPEDKST DIPDRLVDIN QRKTYKRMRF FGKGGFAKCY EIIDVETDDV  50
FAGKIVSKKL MIKHNQKEKT AQEITIHRSL NHPNIVKFHN YFEDSQNIYI 100
VLELCKKRSM MELHKRRKSI TEFECRYYIY QIIQGVKYLH DNRIIHRDLK 150
LGNLFLNDLL HVKIGDFGLA TRIEYEGERK KTLCGTANYI APEILTKKGH 200
SFEVDIWSIG CVMYTLLVGQ PPFETKTLKD TYSKIKKCEY RVPSYLRKPA 250
ADMVIAMLQP NPESRPAIGQ LLNFEFLKGS KVPMFLPSSC LTMAPRIGSN 300
DTIEDSMHRK PLMEMNGIRP DDTRLESTFL KANLHDAITA SAQVCRHSED 350
YRSDIESLYQ QLTNLINGKP RILQGNLGDE NTDPAAQPLF WISKWVDYSD 400
KYGFGYQLCD EGIGVMFNDT TKLILLPNQI NVHFIDKDGK ETYMTTTDYC 450
KSLDKKMKLL SYFKRYMIEH LVKAGANNVN IESDQISRMP HLHSWFRTTC 500
AVVMHLTNGS VQLNFSDHMK LILCPRMSAI TYMDQEKNFR TYRFSTIVEN 550
GVSKDLYQKI RYAQEKLRKM LEKMFT
```

Homologues in other species

In *S. cerevisiae* a homologous kinase is encoded by *CDC5*[3]. The mouse gene Plk encodes a protein kinase related to Polo[4]. Homologues have been identified in *Xenopus laevis* and *S. pombe* (Ohkura and Glover, unpublished).

Physiological substrates and specificity determinants

The enzyme prefers casein to histone H1 as a substrate *in vitro*.

Pattern of expression

In the *D. melanogaster* embryo, the *polo* gene is expressed at developmental stages consistent with a requirement for mitosis and meiosis. The enzyme is maximally active in the mitotic cycle at late anaphase–telophase[5].

References

1 Sunkel, C. and Glover, D. (1988) J. Cell Sci. 89, 25–38.
2 Llamazares, S. et al. (1991) Genes Devel. 5, 2153–2165.
3 Kitada, K. et al. (1993) Mol. Cell. Biol. 13, 4445–4457.
4 Clay, F. et al. (1993) Proc. Natl Acad. Sci. USA 90, 4882–4886.
5 Fenton, B. and Glover, D.M. (1993) Nature 363, 637–639.

Mek	MAPK/Erk kinase (vertebrates) (MAP kinase kinase, MAPKK)

Raymond L. Erikson, Alessandro Alessandrini and Craig Crews
(Harvard University, USA)

Mek is a dual-specificity PK responsible for phosphorylation of MAP kinase (**Erk1/2**) on threonine and tyrosine residues. These phosphorylation events dramatically increase the capacity of MAP kinase to phosphorylate exogenous substrates such as microtubule-associated proteins and myelin basic protein. The MAP kinases encoded by the *ERK1/ERK2* genes are the only substrates of Mek thus far reported.

Subunit structure and isoforms

SUBUNIT	AMINO ACIDS	MOL. WT	SDS-PAGE
Mek1	393	43 470	≈45 kDa
Mek2	401	44 430	≈45 kDa

Two related gene products, Mek1 and Mek2, have been reported in the mouse. Both are apparently monomers.

Genetics[1]
Chromosomal location, (human): 15q21–q22.
Chromosomal locations, (mouse):
   MEK1: 9, middle region, linked to Cyp1a2,d, and Gsta; MEK1 also maps
            near the uncloned recessive neurological mutation, staggerer (sg)
   MEK2: 10, middle region, linked to Bcr, Cola6a-1, Gna-15, and Igf-1.

Domain structure

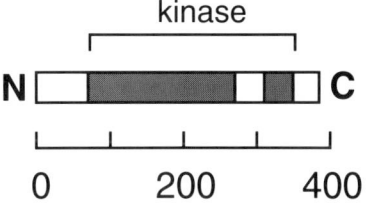

Database accession numbers

		PIR	SWISSPROT	EMBL/GENBANK	REF
Mek1	Mouse			L02526	2
	Xenopus			D13700	3
	Rat			L04485	4
	Human			L05624	5
	Rabbit				6
Mek2	Mouse				7
	Human			L11285	8
	Rat			L14936, D14592	9, 10

Amino acid sequence of mouse Mek

Mek1

```
MPKKKPTPIQ LNPAPDGSAV NGTSSAETNL EALQKKLEEL ELDEQQRKRL  50
EAFLTQKQKV GELKDDDFEK ISELGAGNGG VVFKVSHKPS GLVMARKLIH 100
LEIKPAIRNQ IIRELQVLHE CNSPYIVGFY GAFYSDGEIS ICMEHMDGGS 150
LDQVLKKAGR IPEQILGKVS IAVIKGLTYL REKHKIMHRD VKPSNILVNS 200
RGEIKLCDFG VSGQLIDSMA NSFVGTRSYM SPERLQGTHY SVQSDIWSMG 250
LSLVEMAVGR YPIPPPDAKE LELLFGCHVE GDAAETPPRP RTPGRPLSSY 300
GMDSRPPMAI FELLDYIVNE PPPKLPSGVF SLEFQDFVNK CLIKNPAERA 350
DLKQLMVHAF IKRSDAEEVD FAGWLCSTIG LNQPSTPTHA ASI
```

Mek2

```
MLARRKPVLP ALTINPTIAE GPSPTSEGAS EANLVDLQKK LEELDLDEQQ  50
RKRLEAFLTQ KAKVGELKDD DFERISELGA GNGGVVTKAR HRPSGFIMAR 100
KLIHLEIKPA VRNQIIRELQ VLHECNSPYI VGFYGAFYSD GEISICMEHM 150
DGGSLDQVLK EAKRIPEDIL GKVSIAVLRG LAYLREKHQI MHRDVKPSNI 200
LVNSRGEIKL CDFGVSGQLI DSMANSFVGT RSYMSPERLQ GTHYSVQSDI 250
WSMGLSLVEL AIGRYPIPPP DAKELEASFG RPVVDGADGE PHSVSPRPRP 300
PGRPISVGHG MDSRPAMAIF ELLDYIVNEP PPKLPSGVFS SDFQEFVNKC 350
LIKNPAERAD LKLLMNHAFI KRSEGEEVDF AGWLCRTLRL KQPSTPTRTA 400
V
```

Homologues in other species

Outside of vertebrates, Mek1 is most closely related in size and sequence to *S. pombe* **Byr1**. They are 45% identical in amino acid sequence, although Byr1 has no insert in the kinase domain. Mek1/Mek2 are also related to **Pbs2** and **Ste7** from *S. cerevisiae*.

Physiological substrates and specificity determinants

Mek phosphorylates Thr and Tyr residues in the sequence TEY in MAP kinase. Mutation of the T or Y still permits phosphorylation of the remaining site. Neither denatured MAP kinase nor a synthetic peptide with the TEY sequence appear to be effectively phosphorylated. Mek has no other reported substrates.

Assay

Mek can be assayed using a glutathione-*S*-transferase/Erk fusion protein[6].

Enzyme activators and inhibitors

Mek is regulated by reversible protein phosphorylation on serine or threonine by one or more Mek kinases (MAP kinase kinase kinases), including **Raf-1**.

Pattern of expression

Mek mRNA is widely expressed in many adult rat or mouse tissues.

References

1 Brott, B.K. et al. (1993) Cell Growth Diff. 4, 921–929.
2 Crews, C.M. et al. (1992) Science 258, 478–480.
3 Kosaka, H. et al. (1993) EMBO J. 12, 787–794.
4 Wu, J. et al. (1993) Proc. Natl Acad. Sci. USA 90, 173–177.
5 Seger, R. et al. (1992) J. Biol. Chem. 267, 25628–25631.
6 Ashworth, A. et al. (1992) Oncogene 7, 2555–2556.
7 Alessandrini, A. et al. (1992) Proc. Natl Acad. Sci. USA 89, 8200–8204.
8 Zheng, C.-F. and Guan, K. (1993) J. Biol. Chem. 268, 11435–11439.
9 Wu, et al. (1993) Mol. Cell. Biol. 13, 4539–4548.
10 Otsu, et al. (1993) FEBS Lett. 320, 246–250.

Dsor1 MAP kinase kinase homologue (*D. melanogaster*)

Leo Tsuda and Yasuyoshi Nishida
(Nagoya University, Nagoya, Japan)

Dsor1 (Downstream of Raf1) was genetically identified as a factor acting downstream of **DmRaf** by screening of dominant suppressors of *D-raf*. Dsor1 is involved in the transduction of transmembrane signals from receptor PTKs including **Torso**, **Sev** and **DmEGFR**[1]. Loss of zygotic Dsor1 activity causes reduced rates of cellular proliferation, and maternal Dsor1 activity is necessary for development of both anterior and posterior terminal structures of the embryo[1].

Subunit structure and isoforms

SUBUNIT	AMINO ACIDS	MOL. WT	SDS-PAGE
Dsor1	393	43 542	

Genetics
Chromosomal location:
    1–26.5 on the genetic map
    8D on the salivary gland chromosome map (on the X chromosome)
Mutations:
    Su1 (dominant gain-of-function mutation)
    r1 (loss-of-function mutation, recessive lethal)
    r2 (loss-of-function mutation, recessive lethal)
    Gp158 (loss-of-function mutation, recessive lethal, an insertion of incomplete P at the second exon)

The *Dsor1* gene consists of four exons interrupted with three introns of 644, 193 and 71 bp. The entire coding sequence locates within a 2.7 kb genomic region.

Domain structure

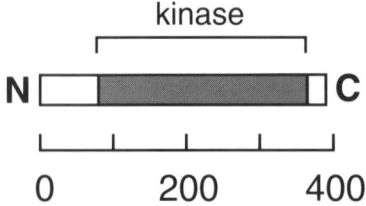

Subdomain X in the kinase domain contains a 20 amino acid insertion (residue 290–309) relative to the yeast homologues, **Ste7**, **Pbs2**, and **Byr1**. The mouse homologue (**Mek**)[2] also contains an insertion of 36 amino acid residues at the same position.

Database accession numbers

	PIR	SWISSPROT	EMBL/GENBANK	REF
Dsor1			D13782	[1]

Amino acid sequence of *D. melanogaster* Dsor1

```
MSKNKLNLVL PPVNTEATVA AATVAPTPPF KTPSGTDLLG KPKTSIDALT  50
ETLEGLDMGD TERKRIKMFL SQKEKIGELS DEDLEKLGEL GSGNGGVVMK 100
VRHTHTHLIM ARKLIHLEVK PAIKKQILRE LKVLHECNFP HIVGFYGAFY 150
SDGEISICME YMDGGSLDLI LKRAGRIPES ILGRITLAVL KGLSYLRDNH 200
AIIHRDVKPS NILVNSSGEI KICDFGVSGQ LIDSMANSFV GTRSYMSPER 250
LQGTHYSVQS DIWSLGLSLV EMAIGMYPIP PPNTATLESI FADNAEESGQ 300
PTDEPRAMAI FELLDYIVNE PPPKLEHKIF STEFKDFVDI CLKKQPDERA 350
DLKTLLSHPW IRKAELEEVD ISGWVCKTMD LPPSTPKRNT SPN
```

Homologues in other species

MAP kinase kinase (**Mek**) has been sequenced from mouse (pre-B cells) and *Xenopus laevis*. Related kinases have been sequenced from *S. cerevisiae* (**Ste7** and **Pbs2**) and *S. pombe* (**Byr1** and **Wis1**).

Pattern of expression

Dsor1 mRNA was detected throughout development at low levels. A large amount of maternal mRNA is stored in the ooplasm[1].

References

[1] Tsuda, L. et al. (1993) Cell 72, 407–414.
[2] Crews, C. M. et al. (1992) Science 258, 478–480.

Beverly Errede (University of North Carolina, USA)

Ste7 (Ste for sterile) functions in pheromone-induced signal transmission in *S. cerevisiae*[1]. The pheromone-induced signal causes haploid cells to arrest cell proliferation and to differentiate into a mating competent state. Five protein kinases Ste20, **Ste11**, Ste7 and a redundant pair of MAP kinase homologues, **Fus3** and **Kss1**, are involved in this signal pathway[2]. Ste7 is a dual-specificity protein kinase that activates Fus3 (and presumably Kss1) by phosphorylation on tyrosine and threonine[3]. Ste7 is itself phosphorylated in response to pheromone[4] and requires a Ste11-dependent phosphorylation for its enzyme activity (unpublished data).

Subunit structure and isoforms

SUBUNIT	AMINO ACIDS	MOL. WT	SDS-PAGE
Ste7	515	57 700	58–67 kDa

Multiple forms on SDS-PAGE are due to different phosphorylated states.

Genetics

Chromosomal location: IV, left arm, linked to *CDC9*.

Domain structure

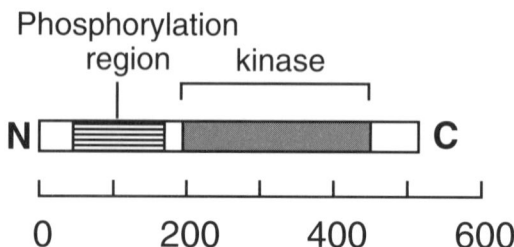

A positive regulatory domain lies within the N-terminal 45 amino acids because deletion of this region results in a stable polypeptide without catalytic activity[5]. The sites associated with Ste7 hyperphosphorylation are located within a Ser-rich region between residues 46–171 (unpublished). A derivative of Ste7 lacking the hyperphosphorylation domain is still enzymatically active[3].

Database accession numbers

	PIR	SWISSPROT	GENBANK/EMBL	REF
Ste7	A25048	P06784	M14097	*1*

Amino acid sequence of *S. cerevisiae* Ste7

```
MFQRKTLQRR NLKGLNLNLH PDVGNNGQLQ EKTETHQGQS RIEGHVMSNI  50
NAIQNNSNLF LRRGIKKKLT LDAFGDDQAI SKPNTVVIQQ PQNEPVLVLS 100
SLSQSPCVSS SSSLSTPCII DAYSNNFGLS PSSTNSTPST IQGLSNIATP 150
VENEHSISLP PLEESLSPAA ADLKDTLSGT SNGNYIQLQD LVQLGKIGAG 200
NSGTVVKALH VPDSKIVAKK TIPVEQNNST IINQLVRELS IVKNVKPHEN 250
IITFYGAYYN QHINNEIIIL MEYSDCGSLD KILSVYKRFV QRGTVSSKKT 300
WFNELTISKI AYGVLNGLDH LYRQYKIIHR DIKPSNVLIN SKGQIKLCDF 350
GVSKKLINSI ADTFVGTSTY MSPERIQGNV YSIKGDVWSL GLMIIELVTG 400
EFPLGGHNDT PDGILDLLQR IVNEPSPRLP KDRIYSKEMT DFVNRCCIKN 450
ERERSSIHEL LHHDLIMKYV SPSKDDKFRH WCRKIKSKIK EDKRIKREAL 500
DRAKLEKKQS ERSTH
```

Homologues in other species

Ste7 is homologous with **Byr1** from *S. pombe*[6], **Pbs2**[7], **Mkk1/2**[8] from *S. cerevisiae*, and **Mek**[9–14] from higher eukaryotes.

Physiological substrates and specificity determinants

The target sequence for phosphorylation by Ste7 is the TEY motif (subdomain VIII) common to Erk/MAP kinase family members[3]. Although phosphorylation *in vitro* has only been shown for **Fus3**, genetic evidence indicates that both **Fus3** and **Kss1** are the physiological substrates of Ste7[3, 15].

Assay

Ste7 enzyme is assayed by transfer of radioactivity from $[\gamma\text{-}^{32}\text{P}]$ATP to a catalytically inactive derivative of **Fus3** (Fus3-R42)[3]. Activity of the enzyme is higher when isolated from pheromone-induced cells compared with uninduced cells. A synthetic peptide encompassing the **Fus3** phosphorylation site is not phosphorylated by Ste7 under standard assay conditions.

Pattern of expression

STE7 mRNA is present in all *S. cerevisiae* cell-types (**a**, α and **a**/α) and does not increase in abundance upon pheromone stimulation (unpublished).

References

[1] Teague, M.A. et al. (1986) Proc. Natl Acad. Sci. USA 83, 7371–7375.
[2] Errede, B. et al. (1993) Curr. Opin. Cell Biol. 5, 254–260.
[3] Errede, B. et al. (1993) Nature 362, 261–264.
[4] Zhou, Z. et al. (1993) Mol. Cell. Biol. 13, 2069–2080.
[5] Zhou, Z. (1993) PhD thesis, University of North Carolina, Chapel Hill.
[6] Nadin-Davis, S.A. and Nasim, A. (1988) EMBO J. 7, 985–993.
[7] Boguslawski, G. and Polazzi J.O. (1987) Proc. Natl Acad. Sci. USA 84, 5842–5848.
[8] Irie, K. et al. (1993) Mol. Cell. Biol. 13, 3076–3083.
[9] Tsuda, L. et al. (1993) Cell 72, 407–414.
[10] Kosaka H. et al. (1993) EMBO J. 12, 787–794.

[11] Yashar, B.M. et al. (1995) submitted.
[12] Crews C.M. et al. (1992) Science 258, 478–480.
[13] Wu J. et al. (1993) Proc. Natl Acad. Sci. USA 90, 173–177.
[14] Seger, R. et al. (1992) J. Biol. Chem. 267, 25628–25631.
[15] Gartner, A. et al. (1992) Genes & Devel. 6, 1280–1292.

| **Pbs2** | **Pbs2 protein kinase (*S. cerevisiae*)** |
| | **(Hog4)** |

George Boguslawski (Indiana University School of Medicine, USA)

An *S. cerevisiae* gene, *PBS2*, is essential for expression of resistance to an antibiotic, polymyxin B[1]. The protein has been expressed in *Escherichia coli* but has not been characterized. To confer resistance, the gene *PBS2* must reside on a high copy plasmid (e.g. YEp24 or similar). The same gene has been found to regulate osmotic sensitivity of yeast cells[2, 3]. Deletion of *PBS2* enables null *ras2* mutants of yeast to grow on non-fermentable carbon sources; overexpression of *PBS2* enhances viability of activated $RAS2^{val19}$ mutant[2]. Disruption of *PBS2* renders cells sensitive to high osmolarity[2, 3]. Pbs2 has been shown to be necessary *in vivo* for tyrosine phosphorylation of Hog1, a yeast homologue of **Erk**/MAP kinase[3]. Consistent with the reported homology between Pbs2 and **Mek**, this observation suggests that Pbs2 is a yeast equivalent of a Mek/MAP kinase kinase that participates in the osmoregulatory signal transduction pathway.

Subunit structure and isoforms

SUBUNIT	*AMINO ACIDS*	*MOL. WT*	*SDS-PAGE*
Pbs2	668	72 714	

Genetics

Chromosomal location: X, left arm, between *INO1* and *ARG3* (18 cM and 29.1 cM); physically immediately upstream of the yeast potassium transporter gene, $TRK1^{1,4}$.

Mutations:
  K389R, inactive *in vivo*
  Deletion of residues 57–142, replacement with RIR, inactive *in vivo*

Domain structure

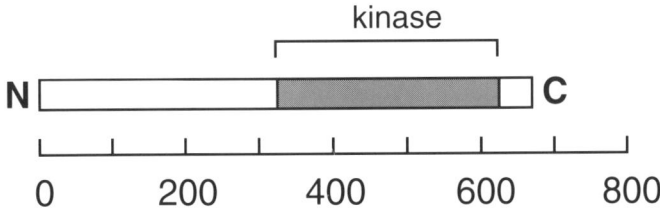

The N-terminal domain exhibits no similarity to any sequence in the GENBANK database, but contains several conspicuous clusters of Ser/Thr residues flanked by charged amino acids.

Database accession numbers

	PIR	*SWISSPROT*	*GENBANK/EMBL*	*REF*
Pbs2			J02946	*1, 2*

Amino acid sequence of *S. cerevisiae* Pbs2

```
MEDKFANLSL HEKTGKSSIQ LNEQTGSDNG SAVKRTSSTS SHYNNINADL  50
HARVKAFQEQ RALKRSASVG SNQSEQDKGS SQSPKHIQQI VNKPLPPLPV 100
AGSSKVSQRM SSQVVQASSK STLKNVLDNQ ETQNITDVNI NIDTTKITAT 150
TIGVNTGLPA TDITPSVSNT ASATHKAQLL NPNRRAPRRP LSTQHPTRPN 200
VAPHKAPAII NTPKQSLSAR RGLKLPPGGM SLKMPTKTAQ QPQQFAPSPS 250
NKKHIETLSN SKVVEGKRSN PGSLINGVQS TSTSSSTEGP HDTVGTTPRT 300
GNSNNSSNSG SSGGGGLFAN FSKYVDIKSG SLNFAGKLSL SSKGIDFSNG 350
SSSRITLDEL EFLDELGHGN YGNVSKVLHK PTNVIMATKE VRLELDEAKF 400
RQILMELEVL HKCNSPYIVD FYGAFFIEGA VYMCMEYMDG GSLDKIYDES 450
SEIGGIDEPQ LAFIANAVIH GLKELKEQHN IIHRDVKPTN ILCSANQGTV 500
KLCDFGVSGN LVASLAKTNI GCQSYMAPER IKSLNPDRAT YTVQSDIWSL 550
GLSILEMALG RYPYPPETYD NIFSQLSAIV DGPPPRLPSD KFSSDAQDFV 600
SLCLQKIPER RPTYAALTEH PWLVKYRNQD VHMSEYITER LERRNKILRE 650
RGENGLSKNV PALHMGGL
```

Homologues in other species

Pbs2 is most closely related to **Wis1**[5] (65% identity within kinase domain) and **Byr1**[6] (43%) from *S. pombe*, and **Mek**[7] (48%) from vertebrates. Within *S. cerevisiae* it is most closely related to **Ste7**[8] (38%).

References

[1] Boguslawski, G. and Polazzi, J.O. (1987) Proc. Natl Acad. Sci. USA 84, 5848–5852.
[2] Boguslawski, G. (1992) J. Gen. Microbiol. 138, 2425–2432.
[3] Brewster, J.L. et al. (1993) Science 259, 1760–1763.
[4] Gaber, R.F. et al. (1988) Mol. Cell. Biol. 8, 2848–2859.
[5] Warbrick, E. and Fantes, P.A. (1991) EMBO J. 10, 4291–4299.
[6] Wang, Y. et al. (1991) Mol. Cell. Biol. 11, 3554–3563.
[7] Crews, C.M. et al. (1992) Science 258, 478–480.
[8] Teague, M.A. et al. (1986) Proc. Natl Acad. Sci. USA 83, 7371–7375.

Mkk1/2

MAP kinase kinase 1/2 (*S. cerevisiae*)

Kenji Irie and Kunihiro Matsumoto (Nagoya University, Japan)

Mkk1 and Mkk2 mediate signalling by the *S. cerevisiae* PKC isozyme **Pkc1**[1]. Loss of *PKC1* function results in cell lysis due to a defect in cell wall construction[2]. Either *MKK1* or *MKK2* on a multicopy plasmid suppresses the cell lysis defect of a temperature-sensitive *pkc1* mutant[3]. Deletion of either *MKK* gene alone does not cause any apparent phenotypic defects, but deletion of both *MKK1* and *MKK2* results in a *ts* cell lysis defect that is suppressed by osmotic stabilizers[3]. Mkk1 and Mkk2 are proposed, based on genetic epistasis experiments, to function within a protein kinase cascade in which Pkc1 activates a protein kinase encoded by **BCK1**[4], which activates Mkk1 and Mkk2, which in turn activate the MAP kinase **Mpk1**[5].

Subunit structure and isoforms

SUBUNIT	AMINO ACIDS	MOL. WT	SDS-PAGE
Mkk1	508	56 719	
Mkk2	506	56 747	

Genetics

Chromosomal location:

MKK1	XV
MKK2	XVI

Domain structure

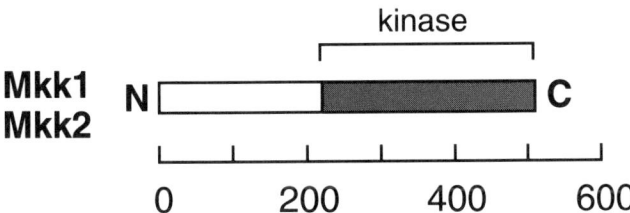

The N-terminal regions are 31% identical and the kinase domains 80% identical.

Database accession numbers

	PIR	SWISSPROT	EMBL/GENBANK	REF
Mkk1			D13001	3
Mkk2			D13785	3

Amino acid sequence of *S. cerevisiae* Mkk1/2

Mkk1

```
MASLFRPPES AKCNPNSPRL KLPLLRNNQV DENNIYLTSN GSSTTAYSSH  50
TPEPLTSSTS TLFSQTRLHP SDSSMTLNTM KKRPAPPSLP SLSINSQSKC 100
KTLPELVPIA DVSDGKHDLG LKQRVIAENE LSGNSDLTPS SMASPFSHTN 150
TSSPYLRNDL SNSVGSDFSN LISAYEQSSS PIKSSSQPKS SSESYIDLNS 200
VRDVDQLDEN GWKYANLKDR IETLGILGEG AGGSVSKCKL KNGSKIFALK 250
VINTLNTDPE YQKQIFRELQ FNRSFQSEYI VRYYGMFTDD ENSSIYIAME 300
YMGGRSLDAI YKNLLERGGR ISEKVLGKIA EAVLRGLSYL HEKKVIHRDI 350
KPQNILLNEN GQVKLCDFGV SGEAVNSLAT TFTGTSFYMA PERIQGQPYS 400
VTSDVWSLGL TILEVANGKF PCSSEKMAAN IAPFELLMWI LTFTPELKDE 450
PESNIIWSPS FKSFIDYCLK KDSRERPSPR QMINHPWIKG QMKKNVNMEK 500
FVRKCWKD
```

Mkk2

```
MASMFRPPES NRSHQKTPKL TLPVNLVQNA KSTNDGQHLN RSPYSSVNES  50
PYSNNSTSAT STTSSMASNS TLLYNRSSTT TIKNRPVPPP LPPLVLTQKK 100
DGIEYRVAGD SQLSERFSNL HVDITYKELL SSAPISTKLS NIDTTFIKKD 150
LDTPEGEDSY PSTLLSAYDF SSSGSNSAPL SANNIISCSN LIQGKDVDQL 200
EEEAWRFGHL KDEITTLGIL GEGAGGSVAK CRLKNGKKVF ALKTINTMNT 250
DSEYQKQIFR ELQFNKSFKS DYIVQYYGMF TDEQSSSIYI AMEYMGGKSL 300
EATYKNLLKR GGRISERVIG KIAESVLRGL SYLHERKVIH RDIKPQNILL 350
NEKGEIKLCD FGVSGEAVNS LAMTFTGTSF YMAPERIQGQ PYSVTCDVWS 400
LGLTLLEVAG GRFPFESDKI TQNVAPIELL TMILTFSPQL KDEPELDISW 450
SKTFRSFIDY CLKKDARERP SPRQMLKHPW IVGQMKKKVN MERFVKKCWE 500
KEKDGI
```

References

1 Errede, B. and Levin, D. E. (1993) Curr. Opin. Cell Biol. 5, 254–260.
2 Levin, D.E. and Bartlett-Heubusch, E. (1992) J. Cell Biol. 116, 1221–1229.
3 Irie, K. et al. (1993) Mol. Cell. Biol. 13, 3076–3083.
4 Lee, K.S. and Levin, D.E. (1992) Mol. Cell. Biol. 12, 172–182.
5 Lee, K.S. et al. (1993) Mol. Cell. Biol. 13, 3067–3075.

Byr1 "Bypass of Ras1" protein kinase (*S. pombe*)

Susan A. Nadin-Davis
(Animal Diseases Research Institute, Ontario, Canada)

The *byr1* gene of fission yeast was isolated as a locus which, in high copy number, can overcome the disruption in sporulation observed in *ras1⁻* strains[1]. This gene, which encodes a putative protein kinase, is allelic to the *ste1* locus, one of a series of at least ten mutations which cause sterility in *S. pombe*[2]. The products of *ras1*, *byr1* and *byr2* (another member of the *ste* gene group which is also predicted to encode a protein kinase[3]) appear to constitute a signal transduction pathway which activates certain genes (i.e. *Pi* and *mei3*) responsible for induction of sexual differentiation in response to nitrogen starvation[2,3].

Subunit structure and isoforms

SUBUNIT	AMINO ACIDS	MOL. WT	SDS-PAGE
Byr1	340	38 186	

Domain structure

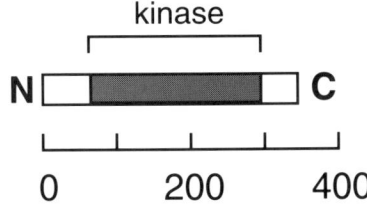

Database accession numbers

	PIR	SWISSPROT	EMBL/GENBANK	REF
Byr1		P10506	X07445	1

Amino acid sequence of *S. pombe* Byr1

```
MFKRRRNPKG LVLNPNASVK SSDNDHKEEL INNQKSFESN VEAFMEQCAH  50
MNRRPAWISD LDNSSLEVVR HLGEGNGGAV SLVKHRNIFM ARKTVYVGSD 100
SKLQKQILRE LGVLHHCRSP YIVGFYGAFQ YKNNISLCME YMDCGSLDAI 150
LREGGPIPLD ILGKIINSMV KGLIYLYNVL HIIHRDLKPS NVVVNSRGEI 200
KLCDFGVSGE LVNSVAQTFV GTSTYMSPER IRGGKYTVKS DIWSLGISII 250
ELATQELPWS FSNIDDSIGI LDLLHCIVQE EPPRLPSSFP EDLRLFVDAC 300
LHKDPTLRAS PQQLCAMPYF QQALMINVDL ASWASNFRSS
```

Homologues in other species

Although not demonstrated to be true functional homologues, **Mek** (MAP kinase kinase) from mouse, rat and *D. melanogaster* exhibit substantial sequence similarity to Byr1, as does **Ste7** of *S. cerevisiae*. Several of these proteins also exhibit serological crossreactivity.

Pattern of expression

The *byr1* gene appears to be constitutively transcribed in yeast cells. Protein expression has not been examined.

References

[1] Nadin-Davis, S.A. and Nasim, A. (1988) EMBO J. 7, 985–993.
[2] Nadin-Davis, S.A. and Nasim, A. (1990) Mol. Cell. Biol. 10, 549–560.
[3] Wang, Y. et al. (1991) Mol. Cell. Biol. 11, 3554–3563.

Wis1

Wis 1 protein kinase (*S. pombe*)

Emma Warbrick, Shaun Mackie and Peter Fantes
(University of Edinburgh, UK)

The Wis1 protein kinase in *S. pombe* is involved in regulating the G2–M transition[1,2]. Overexpression of the *wis1*[+] gene advances mitosis in a dosage-dependent manner. Strains deleted for *wis1*[+] are viable but the cells are elongated, indicating delayed mitosis. *wis1*[+] deletion strains show very poor viability in stationary phase and altered conjugation activity, implicating *wis1*[+] in several aspects of the *S. pombe* life cycle.

Subunit structure and isoforms

SUBUNIT	AMINO ACIDS	MOL. WT	SDS-PAGE
Wis1	605	64 700	68 kDa

Domain structure

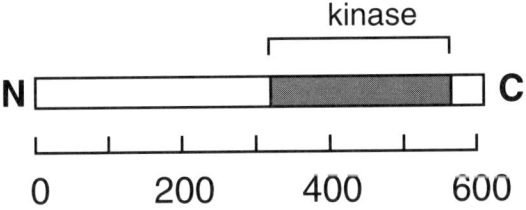

The N-terminal domain shows no overall similarity with proteins in current databases, although it is rich in Ser (particularly in the 100 most N-terminal residues), Thr, Pro and Glu. Clusters of these amino acids (PEST-rich sequences) are characteristic of some unstable proteins.

Database accession numbers

	PIR	SWISSPROT	GENBANK/EMBL	REF
Wis1			X62631	1

Amino acid sequence of *S. pombe* Wis1

```
MSSPNNQPLS CSLRQLSISP TAPPGDVGTP GSLLSLSSSS SSNTDSSGSS  50
LGSLSLNSNS SGSDNDSKVS SPSREIPSDP PLPRAVPTVR LGRSTSSRSR 100
NSLNLDMKDP SEKPRRSLPT AAGQNNIGSP PTPPGPFPGG LSTDIQEKLK 150
AFHASRSKSM PEVVNKISSP TTPIVGMGQR GSYPLPNSQL AGRLSNSPVK 200
SPNMPESGLA KSLAAARNPL LNRPTSFNRQ TRIRRAPPGK LDLSNSNPTS 250
PVSPSSMASR RGLNIPPTLK QAVSETPFST FSDILDAKSG TLNFKNKAVL 300
NSEGVNFSSG SSFRINMSEI IKLEELGKGN YGVVYKALHQ PTGVTMALKE 350
IRLSLEEATF NQIIMELDIL HKAVSPYIVD FYGAFFVEGS VFICMEYMDA 400
GSMDKLYAGG IKDEGVLART AYAVVQGLKT LKEEHNIIHR DVKPTNVLVN 450
SNGQVKLCDF GVSGNLVASI SKTNIGCQSY MAPERIRVGG PTNGVLTYTV 500
QADVWSLGLT ILEMALGAYP YPPESYTSIF AQLSAICDGD PPSLPDSFSP 550
EARDFVNKCL NKNPSLRPDY HELANHPWLL KYQNADVDMA SWAKGALKEK 600
GEKRS
```

Homologues in other species

Wis1 is highly similar in sequence throughout its catalytic domain to **Pbs2** and **Ste7** of *S. cerevisiae*, **Byr1** of *S. pombe*, and **Mek**/MAP kinase kinases of vertebrates, all of which are elements or presumptive elements of signal transduction pathways. Wis1 is most similar to **Pbs2**, where the similarity extends beyond the highly conserved protein kinase domains toward both the N- and C-termini (residues 278–589 of Wis1). It is not however known whether Wis1 and Pbs2 are functional homologues.

References

[1] Warbrick, E. and Fantes, P.A. (1991). EMBO J. 10, 4291–4299.
[2] Warbrick, E. and Fantes, P.A. (1992). Mol. Gen. Genet. 232, 440–446.

Ste11 — Ste11 protein kinase (*S. cerevisiae*)

Beverly Errede (University of North Carolina, Chapel Hill, USA)

Ste11 (Ste for sterile) functions in pheromone-induced signal transmission in *S. cerevisiae*[1]. The pheromone-induced signal causes haploid cells to arrest cell proliferation and to differentiate into a mating competent state. Five protein kinases, Ste20, **Ste11**, **Ste7** and a redundant pair of MAP kinase homologues, designated **Fus3** and **Kss1**, are part of this signal pathway[2]. Based on genetic analyses, Ste11 functions after Ste20 and before **Ste7** and **Fus3/Kss1**[3, 4]. It is not yet clear whether Ste20 directly activates **Ste11** and whether Ste11 directly activates **Ste7**.

Subunit structure and isoforms

SUBUNIT	AMINO ACIDS	MOL. WT	SDS-PAGE
Ste11	717	80 700	80 kDa

Genetics
Chromosomal location: XII, right arm, linked to *ILV5*.

Domain structure

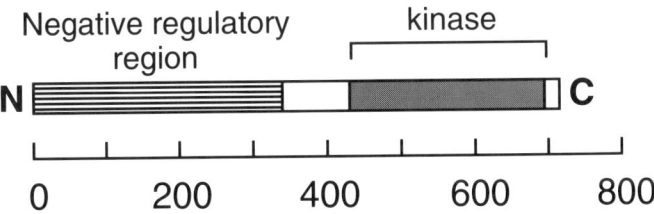

The constitutive activity associated with expression of a Ste11 polypeptide encompassing the kinase domain (342–717) is consistent with the N-terminal domain having a negative effect on Ste11 activity[5]. Also consistent with a negative role for the N-terminus is the finding that overexpression of a Ste11 N-terminal polypeptide 1–209) interferes with pheromone response[6]. Mutations in either the regulatory domain (P279S) or the catalytic domain (T596I) converts Ste11 to a hyperactive form[3].

Database accession numbers

	PIR	SWISSPROT	GENBANK/EMBL	REF
Ste11	A36456	P23561	X53431	*1*

Amino acid sequence of *S. cerevisiae* Ste11

```
MEQTQTAEGT DLLIGDEKTN DLPFVQLFLE EIGCTQYLDS FIQCNLVTEE  50
EIKYLDKDIL IALGVNKIGD RLKILRKSKS FQRDKRIEQV NRLKNLMEKV 100
SSLSTATLSM NSELIPEKHC VIFILNDGSA KKVNVNGCFN ADSIKKRLIR 150
RLPHELLATN SNGEVTKMVQ DYDVFVLDYT KNVLHLLYDV ELVTICHAND 200
RVEKNRLIFV SKDQTPSDKA ISTSKKLYLR TLSALSQVGP SSSNLLAQNK 250
GISHNNAEGK LRIDNTEKDR IRQIFNQRPP SEFISTNLAG YFPHTDMKRL 300
QKTMRESFRH SARLSIAQRR PLSAESNNIG DILLKHSNAV DMALLQGLDQ 350
TRLSSKLDTT KIPKLAHKRP EDNDAISNQL ELLSVESGEE EDHDFFGEDS 400
DIVSLPTKIA TPKNWLKGAC IGSGSFGSVY LGMNAHTGEL MAVKQVEIKN 450
NNIGVPTDNN KQANSDENNE QEEQQEKIED VGAVSHPKTN QNIHRKMVDA 500
LQHEMNLLKE LHHENIVTYY GASQEGGNLN IFLEYVPGGS VSSMLNNYGP 550
FEESLITNFT RQILIGVAYL HKKNIIHRDI KGANILIDIK GCVKITDFGI 600
SKKLSPLNKK QNKRASLQGS VFWMSPEVVK QTATTAKADI WSTGCVVIEM 650
FTGKHPFPDF SQMQAIFKIG TNTTPEIPSW ATSEGKNFLR KAFELDYQYR 700
PSALELLQHP WLDAHII
```

Homologues in other species

Ste11 is most closely related to **Bck1** (*S. cerevisiae*)[7], **Byr2** (*S. pombe*)[8] and **Mek** (mouse)[9].

Assay

Ste11 is assayed in immune complexes by transfer of radioactivity from $[\gamma\text{-}^{32}P]ATP$ to p78. p78 is a polypeptide of unknown identity that co-precipitates with Ste11[1].

Pattern of expression

STE11 mRNA is present in all *S. cerevisiae* cell types (**a**, α and **a**/α) and does not increase in abundance upon pheromone stimulation[1].

References

1   Rhodes, N. et al. (1990) Genes Devel. 4, 1862–1874.
2   Errede, B. and Levin, D.E. (1993) Curr. Opin. Cell Biol. 5, 254–260.
3   Stevenson, B.J. et al. (1992) Genes Devel. 6, 1293–1304.
4   Leberer E. et al. (1992) EMBO J. 11, 4815–4824.
5   Cairns, B.R. et al. (1992) Genes Devel. 6, 1305–1318.
6   Rhodes, N.R. (1991) PhD thesis, University of North Carolina, Chapel Hill.
7   Lee, K.S. and Levin, D.E. (1992) Mol. Cell. Biol. 12, 172–182.
8   Wang, Y. et al. (1991) Mol. Cell. Biol. 11, 3554–3563.
9   Lange-Carter, C.A. et al. (1993) Science 260, 315–319.

| **Bck1** | Bypass of C-kinase (*S. cerevisiae*)
(Slk1, Ssp31) |

David E. Levin (Johns Hopkins University, USA)

Bck1 mediates signalling by the protein kinase C isozyme in *S. cerevisiae* **ScPKC**[1]. Loss of *PKC1* function results in cell lysis due to a defect in cell wall construction[2,3]. Dominant mutations in *BCK1* suppress the *pkc1*-associated lysis defect[4]. Genetic analysis of *BCK1* revealed that it also plays a role in cell wall construction. Deletion mutants display a temperature-dependent cell lysis defect[4]. Bck1 is proposed, based on genetic epistasis experiments, to function within a protein kinase cascade in which **ScPKC** activates Bck1, which activates a redundant pair of MAP kinase kinases (**Mkk1/2**)[5], which activate the MAP kinase encoded by *MPK1*[6]. Bck1 is phosphorylated on serine and threonine *in vitro* by Pkc1. The *BCK1* gene was also isolated in two other genetic screens: (1) recessive mutations in *BCK1 (SLK1)* are synthetically lethal with a deletion mutant of *SPA2*[7]; (2) *BCK1 (SSP31)* is a dosage-dependent suppressor of a conditional mutation in *SMP3*[8].

Subunit structure and isoforms

SUBUNIT	AMINO ACIDS	MOL. WT	SDS-PAGE
Bck1	1478	163 000	190 kDa

Genetics
Chromosomal location: X, tightly linked to *ARG3*[4].
Gene structure: single uninterrupted ORF of 1478 amino acids
Mutations:
  K1204R (inactive)
  T1119P, I1120K, I1120T, G1146V, A1174P (constitutively active)

Domain structure

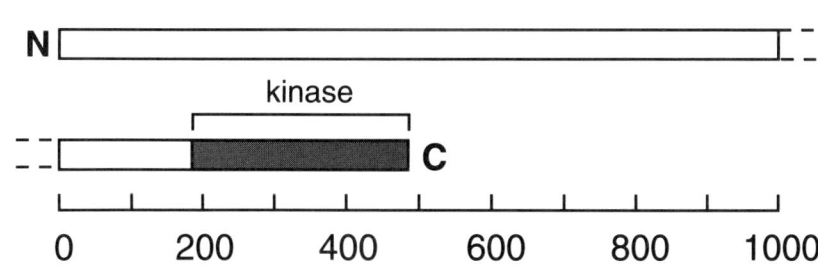

Database accession numbers

	PIR	SWISSPROT	GENBANK/EMBL	REF
Bck1			X60227, M84389, D90446	4, 7, 8

Amino acid sequence of *S. cerevisiae* Bck1

```
MPFLRKIAGT AHTHSRSDSN SSVKFGHQPT SSVASTKSSS KSPRATSRKS   50
IYDDIRSQIP NLTPNSTSSQ FYESTPVIEQ SFNWTTDDHI SAGTLENPTS  100
FTNSSYKNDN GPSSLSDSRK SSGGNSVNSL SFDKLILSWD PTDPDEWTMH  150
RVTSWFKFHD FPESWILFFK KHQLFGHRFI KLLAYDNFAV YEKYLPQTKT  200
ASYTRFQQLL KKTMTKNVTN SHIRQKSASK LKSSRSSSES IKSKLKNSKS  250
QEDISNSRST SESALSPTKS GPSKTDEKNF LHSTSTHQKT KSASSLYRRS  300
FISLRGSSSS NASSAKSPSN IKLSIPARPH SIIESNSTLT KSASPPASPS  350
YPSIFRRHHK SSSSESSLLN SLFGSGIGEE APTKPNPQGH SLSSENLAKG  400
KSKHYETNVS SPLKQSSLPT SDDKGNLWNK FKRKSQIGVP SPNTVAYVTS  450
QETPSLKSNS STATLTVQTA DVNIPSPSSS PPPIPKTANR SLEVISTEDT  500
PKISSTTASF KETYPDCINP DKTVPVPVNN QKYSVKNFLL DQKFYPLKKT  550
GLNDSENKYI LVTKDNVSFV PLNLKSVAKL SSFKESALTK LGINHKNVTF  600
HMTDFDCDIG AAIPDDTLEF LKKSLFLNTS GKIYIKDQMK LQQKPKPAPL  650
TSENNVPLKS VKSKSSMRSG TSSLIASTDD VSIVTSSSDI TSFDEHASGS  700
GRRYPQTPSY YYDRVSNTNP TEELNYWNIK EVLSHEENAP KMVFKTSPKL  750
ELNLPDKGSK LNIPTPITEN ESKSSFQVLR KDEGTEIDFN HRRESPYTKP  800
ELAPKREAPK PPANTSPQRT LSTSKQNKPI RLVRASTKIS RSKRSKPLPP  850
QLLSSPIEAS SSSPDSLTSS YTPASTHVLI PQPYKGANDV MRRLKTDQDS  900
TSTSPSLKMK QKVNRSNSTV STSNSIFYSP SPLLKRGNSK RVVSSTSAAD  950
IFEENDITFA DAPPMFDSDD SDDDSSSSDD IIWSKKKTAP ETNNENKKDE 1000
KSDNSSTHSD EIFYDSQTQD KMERKMTFRP SPEVVYQNLE KFFPRANLDK 1050
PITEGIASPT SPKSLDSLLS PKNVASSRTE PSTPSRPVPP DSSYEFIQDG 1100
LNGKNKPLNQ AKTPKRTKTI RTIAHEASLA RKNSVKLKRQ NTKMWGTRMV 1150
EVTENHMVSI NKAKNSKGEY KEFAWMKGEM IGKGSFGAVY LCLNVTTGEM 1200
MAVKQVEVPK YSSQNEAILS TVEALRSEVS TLKDLDHLNI VQYLGFENKN 1250
NIYSLFLEYV AGGSVGSLIR MYGRFDEPLI KHLTTQVLKG LAYLHSKGIL 1300
HRDMKADNLL LDQDGICKIS DFGISRKSKD IYSNSDMTMR GTVFWMAPEM 1350
VDTKQGYSAK VDIWSLGCIV LEMFAGKRPW SNLEVVAAMF KIGKSKSAPP 1400
IPEDTLPLIS QIGRNFLDAC FEINPEKRPT ANELLSHPFS EVNETFNFKS 1450
TRLAKFIKSN DKLNSSKLRI TSQENKTE
```

Homologues in other species

Bck1 is related to **Ste11** of *S. cerevisiae*, **Byr2** of *S. pombe*, and mouse Mek kinase[5].

References

[1] Errede, B. and Levin, D.E. (1993) Curr. Opin. Cell Biol. 5, 254–260.
[2] Levin, D.E. and Bartlett-Heubusch, E. (1992) J. Cell Biol. 116, 1221–1229.
[3] Paravicini, G. et al. (1992) Mol. Cell. Biol. 12, 4896–4905.
[4] Lee, K.S. and Levin, D.E. (1992) Mol Cell. Biol. 12, 172–182.
[5] Irie, K. et al. (1993) Mol. Cell. Biol. 13, 3076–3083.
[6] Lee, K.S. et al. (1993) Mol. Cell. Biol. 13, 3067–3075.
[7] Costigan, C. et al. (1992) Mol. Cell. Biol. 12, 1162–1178.
[8] Irie, K. et al. (1991) Gene 108, 139–144.
[9] Lange-Carter, C.A. (1993) Science 260, 315–319.

Byr2

Byr2 protein kinase (*S. pombe*)

Stevan Marcus (University of Texas M.D. Anderson Cancer Center, USA)

Byr2 is a member of the growing family of MAP kinase kinase kinases, proteins that phosphorylate and activate MAP kinase kinases[1]. Byr2 is required for sexual differentiation in *S. pombe*[2]. Genetic and molecular studies have demonstrated that the kinase functions downstream of both Ras1 (a Ras oncoprotein homologue) and Gpa1 (a homologue of mammalian heterotrimeric G protein α subunits), and upstream of **Byr1** (an *S. pombe* MAP kinase kinase **Mek** homologue)[2–4]. Ras1 and Byr2 interact physically, suggesting that Byr2 is an effector of Ras1 signalling in *S. pombe*[5].

Subunit structure and isoforms

SUBUNIT	AMINO ACIDS	MOL. WT	SDS-PAGE
Byr2	659	73 632	

Genetics

Chromosomal location: 2, right arm (allelic to the *ste8* gene).
Site-directed mutants:
  G534D (dominant interfering)
  P209S (hyperactive)

Domain structure

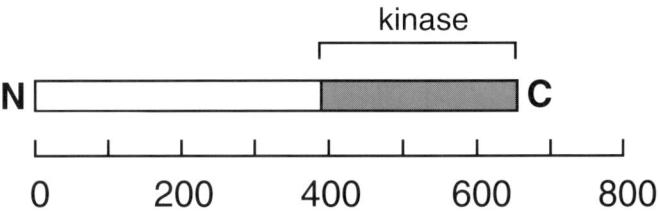

The N-terminal region is believed to constitute a regulatory domain, because expression of a truncated mutant (lacking 389–606) inhibits the sexual response in *S. pombe*. Deletion of 1–320 results in a hyperactive mutant. The sequence QRPPSE (207–212) is conserved in the N-terminal region of *S. cerevisiae* Ste11, which is structurally and functionally related to Byr2. Substitution of P209 (and its equivalent in Ste11) results in hyper-activation.

Database accession numbers

	PIR	SWISSPROT	EMBL/GENBANK	REF
Byr2	A39723	P28829	M74293	2

Amino acid sequence of *S. pombe* Byr2

```
MEYYTSKEVA EWLKSIGLEK YIEQFSQNNI EGRHLNHLTL PLLKDLGIEN  50
TAKGKQFLKQ RDYLREFPRP CILRFIACNG QTRAVQSRGD YQKTLAIALK 100
KFSLEDASKF IVCVSQSSRI KLITEEEFKQ ICFNSSSPER DRLIIVPKEK 150
PCPSFEDLRR SWEIELAQPA ALSSQSSLSP KLSSVLPTST QKRSVRSNNA 200
KPFESYQRPP SELINSRISD FFPDHQPKLL EKTISNSLRR NLSIRTSQGH 250
NLGNFGQEIL PRSSRRARPS ELVCPLSSLR ISVAEDVNRL PRIDRGFDPP 300
LTVSSTQRIS RPPSLQKSIT MVGVEPLYQS NGNEKSSKYN VFSESAHGNH 350
QVLSFSPGSS PSFIEQPSPI SPTSTTSEDT NTLEEDTDDQ SIKWIRGALI 400
GSGSFGQVYL GMNASSGELM AVKQVILDSV SESKDRHAKL LDALAGEIAL 450
LQELSHEHIV QYLGSNLNSD HLNIFLEYVP GGSVAGLLTM YGSFEETLVK 500
NFIKQTLKGL EYLHSRGIVH RDIKGANILV DNKGKIKISD FGISKKLELN 550
STSTKTGGAR PSFQGSSFWM APEVVKQTMH TEKTDIWSLG CLVIEMLTSK 600
HPYPNCDQMQ AIFRIGENIL PEFPSNISSS AIDFLEKTFA IDCNLRPTAS 650
ELLSHPFVS
```

Homologues in other species

Byr2 is most closely related both structurally and functionally[3] to *S. cerevisiae* **Ste11**. It is also related to *S. cerevisiae* **Bck1**, Mek kinase (MEKK) from mouse[6], and NPK1 from tobacco[7].

Physiological substrates and specificity determinants

Based on genetic analysis, and by analogy to the metazoan homologues, it is presumed that **Byr1** is a substrate of Byr2 in *S. pombe*.

References
1 Pelech, S.L. (1993) Curr. Biol. 3, 513–515.
2 Wang, Y. et al. (1991) Mol. Cell. Biol. 11, 3554–3563.
3 Neiman, A.M. et al. (1993) Mol. Biol. Cell 4, 107–120.
4 Xu, H.-P. et al. (1994) Mol. Cell. Biol. 14, 50–58.
5 Van Aelst, L. et al. (1993) Proc. Natl Acad. Sci. USA, 6213–6217.
6 Lange-Carter, C.A. et al. (1993) Science 260, 315–319.
7 Banno, H. et al. (1993) Mol. Cell. Biol. 13, 4745–4752.

Nek1   NimA-related kinase (vertebrates)

Karen Colwill, Vladimir Lhotak and Tony Pawson
(Mount Sinai Hospital, Canada)

Nek1 is a NimA-related PK cloned in mouse. Although its function is unknown, its high-level RNA expression in meiotic cells implies that it may be required for survival and development of germ cells. After expression in bacteria, Nek1 has been shown to be a dual-specificity kinase which autophosphorylates preferentially on serine and threonine, and to a lesser extent on tyrosine.

Subunit structure and isoforms

SUBUNIT	AMINO ACIDS	MOL. WT	SDS-PAGE
Nek1	774	88 427	

There may be at least two isoforms of Nek1, as there are two RNA transcripts of 6.5 kb and 4.4 kb in mouse tissues.

Domain structure

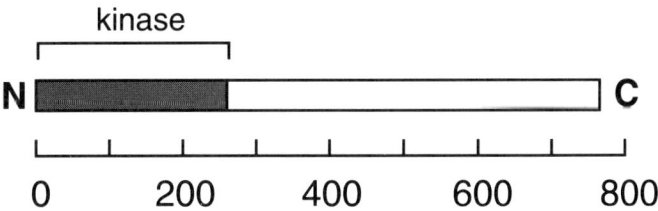

The catalytic domain is followed by a long basic tail. Within the tail are four clusters of basic residues which may represent bipartite nuclear targeting motifs.

Database accession numbers

	PIR	SWISS-PROT	EMBL/GENBANK	REF
Nek1			S45828, S114983	1

Amino acid sequence of mouse Nek1

```
MEKYVRLQKI GEGSFGKAVL VKSTEDGRHY VIKEINISRM SDKERQESRR  50
EVAVLANMKH PNIVQYKESF EENGSLYIVM DYCEGGDLFK RINAQKGALF 100
QEDQILDWFV QICLALKHVH DRKILHRDIK SQNIFLTKDG TVQLGDFGIA 150
RVLNSTVELA RTCIGTPYYL SPEICENKPY NNKSDIWALG CVLYELCTLK 200
HAFEAGNMKN LVLKIISGSF PPVSPHYSYD LRSLLSQLFK RNPRDRPSVN 250
SILEKGFIAK RIEKFLSPQL IAEEFCLKTL SKFGPQPLPG KRPASGQGVS 300
SFVPAQKITK PAAKYGVPLT YKKYGDKKLL EKKPPPKHKQ AHQIPVKKMN 350
SGEERKKMSE EAAKKRRLEF IEKEKKQKDQ IRFLKAEQMK RQEKQRLERI 400
NRAREQGWRN VLRAGGSGEV KASFFGIGGA VSPSPCSPRG QYEHYHAIFD 450
QMQRLRAEDN EARWKGGIYG RWLPERQKGH LAVERANQVE EFLQRKREAM 500
QNKARAEGHV VYLARLRQIR LQNFNERQQI KAKLRGENKE ADGTKGQEAT 550
EETDMRLKKM ESLKAQTNAR AAVLKEQLER KRKEAYEREK KVWEEHLVAR 600
```

Amino acid sequence of mouse Nek1 *continued*

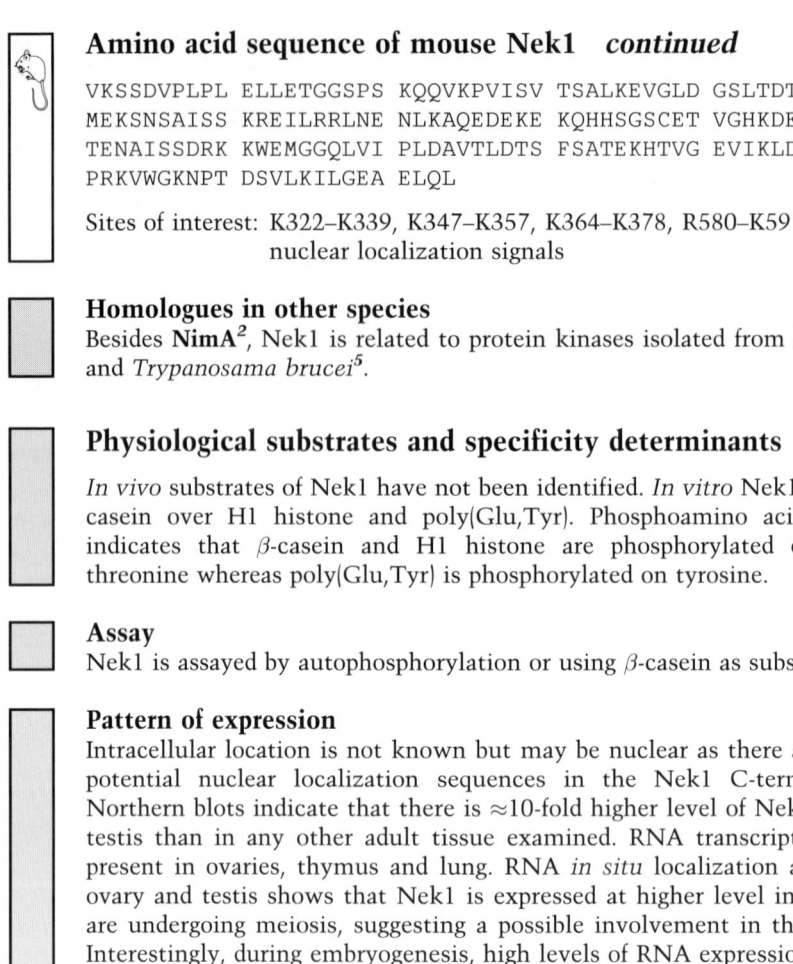

```
VKSSDVPLPL ELLETGGSPS KQQVKPVISV TSALKEVGLD GSLTDTQEEE 650
MEKSNSAISS KREILRRLNE NLKAQEDEKE KQHHSGSCET VGHKDEREYE 700
TENAISSDRK KWEMGGQLVI PLDAVTLDTS FSATEKHTVG EVIKLDSNGS 750
PRKVWGKNPT DSVLKILGEA ELQL
```

Sites of interest: K322–K339, K347–K357, K364–K378, R580–K591, putative nuclear localization signals

Homologues in other species

Besides **NimA**[2], Nek1 is related to protein kinases isolated from humans[3, 4] and *Trypanosama brucei*[5].

Physiological substrates and specificity determinants

In vivo substrates of Nek1 have not been identified. *In vitro* Nek1 prefers β-casein over H1 histone and poly(Glu,Tyr). Phosphoamino acid analysis indicates that β-casein and H1 histone are phosphorylated on serine/threonine whereas poly(Glu,Tyr) is phosphorylated on tyrosine.

Assay

Nek1 is assayed by autophosphorylation or using β-casein as substrate.

Pattern of expression

Intracellular location is not known but may be nuclear as there are several potential nuclear localization sequences in the Nek1 C-terminal tail. Northern blots indicate that there is ≈10-fold higher level of Nek1 RNA in testis than in any other adult tissue examined. RNA transcripts are also present in ovaries, thymus and lung. RNA *in situ* localization analysis in ovary and testis shows that Nek1 is expressed at higher level in cells that are undergoing meiosis, suggesting a possible involvement in this process. Interestingly, during embryogenesis, high levels of RNA expression are seen in the peripheral nervous system including sympathetic, parasympathetic and sensory neurones as well as motor neurones (Benny Motro, personal communication).

References
1   Letwin, K. et al. (1992) EMBO J. 11, 3521–3531.
2   Osmani, S.A. et al. (1988) Cell 53, 237–244.
3   Schultz, S.J. and Nigg, E.A. (1993) Cell Growth Differentiation 4, 821–830.
4   Levedakon, E.N. et al. (1994) Oncogene 9, 1977–1988.
5   Gale. M. and Parsons, M. (1993) Mol. Biochem. Parasitol. 59, 111–121.

 **NimA protein kinase (*Aspergillus nidulans*)**

Stephen A. Osmani (Geisinger Clinic, Weis Center for Research, USA)

The *nimA* (never in mitosis) gene was identified as *ts* mutations that would reversibly arrest cells in G2 of the cell cycle. The *nimA* gene was cloned by complementation of the *ts nimA5* mutation, and sequence analysis indicated that it encoded a serine/threonine protein kinase designated NimA. Subsequent biochemical analysis demonstrated that NimA is a cell cycle regulated serine/threonine PK that is required, in addition to the universal mitotic regulator Cdc2, for entry into mitosis in *A. nidulans*.

Subunit structure and isoforms

SUBUNIT	AMINO ACIDS	MOL. WT	SDS-PAGE
NimA	699	79 000	

Genetics
Chromosomal location: II[1].

Mutants:	nimA1, A5, A7, are all recessive *ts* mutations that arrest cells at the G2/M boundary[2].
Site-directed mutant:	K40M (inactive)[3].

Domain structure

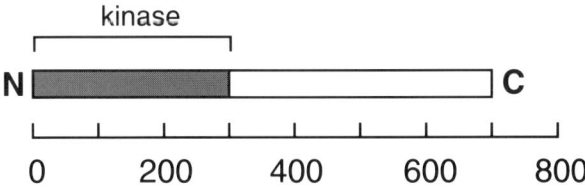

Database accession numbers

	PIR	SWISS-PROT	EMBL/GENBANK	REF
NimA		P11837	M20249	4

 ## Amino acid sequence of *A. nidulans* NimA

```
MAIALAEADK YEVLEKIGCG SFGIIRKVKR KSDGFILCRK EINYIKMSTK  50
EREQLTAEFN ILSSLRHPNI VAYYHREHLK ASQDLYLYME YCGGGDLSMV 100
IKNLKRTNKY AEEDFVWRIL SQLVTALYRC HYGTDPAEVG SNLLGPAPKP 150
SGLKGKQAQM TILHRDLKPE NIFLGSDNTV KLGDFGLSKL MHSHDFASTY 200
VGTPFYMSPE ICAAEKYTLR SDIWAVGCIM YELCQREPPF NARTHIQLVQ 250
KIREGKFAPL PDFYSSELKN VIASCLRVNP DHRPDTATLI NTPVIRLMRR 300
EVELNNLSRA ARKREEATMQ KAKDVEQAFA KLEKEKQQIR SELENSIRRE 350
WEVKARLEID RQVQNELDKL RKRFECEVQD RVAQEVEKQR RNANYREDAS 400
LRSSGHSSQM SSSNSEDSDF PSSTDISQLS LESPTNKAAK LPKKESRTPF 450
TRSKTVVDSP MDIQMAEPSP ISIASLSLSP RRTSATYSGK NIFAEGERKR 5 ) 
PKFEPTLAYS DDEDDTPELP SPTRPKVKPD PFKAPSRPLL RQNTTALMQK 550
LSTQPPIFPA NPSRLPQMSA PDVRESKSRS PHRRLSKIPS SANLAADAGS 600
PTRKNGVKSS PSKMNGGDEM FKAVMQRNMG GRTLVELAQA RAGGRPIDEV 650
KRCASDSRSG CSVPMKSADR DPPAVWDPER DEMPSPFLAR GRKVIRNLR
```

Homologues in other species

A functional homologue from *Neurospora crassa* that complements the *nimA5* mutation of *A. nidulans* has been cloned[5]. Protein kinases with similarity to NimA have been cloned from mouse (**Nek1**)[6] and *S. cerevisiae* (**Kin3**)[7], but neither has been shown to be a functional homologue. NimA is the founder member of a class of PKs that have an N-terminal catalytic domain, a basic C-terminal extension and an overall predicted isoelectric point of 9 or greater[6].

Assay

NimA can be assayed in immunoprecipitates by transfer of radioactivity from [γ-^{32}P]ATP to partially dephosphorylated β-casein[8]. The original assay conditions have been optimized[3].

Pattern of expression

Levels of NimA kinase activity are regulated through the cell cycle such that maximal levels are detected during G2 and mitosis[8].

References

1   Morris, N.R. (1976) Genet. Res. 26, 237–254.
2   Bergen, L.G. (1984) J. Bact. 159, 114–119.
3   Lu, K.P. et al. (1993) J. Biol. Chem. 268, 8769–8776.
4   Osmani, S.A. et al. (1988) Cell 53, 237–244.
5   Pu, R.T. and Osmani, S.A. (1995), in preparation.
6   Letwin, K. et al. (1992) EMBO J. 11, 3521–3531.
7   Jones, D.G. and Rosamond, J. (1990) Gene 90, 87–92.
8   Osmani, A.H. et al. (1991) EMBO J. 10, 2669–2679.

Fused

fused gene product (*D. melanogaster*)

Denise Busson, Bernadette Limbourg-Bouchon, Thomas Préat,
Pascal Thérond, Hervé Tricoire and Claudie Lamour-Isnard
(Institut Jacques Monod, Paris and Centre de Génétique
Moléculaire, Gif/Yvette, France)

The *fused* gene belongs to the segment-polarity class of *D. melanogaster* segmentation genes. It is required for pattern formation within embryonic segments and imaginal discs and during oogenesis. As deduced from the DNA sequence, the Fused protein is a putative serine/threonine PK, thought to be involved in signal transduction pathways at different stages of development[1,2].

Subunit structure and isoforms

SUBUNIT	AMINO ACIDS	MOL. WT	SDS-PAGE
Fused	805	90 300	

Genetics

Chromosomal location: X, 1–59.5, 17C3–D2[3].

Domain structure

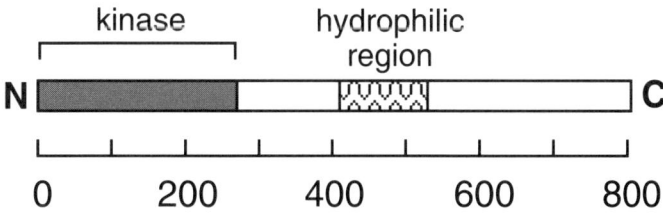

The function of the C-terminal region is not known but it appears to be essential for Fused activity and/or regulation[1,2,4]. A hydrophilic region in the centre could correspond to a hinge region between catalytic and regulatory domains.

Database accession numbers

	PIR	SWISSPROT	EMBL/GENBANK	REF
Fused	S11380	P23647	X55759	1

Amino acid sequence of *D. melanogaster* Fused

```
MNRYAVSSLV GQGSFGCVYK ATRKDDSKVV AIKVISKRGR ATKELKNLRR  50
ECDIQARLKH PHVIEMIESF ESKTDLFVVT EFALMDLHRY LSYNGAMGEE 100
PARRVTGHLV SALYYLHSNR ILHRDLKPQN VLLDKNMHAK LCDFGLARNM 150
TLGTHVLTSI KGTPLYMAPE LLADEPYDHH ADMWSLGCIA YESMAGQPPF 200
CASSILHLVK MIKHEDVKWP STLTCECRSF LQGLLEKDPG LRISWTQLLC 250
HPFVEGRIFI AETQAEAAKE SPFTNPEAKV KSSKQSDPEV GDLDEALAAL 300
DFGESRQENL TTSRDSINAI APSDVDHLET DVEDNMQRVV VPFADLSYRD 350
LSGVRAMPMV HQPVINSHTC FVSGNSNMIL NHMNDNFDFQ ASLRGGVVAA 400
```

Amino acid sequence of *D. melanogaster* Fused *continued*

```
KPIVAPTVRQ SRSKDLEKRK LSQNLDNFSV RLGHSVDHEA QRKATEIATQ 450
EKHNQENKPP AEAISYANSQ PPQQQPQQLK HSMHSTNEEK LSSDNTPPCL 500
LPGWDSCDES QSPPIENDEW LAFLNRSVQE LLDGELDSLK QHNLVSIIVA 550
PLRNSKAIPR VLKSVAQLLS LPFVLVDPVL IVDLELIRNV YVDVKLVPNL 600
MYACKLLLSH KQLSDSAASA PLTTGSLSRT LRSIPELTVE ELETACSLYE 650
LVCHLVHLQQ QFLTQFCDAV AILAASDLFL NFLTHDFRQS DSDAASVRLA 700
GCMLALMSCV LRELPENAEL VERIVFNPRL NFVSLLQSRH HLLRQRSCQL 750
LRLLARFSLR GVQRIWNGEL RFALQQLSEH HSYPALRGEA AQTLDEISHF 800
TFFVT
```

Pattern of expression
Hypothetical, as deduced from mRNA expression: highest levels in adult females (present in ovaries, in nurse cells and oocyte) and early embryonic stages; lowest expression in larvae (present in imaginal discs), pupae and adult males[2].

References
1. Préat, T. et al. (1990) Nature 347, 87–89.
2. Thérond, P. et al. (1993) Mech. Devel. 44, 65–80.
3. Busson, D. et al. (1988) Roux's Arch. Devel. Biol. 197, 221–230.
4. Préat, T. et al. (1993) Genetics 135, 1047–1062.

Kin3 — Kin3 protein kinase (*S. cerevisiae*)

John Rosamond (University of Manchester, UK)

The *KIN3* gene was isolated from a library of *S. cerevisiae* genomic DNA during a screen for novel protein kinases, using a 32-fold degenerate synthetic oligonucleotide corresponding to a consensus sequence for subdomain VIb as the primary probe. Candidate clones were rescreened with a second 32-fold degenerate oligonucleotide for subdomain I[1]. The function of Kin3 is unclear since cells that carry disrupted copies of the gene undergo normal mitosis, conjugation and meiosis. No phenotype has been established for the null mutant.

Subunit structure and isoforms

SUBUNIT	AMINO ACIDS	MOL. WT	SDS-PAGE
Kin3	369	43 435	

Genetics
Chromosomal location: I.

Domain structure

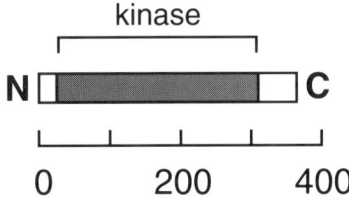

Database accession numbers

	PIR	SWISSPROT	EMBL/GENBANK	REF
Kin3	S23580	P22209	M53416	1

Amino acid sequence of *S. cerevisiae* Kin3

```
MHRRQFFQEY RSPQQQQGHP PRSEYQVLEE IGRGSFGSVR KVIHIPTKKL  50
LVRKDIKYGH MNSKERQQLI AECSILSQLK HENIVEFYNW DFDEQKELLY 100
LYMEYCSRGD LSQMIKHYKQ QHKYIPEKII WGILAQLLTA LYKCHYGVEL 150
PTLTTIYDRM KPPVKGKNIV IHRDLKPGNI FLSYDDSDYN INEQVDGHEE 200
VNSNYYRDHR VNSGKRGSPM DYSQVVVKLG DFGLSQISET SIQFATTYVG 250
TPYYMSPEVL MDQPYSPLSD IWSLGCVIFE MCSLHPPFQA KNYLELQTKI 300
KNGKCDTVPE YYSRGLNAII HSMIDVNLRT RPSTFELLQD IQIRTARKSL 350
QLERFERRLI DYENELTNI
```

Homologues in other species
Kin3 has significant homology to **NimA** from *A. nidulans*, although the functional significance of this is unclear given the absence of a mitotic phenotype in *KIN3* null mutants.

Pattern of expression

The *KIN3* transcript is present constitutively in cells undergoing mitosis; transcript levels during meiosis and protein levels during mitosis and meiosis have not been examined.

Reference

[1] Jones, D.G.L. and Rosamond, J. (1990) Gene 90, 87–92.

Wee1 — Wee1 protein kinase (vertebrates)

Hiroto Okayama, Makato Igarashi and Akihisa Nagata
(University of Tokyo and ERATO, JRDC, Japan)

In eukaryotes, mitosis is initiated by the activation of the cyclin-associated Cdc2 kinase. In the S and G2phases of the cell cycle, this kinase is inactivated by phosphorylation at Tyr15. Wee1 kinase catalyses this phosphorylation[1,2]. At the end of G2 phase, Cdc2 undergoes dephosphorylation and concomitant activation, which is catalysed by specific phosphatase(s), Cdc25. Wee1 resembles **SpWee1** and complements the *S. pombe wee1* mutant[1]. Structurally, this kinase belongs to the serine/threonine PK family, yet possesses tyrosine kinase activity[1,2].

Subunit structure and isoforms

SUBUNIT	AMINO ACIDS	MOL. WT	SDS-PAGE
Wee1	646	72 000	100 kDa

Genetics
Chromosomal location (human): 11p 15.1–15.3[3].

Domain structure

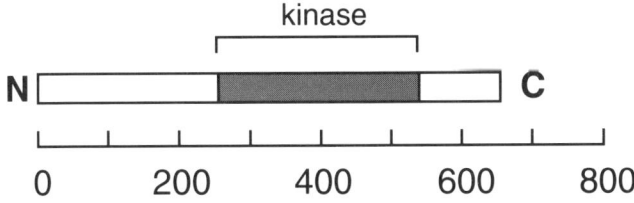

The region N-terminal to the catalytic domain is dispensable for Wee1 function[1].

Database accession numbers

	PIR	SWISSPROT	EMBL/GENBANK	REF
Human			X62048	1

Amino acid sequence of human Wee1

```
MSFLSRQQPP PPRRAGAACT LRQKLIFSPC SDCEEEEEE EEEGSGHSTG  50
EDSAFQEPDS PLPPERSPTE PGPERRRSPG PAPGSPGELE EDLLLPGACP 100
GADEAGGGAE GDSWEEEGFG SSSPVKSPAA PYFLGSSFSP VRCGGPGDAS 150
PRGCGARRAG EGRRSPRPDH PGTPPHKTFR KLRLFDTPHT PKSLLSKARG 200
IDSSSVKLRG SSLFMDTEKS GKREFDVRQT PQVNINPFTP DSLLLHSSGQ 250
CRRRKRTYWN DSCGEDMEAS DYELEDETRP AKRITITESN MKSRYTTEFH 300
ELEKIGSGEF GSVFKCVKRL DGCIYAIKRS KKPLAGSVDE QNALREVYAH 350
AVLGQHSHVV RYFSAWAEDD HMLIQNEYCN GGSLADAISE NYRIMSYFKE 400
AELKDLLLQV GRGLRYIHSM SLVHMDIKPS NIFISRTSIP NAASEEGDED 450
DWASNKVMFK IGDLGHVTRI SSPQVEEGDS RFLANEVLQE NYTHLPKADI 500
FALALTVVCA AGAEPLPRNG DQWHEIRQGR LPRIPQVLSQ EFTELLKVMI 550
HPDPERRPSA MALVKHSVLL SASRKSAEQL RIELNAEKFK NSLLQKELKK 600
AQMAKAAAEE RALFTDRMAT RSTTQSNRTS RLIGKKMNRS VSLTIY
```

Homologues in other species

Human and rat Wee1 kinase cDNAs have been cloned and sequenced.

Physiological substrate and specificity determinants

The only known physiological substrate is **Cdc2** associated with cyclin B[2].

Assay

The enzyme is assayed by transfer of radioactivity from $[\gamma\text{-}^{32}\text{P}]$ATP to cyclin B-associated Cdc2[2].

Pattern of expression

The *Wee1* gene is expressed in a cell cycle-dependent manner: the transcript begins to appear in S phase, peaks at G2 phase, and decreases thereafter. The enzyme is localized in the nucleus[4].

References

1 Igarashi, M. et al. (1991) Nature 353, 80–83.
2 Parker, L.L and Piwnica-Worms, H. (1992) Science 257, 1955–1957.
3 Tariaux, S.A. et al. (1993) Genomics 15, 194–196.
4 Heald, R. et al. (1993) Cell 74, 1–20.

SpWee1 — Wee1 protein kinase (*S. pombe*)

Paul Russell (Scripps Research Institute, La Jolla, California, USA)

Wee1 functions to delay the onset of mitosis until cells have grown to a sufficient size[1]. On the basis of sequence comparisons Wee1 was predicted to be a protein-serine/threonine kinase, but it inhibits mitosis by phosphorylating the **SpCdc2** protein kinase on Tyr15[2,3].

Subunit structure and isoforms

SUBUNIT	AMINO ACIDS	MOL. WT	SDS-PAGE
Wee1	877	96 154	107 kDa

Wee1 is active as a monomer[4].

Genetics

Chromosomal location: 3, short arm.
Mutation: *wee1*-50 *(ts)*.

Domain structure

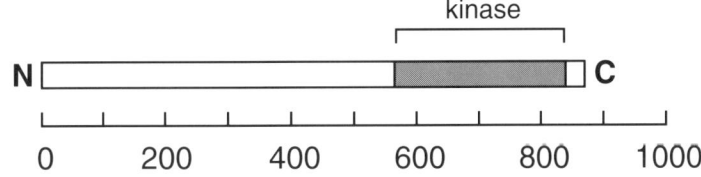

Database accession numbers

	PIR	SWISS-PROT	EMBL/GENBANK	REF
Wee1	A25962	P07527	M16508	1

Amino acid sequence of *S. pombe* Wee1

```
MSSSSNTSSH RSYGLRRSQR SMNLNRATLL APPTPSSLYD ANNSTSSTSS  50
QKPNTSFTSL FGPRKQTTSS PSFSHAAPLH PLSPPSFTHS QPQIQAQPVP 100
RRPSLFDRPN LVSRSSSRLG DSPSLSPVAQ VANPIHHTAP SPSDVRAFPI 150
HKNASTGVKR SFFSSSMSNG AMSPPSHSPS PFLQSSQHIP PSTPAQKLRK 200
KNNFDSFRIS NSHISPFASG SFSPFATSSP NFLSTSTPAP PNSNNANPST 250
LFSSIPSSRH TTSNHFPSNS AQSSLFSPTA RPLTARKLGF ASSQTKSAVS 300
NNHSRNSSKD ASFMMKSFIP SNRSHPQTQQ NESSLFSDNS MVNSSSNSFS 350
LFPNATLPNP PSSELLTTPF QQIKPPSQVF MSTGLLSKQH RPRKNINFTP 400
LPPSTPSKPS TFVRPHSSST DSPPSPSTPS NTQTDSYFIQ RENTPTNHNS 450
IPTIQLEKSS MDFLRFDPPP SAVKTSHNYG LPFLSNQRCP ATPTRNPFAF 500
ENTVSIHMDG RQPSPIKSRN NNQMSFAMEE EADVSQPSSS SFTLSFPSAL 550
TSSKVSSSTS HLLTRFRNVT LLGSGEFSEV FQVEDPVEKT LKYAVKKLKV 600
KFSGPKERNR LLQEVSIQRA LKGHDHIVEL MDSWEHGGFL YMQVELCENG 650
SLDRFLEEQG QLSRLDEFRV WKILVEVALG LQFIHHKNYV HLDLKPANVM 700
ITFEGTLKIG DFGMASVWPV PRGMEREGDC EYIAPEVLAN HLYDKPADIF 750
SLGITVFEAA ANIVLPDNGQ SWQKLRSGDL SDAPRLSSTD NGSSLTSSSR 800
ETPANSIIGQ GGLDRVVEWM LSPEPRNRPT IDQILATDEV CWVEMRRKAG 850
AIIYEGIHGS SSNPQGDQMM EDWQVNV
```

Homologues in other species

A human homologue (**Wee1**)[3, 5] and *Saccharomycer cerevisiae* (Swe1)[6] homologues have been described.

Physiological substrates and specificity determinants

Phosphorylates **SpCdc2** kinase on Tyr15[2, 3].

Assay

The enzyme is most conveniently assayed by transfer of radioactivity from $[\gamma\text{-}^{32}\text{P}]$ATP to angiotensin II[4] or Cdc2[2].

Enzyme activators and inhibitors

Wee1 is inactivated by phosphorylation carried out by **Nim1**.

References

1 **Russell, P. and Nurse, P. (1987) Cell 49, 559–567.**
2 Parker, L.L. et al. (1992) Proc. Natl Acad. Sci. USA 89, 2917–2921.
3 McGowan, C.H. and Russell, P. (1993). EMBO J. 12, 75–85.
4 Featherstone, C. and Russell, P. (1991) Nature 349, 808–811.
5 Igarashi, M. et al. (1991) Nature 353, 80–83.
6 Booher, R.N. et al. (1993) EMBO J. 12, 3417–3426.

Mik1

Mik1 protein kinase (*S. pombe*)

Karen Lundgren (Cold Spring Harbor Laboratory, USA)

Mik1 was characterized via a genetic screen designed to identify mitotic inhibitors in the fission yeast, *S. pombe*. The *mik1* gene complements the mitotic catastrophe defect in the double mutant strain, *cdc2-3w wee1-50*[1]. It was sequenced and shown to be most homologous to the **SpWee1** protein kinase[2] which also complements *cdc2-3w wee1-50*. The *mik1* gene disruption strain shows no apparent phenotype, but, the *mik1::ura4 wee1-50* double mutant is inviable, suggesting that the two kinases have a redundant function in *S. pombe*: phosphorylation of the **SpCdc2** protein kinase on Tyr15.

Subunit structure and isoforms

SUBUNIT	AMINO ACIDS	MOL. WT	SDS-PAGE
Mik1	581	66 000	

Domain structure

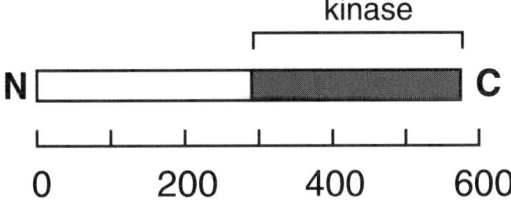

The ATP-binding site as defined by Hanks *et al.* is not completely conserved in Mik1. Instead of the GXGXXG motif, Mik1 has HXSXXS. However, mutation of the highly conserved lysine (K320A) results in a Mik1 that is unable to complement the *mik1::ura wee1-50* double mutant (K. Lundgren, unpublished). The wildtype *mik1* and *wee1* genes are able to complement the double mutant.

Database accession numbers

	PIR	SWISSPROT	GENBANK/EMBL	REF
Mik1			M60834	1

Amino acid sequence of *S. pombe* Mik1

```
MDSSTTIPIT PTRTPCFFNI SSSFNEHSPL NFYDEPIYNF SSGHEENQSH  50
KSSKLTFFKP SNTKRSPHTP MQNNAKAIRL STTVRHGIFK NSDLDGCSKP 100
FAFSSGLKLS KKIVDASTPI DLKRKRAVTS LSTGLLSKRE KWSLWEGNLT 150
NPRSEQPHTP CKKGTKIKLK PPQSPLSPTT SLLARKCKHI DLDTFSRLDH 200
PNSDSSDETF EMEELPSLSY GSEDLLEFCE TPCKSQPIFL SSSHVNNWDE 250
KDVPSSLSWT PTSPIFLNIN SADDYEEEED WTSDLRIRFQ QVKPIHESDF 300
SFVYHVSSIN PPTETVYVVK MLKKNAAKFT GKERHLQEVS ILQRLQACPF 350
VVNLVNVWSY NDNIFLQLDY CENGDLSLFL SELGLLQVMD PFRVWKMLFQ 400
LTQALNFIHL LEFVHLDVKP SNVLITRDGN LKLGDFGLAT SLPVSSMVDL 450
```

Amino acid sequence of *S. pombe* Mik1   *continued*

```
EGDRVYIAPE ILASHNYGKP ADVYSLGLSM IEAATNVVLP ENGVEWQRLR 500
SGDYSNLPNL KDLLLLSKEKV QINKVRCAES LQCLLQRMTH PYVDCRPTTQ 550
DLLAMPEMIF ISEHSQKAAI IYEDHNSWLE T
```

Homologues in other species
A human gene homologous to Mik1 and SpWee1, **Wee1**, has been sequenced[3].

Assay
Genetic analysis predicts that the substrate for Mik1 (and Wee1) is the Tyr15 residue of **SpCdc2**. Using antiphosphotyrosine antibodies, it was shown that in the *mik1::ura wee1-50* double mutant, Cdc2 is not phosphorylated on tyrosine[1]. The sequence of Mik1 is more similar to serine/threonine PKs than to TPKs but, like Wee1, Mik1 may be a kinase with specificity for tyrosine and serine[4]. This is currently being tested.

References
[1]   Lundgren, K. et al. (1991) Cell 64, 1111–1122.
[2]   Russell, P. and Nurse, P. (1987) Cell 49, 559–567.
[3]   Igarashi, M. et al. (1991) Nature 353, 80–83.
[4]   Featherstone, C. and Russell, P. (1991) Nature 349, 806–811.

Haem-regulated eIF-2α kinase (vertebrates)
(Haem-regulated inhibitor, haem-controlled repressor, HCR)

Jane-Jane Chen and Irving M. London
(Massachusetts Institute of Technology, USA)

HRI, a haem-regulated translational inhibitor, is activated in haem deficiency of reticulocytes and immature erythroid cells. When activated it phosphorylates the α subunit of eukaryotic initiation factor 2 (eIF-2α)[1,2]. Phosphorylation of eIF-2α results in the binding and sequestration of eIF-2B, which is required for the recycling of eIF-2 and reinitiation of protein synthesis. The unavailability of eIF-2B results in the cessation of protein synthesis[1-3]. The main physiological function of HRI in erythroid cells appears to be the balanced syntheses of globin chains and haem.

Subunit structure and isoforms

SUBUNIT	AMINO ACIDS	MOL. WT	SDS-PAGE
HRI	626	70 300	90 kDa

Purified HRI is a homodimer of 180 kDa.

Domain structure

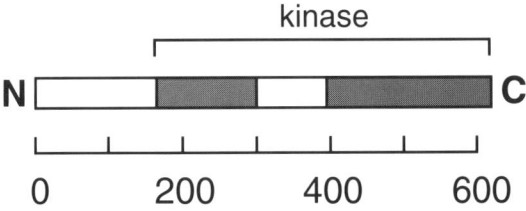

HRI contains a large insert of ≈100 residues between domains V and VI. The N-terminal region and kinase insert are unique to HRI and may be involved in haem-regulation.

Database accession numbers

	PIR	SWISSPROT	EMBL/GENBANK	REF
Rabbit			M69035	4

Amino acid sequence of rabbit HRI

```
MLGGSAGTRG GEAEGDGAGA VGAVAPPPAI DFPAEVSDPK YDESDVPAEL  50
QVLKEPLQQP AFPFAVANQL LLVSLLEHLS HVHEPNPLRS RQVFKLLCQT 100
FIKMGLLSSF TCSDEFSSLR LHHNRAITHL MRSARERVRQ DPCADNSHIQ 150
KIRSREVALE AQTSRYLNEF EELSILGKGG YGRVYKVRNK LDGQYYAIKK 200
ILIKGATKTD CMKVLREVKV LAGLQHPNIV GYHTAWIEHV HVHVQADRVP 250
IQLPSLEVLS DQEEDRDQYG VKNDASSSSS IIFAEFSPEK EKSSHECAVE 300
SQNNKLVNYT TNLVVRDTGE FESSTERQEN GSIVERQLLF GHNSDVEEDF 350
TSAEESSEED LSALRHTEVQ YHLMLHIQMQ LCELALWDWI AERNRRSREC 400
```

Amino acid sequence of rabbit HRI *continued*

```
VDESACPYVM VSVATKIFQE LVEGVFYIHN MGIVHRDLKP RNIFLHGPDQ 450
QVKIGDFGLA CADIIQKNTA RTSRNGERAP THTSRVGTCL YASPEQLEGS 500
EYDAKSDMYS VGVILLELFQ PFGTEMERAE VLTGVRAGRI PDSLSKRCPA 550
QAKYVQLLTR RNASQRPSAL QLLQSELFQN SAHVNLTLQM KIIEQEREIE 600
ELKKQLSLLS QARGVRSDRR DGELPA
```

Homologues in other species

Yeast eIF-2α kinase (**Gcn2**) shares extensive sequence similarity with HRI. The similarity is greater than that between HRI and **PKR**, the other known mammalian eIF-2α kinase. The mechanisms of activation of these three eIF-2α kinases are different.

Physiological substrates and specificity determinants

The only known physiological substrate of HRI is the α subunit of eIF-2. The site of the phosphorylation is Ser51. The sequence surrounding Ser51 is LSELSRRRIR. The consensus sequence for phosphorylation by HRI appears to be **SXX<u>R</u>** or **SXXX<u>R</u>**[5].

Assay

The enzyme is usually assayed by the phosphorylation of the α subunit of eIF-2[6]. In addition, HRI undergoes autophosphorylation, which can be used as an assay.

Enzyme inhibitor and activators

HRI is inactivated by haem. Autophosphorylation, which is accompanied by the activation of HRI, and eIF-2α phosphorylation are inhibited by haemin with an apparent K_i of about 1 μM. Haemin inhibits ATP binding to HRI. Haemin also promotes intersubunit disulphide bond formation in HRI, and produces a disulphide linked homodimer. HRI in reticulocyte lysates becomes haem-irreversible when it is activated by prolonged heat treatment. The mechanism for this activation is not clear. In addition, HRI in haemin-supplemented reticulocyte lysates is activated by treatment with heat shock, N-ethylmaleimide, oxidized glutathione and heavy metal ions. Heat shock proteins may be involved in the activation of HRI in haemin-supplemented lysates[2].

Pattern of expression

Erythroid-specific; most abundant in reticulocytes, detectable in bone marrow, not detectable in mature erythrocytes or any other tissues[7-9].

References

1   London, I.M. et al. (1987) In The Enzymes, Vol. XVII, 3rd edn, Boyer, P.D. and Krebs, E.G., eds, Academic Press, New York, pp. 359–380.
2   Chen, J.-J. (1993) In Translational Control of Gene Expression 2, Ilan, J., ed., pp. 349–372.
3   Hershey, J.W.B. (1991) Annu. Rev. Biochem. 60, 717–755.

[4] Chen, J.-J. et al. (1991) Proc. Natl Acad. Sci. USA 88, 7729–7733.

[5] Proud, C.G. et al. (1991) Eur. J. Biochem. 195, 771–779.

[6] Chen, J.-J. et al. (1989) J. Biol. Chem. 264, 9559–9564.

[7] Petryshyn, R. et al. (1984) Biochem. Biophys. Res. Commun. 119, 891–898.

[8] Pal, J.K. et al. (1991) Biochemistry, 30, 2555–2562.

[9] Crosby, J.S. et al. (1994) Mol. Cell. Biol. 14, 3906–3914.

PKR | Protein kinase (RNA dependent) (vertebrates) (dsRNA-activated PK, P1/eIF2 kinase, dsRNA-activated inhibitor (DAI), p68, p65)

Ara G. Hovanessian (Institut Pasteur, Paris, France)

PKR is induced transcriptionally in mouse and human cells upon treatment with interferon. An analogous enzyme has been described in rabbit reticulocytes, different mouse tissues, and human peripheral blood mononuclear cells[1]. PKR is mostly associated with the microsomal fraction. This serine/threonine PK is responsible for the phosphorylation of the α subunit of the protein synthesis initiation factor eIF2, thus mediating inhibition of protein synthesis. Direct evidence for this has been provided by the expression of the recombinant human PKR in murine cells resulting in an antiviral effect[2], and in *S. cerevisiae* resulting in inhibition of growth[3]. PKR has also been reported to manifest a potential tumour suppressor function[4, 5] by an unknown mechanism.

Subunit structure and isoforms

SUBUNIT	AMINO ACIDS	MOL. WT	SDS-PAGE
PKR	551	62 100	68 kDa

Genetics
Chromosomal location:
  Human   2p21–22[6, 7]
  Mouse   17E2[7]
Mutations: K296R, K296P (inactive)[8].

Domain structure

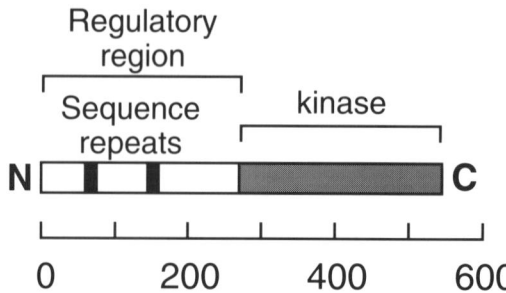

The regulatory region interacting with RNA has been localized at the N-terminal end of PKR, upstream of the protein kinase catalytic subdomains[1]. Mutations of K296 to R or P inactivate PKR[8]. Residues 362–371 are similar between human and murine PKRs, **HRI** and **Gcn2**, and may be implicated as a substrate recognition site[3, 9]. Residues 58–75 and 148–166 manifest strong identity with each other[10] and might represent an important motif for binding to RNA[11].

Database accession numbers

	PIR	SWISSPROT	EMBL/GENBANK	REF
Human			M35663	12
Mouse			M93567	9

Amino acid sequence of human PKR

```
MAGDLSAGFF MEELNTYRQK QGVVLKYQEL PNSGPPHDRR FTFQVIIDGR  50
EFPEGEGRSK KEAKNAAAKL AVEILNKEKK AVSPLLLTTT NSSEGLSMGN 100
YIGLINRIAQ KKRLTVNYEQ CASGVHGPEG FHYKCKMGQK EYSIGTGSTK 150
QEAKQLAAKL AYLQILSEET SVKSDYLSSG SFATTCESQS NSLVTSTLAS 200
ESSSEGDFSA DTSEINSNSD SLNSSSLLMN GLRNNQRKAK RSLAPRFDLP 250
DMKETKYTVD KRFGMDFKEI ELIGSGGFGQ VFKAKHRIDG KTYVIKRVKY 300
NNEKAEREVK ALAKLDHVNI VHYNGCWDGF DYDPETSDDS LESSDYDPEN 350
SKNSSRSKTK CLFIQMEFCD KGTLEQWIEK RRGEKLDKVL ALELFEQITK 400
GVDYIHSKKL IHRDLKPSNI FLVDTKQVKI GDFGLVTSLK NDGKRTRSKG 450
TLRYMSPEQI SSQDYGKEVD LYALGLILAE LLHVCDTAFE TSKFFTDLRD 500
GIISDIFDKK EKTLLQKLLS KKPEDRPNTS EILRTLTVWK KSPEKNERHT 550
C
```

Homologues in other species

Human and mouse PKR are 61% identical[9, 12]. Murine PKR[9] is identical to murine **Tik**[13]. PKR shows 38% sequence identity with yeast **Gcn2**.

Assay

PKR can be assayed in non-ionic detergent extracts of cells by incubation with poly(I).poly(C), Mg^{2+}, Mn^{2+} and $[\gamma-^{32}P]ATP$. Once autophosphorylated, it can be assayed for by phosphorylation of eIF2 or histone H2B. Monoclonal and polyclonal antibodies have been raised against human PKR.

Enzyme activators and inhibitors

PKR is activated by binding to ds (double stranded) RNAs; however, some ssRNAs, by virtue of stem-loop structures, can bind and trigger activation. Other polyanions such as heparin can activate PKR[1]. PKR is activated by low concentrations of RNA, whereas high concentrations inhibit its activation[14]. Besides the existence of some cellular inhibitors, different viruses have developed specific mechanisms to downregulate PKR activity[15]. PKR is inhibited by 2-aminopurine.

References

1 Hovanessian, A.G. (1993) Semin. Virol. 4, 237–245.
2 Meurs, E. et al. (1992) J. Virol. 66, 5805–5814.
3 Chong, K.L. et al. (1992) EMBO J. 11, 1553–1562.
4 Koromilas, A.E. et al. (1992) Science 257, 1685–1689.
5 Meurs, E. et al. (1993) Proc. Natl Acad. Sci. USA. 90, 232–236.
6 Barber, G.N. et al. (1993) Genomics 16, 765–767.
7 Squire, J. et al. (1993) Genomics 16, 768–770.
8 Katze, M. et al. (1991) Mol. Cell. Biol. 11, 5497–5505.
9 Fen, G.S. (1992) Proc. Natl Acad. Sci. USA 89, 5447–5451.
10 McCormack, S.J. et al. (1992) Virol. 188, 47–56.
11 Green, S.R. and Mathews, M.B. (1992) Genes Devel. 6, 2478–2490.
12 Meurs, E. et al. (1990) Cell 62, 379–390.

13 Icely, P.L. et al. (1991) J. Biol. Chem. 266, 16073–16077.
14 Galabru et al. (1989) Eur. J. Biochem. 178, 581–589.
15 Katze, M.G. (1993) Semin. Virol. 4, 259–268.

Tik

Protein kinase (RNA dependent) (vertebrates)
(p65, dsRNA-dependent PK)

Ninan Abraham and John C. Bell (University of Ottawa, Canada)

TIK was cloned by screening of a cDNA expression library with an anti-P-Tyr antibody[1]. Paradoxically, both the human and murine Tik kinases phosphorylate serine and threonine *in vitro* and *in vivo* and therefore have been placed in the family of dual-specificity PKs. Reactivity with anti-P-Tyr antibodies is intrinsic to the enzyme since phosphotransfer mutants lack such immunoreactivity. It is now clear that Tik is the murine homologue of human **PKR**.

Subunit structure and isoforms

SUBUNIT	AMINO ACIDS	MOL. WT	SDS-PAGE
Tik	515	58 570	65 kDa

Domain structure

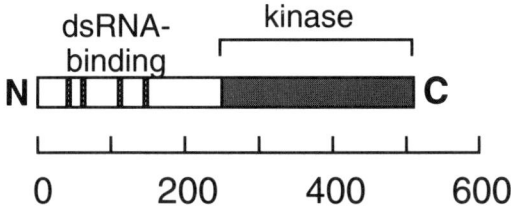

Database accession numbers

	PIR	SWISSPROT	EMBL/GENBANK	REF
Mouse			M65029	1

Amino acid sequence of mouse Tik

```
MASDTPGFYM DKLNKYRQMH GVAITYKELS TSGPPHDRRF TFQVLIDEKE  50
FPEAKGRSKQ EARNAAAKLA VDILDNENKV DCHTSASEQG LFVGNYIGLV 100
NSFAQKKKLS VNYEQCEPNS ELPQRFICKC KIGQTMYGTG SGVTKQEAKQ 150
LAAKEAYQKL LKSPPKTAGT SSSVVTSTFS GFSSSSSMTS NGVSQSAPGS 200
FSSENVFTNG LGENKRKSGV KVSPDDVQRN KYTLDARFNS DFEDIEEIGL 250
GGFGQVFKAK HRIDGKRYAI KRVKYNTEKA EHEVQALAEL NHVNIVQYHS 300
CWEGVDYDPE HSMSDTSRYK TRCLFIQMEF CDKGTLEQWM RNRNQSKVDK 350
ALILDLYEQI VTGVEYIHSK GLIHRDLKPG NIFLVDERHI KIGDFGLATA 400
LENDGKSRTR RTGTLQYMSP EQLFLKHYGK EVDIFALGLI LAELLHTRFT 450
ESEKIKFFES LRKGDFSNDI FDNKEKSLLK KLLSEKPKDR PETSEILKTL 500
AEWRNISEKK KRNTC
```

Homologues in other species

Tik is now known to be the mouse homologue of **PKR**, and is also related to *S. cerevisiae* **Gcn2**. *TIK* has been cloned from a 70Z/3 pre-B cell expression library[1] and from an FM3A mammary carcinoma expression library[2].

Physiological substrates and specificity determinants

Tik phosphorylates the translation factor eIF2α, and inhibits its ability to mediate reinitiation of translation.

Assay

Tik can be assayed by autophosphorylation in the presence of dsRNA and [γ-^{32}P]ATP, either directly in an extract (as a 65 kDa band[3]), or in an immune complex.

Enzyme activators and inhibitors

The activity of Tik is increased in the presence of dsRNA, including viral replication intermediates, and the TAR region of HIV. Heparin and poly(I):poly(C) are also effective *in vitro* activators. Viral proteins such as Tat, and viral RNA products such as the VAI RNA of adenovirus, block Tik binding to RNA. 2-Aminopurine is an effective *in vitro* inhibitor of Tik activity.

Pattern of expression

Tik is found predominantly as a soluble cytoplasmic enzyme, or associated with polysomes. It is expressed in all adult tissues. Expression of the *TIK* gene can be dramatically increased following interferon treatment of tissue culture cells.

References

1 Icely, P.L. et al. (1991) J. Biol. Chem. 266, 16073–16077.
2 Feng, G. et al. (1992) Proc. Natl Acad. Sci. USA 89, 5447–5451.
3 Roy, S. et al. (1990) Science 247, 1216–1219.

Alan G. Hinnebusch (National Institutes of Health, Bethesda, MD, USA)

Gcn2 regulates protein synthesis at the initiation step by phosphorylating the α subunit of eIF2 on Ser51. Activation of the wildtype enzyme in response to amino acid starvation produces a low level of eIF2α phosphorylation that stimulates the translation of *GCN4* mRNA by a unique reinitiation mechanism. Because Gcn4 is a transcriptional activator of amino acid biosynthetic genes, Gcn2 is required for growth when amino acids are limiting[1]. Mutationally activated forms of Gcn2 produce a high level of eIF2α phosphorylation that inhibits total protein synthesis[2,3], similar to what occurs in mammalian cells upon activation of the eIF2α kinases **PKR** or **HRI**[4].

Subunit structure and isoforms

SUBUNIT	AMINO ACIDS	MOL. WT	SDS-PAGE
Gcn2	1590	182 000	180 kDa

Genetics

Chromosomal location: IV, right arm[1].

Mutations:

K559R: kinase inactive *in vivo* and *in vitro*

Various deletions and insertions in regions flanking the kinase domain (e.g. Δ15–421, 437S,S) prevent induction of *GCN4* translation *in vivo*, but the kinase remains active *in vitro*[5,6]

Mutations causing constitutive activation of *GCN4* translation (e.g. E532K) map in the kinase domain, tRNA synthetase domain or the C-terminal region[3,5]

Domain structure[6,7]

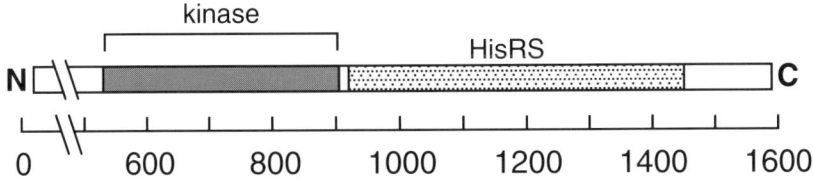

The HisRS domain is similar in sequence to histidyl-tRNA synthetase[6]. The C-terminal domain is critical for the ribosome association[8].

Database accession numbers

	PIR	SWISSPROT	EMBL/GENBANK	REF
Gcn2	OKBYN2	P15442	M27082	6

Amino acid sequence of *S. cerevisiae* Gcn2

```
MTPMYPYTAP EIEFKNVQNV MDSQLQMLKS EFKKIHNTSR GQEIIFEITS   50
FTQEKLDEFQ NVVNTQSLED DRLQRIKETK EQLEKEEREK QQETIKKRSD  100
EQRRIDEIVQ RELEKRQDDD DDLLFNRTTQ LDLQPPSEWV ASGEAIVFSK  150
TIKAKLPNNS MFKFKAVVNP KPIKLTSDIF SFSKQFLVKP YIPPESPLAD  200
FLMSSEMMEN FYYLLSEIEL DNSYFNTSNG KKEIANLEKE LETVLKAKHD  250
NVNRLFGYTV ERMGRNNATF VWKIRLLTEY CNYYPLGDLI QSVGFVNLAT  300
ARIWMIRLLE GLEAIHKLGI VHKCINLETV ILVKDADFGS TIPKLVHSTY  350
GYTVLNMLSR YPNKNGSSVE LSPSTWIAPE LLKFNNAKPQ RLTDIWQLGV  400
LFIQIISGSD IVMNFETPQE FLDSTSMDET LYDLLSKMLN NDPKKRLGTL  450
ELLPMKFLRT NIDSTINRFN LVSESVNSNS LELTPGDTIT VRGNGGRTLS  500
QSSIRRRSFN VGSRFSSINP ATRSRYASDF EEIAVLGQGA FGQVVKARNA  550
LDSRYYAIKK IRHTEEKLST ILSEVMLLAS LNHQYVVRYY AAWLEEDSMD  600
ENVFESTDEE SDLSESSSDF EENDLLDQSS IFKNRTNHDL DNSNWDFISG  650
SGYPDIVFEN SSRDDENEDL DHDTSSTSSS ESQDDTDKES KSIQNVPRRR  700
NFVKPMTAVK KKSTLFIQME YCENRTLYDL IHSENLNQQR DEYWRLFRQI  750
LEALSYIHSQ GIIHRDLKPM NIFIDESRNV KIGDFGLAKN VHRSLDILKL  800
DSQNLPGSSD NLTSAIGTAM YVATEVLDGT GHYNEKIDMY SLGIIFFEMI  850
YPFSTGMERV NILKKLRSVS IEFPPDFDDN KMKVEKKIIR LLIDHDPNKR  900
PGARTLLNSG WLPVKHQDEV IKEALKSLSN PSSPWQQQVR ESLFNQSYSL  950
TNDILFDNSV PTSTPFANIL RSQMTEEVVK IFRKHGGIEN NAPPRIFPKA 1000
PIYGTQNVYE VLDKGGTVLQ LQYDLTYPMA RYLSKNPSLI SKQYRMQHVY 1050
RPPDHSRSSL EPRKFGEIDF DIISKSSSES GFYDAESLKI IDEILTVFPV 1100
FEKTNTFFIL NHADILESVF NFTNIDKAQR PLVSRMLSQV GFARSFKEVK 1150
NELKAQLNIS STALNDLELF DFRLDFEAAK KRLYKLMIDS PHLKKIEDSL 1200
SHISKVLSYL KPLEVARNVV ISPLSNYNSA FYKGGIMFHA VYDDGSSRNM 1250
IAAGGRYDTL ISFFARPSGK KSSNTRKAVG FNLAWETIFG IAQNYFKLAS 1300
GNRIKKRNRF LKDTAVDWKP SRCDVLISSF SNSLLDTIGV TILNTLWKQN 1350
IKADMLRDCS SVDDVVTGAQ QDGIDWILLI KQQAYPLTNH KRKYKPLKIK 1400
KLSTNVDIDL DLDEFLTLYQ QETGNKSLIN DSLTLGDKAD EFKRWDENSS 1450
AGSSQEGDID DVVAGSTNNQ KVIYVPNMAT RSKKANKREK WVYEDAARNS 1500
SNMILHNLSN APIITVDALR DETLEIISIT SLAQKEEWLR KVFGSGNNST 1550
PRSFATSIYN NLSKEAHKGN RWAILYCHKT GKSSVIDLQR
```

Homologues in other species

The kinase domain shows 36 and 42% sequence identity with **HRI** and human **PKR**, respectively. In addition, these three eIF2α kinases contain inserts in subdomains IV and VII; the inserts in subdomain IV of Gcn2 and HRI are quite large and share a short stretch of sequence similarity. Based on these structural similarities, it was proposed that Gcn2, PKR and HRI are members of a kinase subfamily[3, 9, 10]. In fact, expression of PKR or HRI in yeast cells can substitute for Gcn2 in the regulation of *GCN4*-specific and total mRNA translation[10, 11].

Physiogical substrates and specificity determinants

eIF2α is the only known substrate for Gcn2. A sequence of 18 amino acids (NIEGMILLSELSRRRIRSI) surrounding S51 is identical between yeast and human eIF2α[12], and Gcn2 phosphorylates rabbit eIF2α *in vitro*[2]. These considerations make it likely that Gcn2 closely resembles the mammalian kinases regarding specificity determinants.

Assay

The enzyme has been assayed in immune complexes by transfer of radio-activity from $[\gamma\text{-}^{32}\text{P}]$ATP to purified *S. cerevisiae* or rabbit eIF2[2]. Autophos-phorylation can also be assayed[5]. The specific activity of Gcn2 in these reactions is not increased by growing cells under starvation conditions prior to isolating the enzyme, nor by most mutations (*GCN2ᶜ*) that activate Gcn2 function *in vivo*[5] (see ref. 16 for an exception to the latter). Similarly, it is not diminished by *gcn1* mutations which impair function *in vivo*[13].

Enzyme activators and inhibitors

Gcn2 exhibits a low basal activity when cells are grown under non-starvation conditions and its ability to phosphorylate eIF2α is stimulated by limiting cells for an amino acid[2]. Because mutations that reduce aminoacyl-tRNA-synthetase activity lead to high-levels of *GCN4* expression that are dependent on Gcn2, uncharged tRNA is thought to be a physiological activator of Gcn2 kinase function[1, 14]. The large C-terminal regulatory region that is similar in sequence to histidyl-tRNA synthetases could have a role in detecting uncharged tRNA in amino acid-starved cells and activating the adjacent kinase domain[3, 5, 6]. Gcn2 is ribosome-associated[8]; thus, activation of its kinase function by uncharged tRNA may be coupled to the translation process. Gcn2 is also activated by purine starvation[15], but the activator in this situation is unknown. The Gcn1 protein is required *in vivo* for activating Gcn2 kinase function in response to amino acid or purine starvation, and to various *GCN2ᶜ* mutations[13, 15]; however, the exact role of Gcn1 in the activation mechanism is unknown.

Pattern of expression

There is evidence that transcription of the *GCN2* gene increases under starvation conditions where it functions to stimulate the translation of *GCN4*[7]; however, an increase in the level of Gcn2 protein has not been observed in these circumstances[5]. Moreover, *GCN2ᶜ* mutations elevate the level of Gcn2 kinase function without increasing the level of the protein; in fact, the steady-state levels of the most highly activated Gcn2ᶜ proteins appear to be less than that of wildtype Gcn2[3, 5].

References

1   Hinnebusch, A.G. (1992) In The Molecular and Cellular Biology of the Yeast Saccharomyces: Gene Expression, Jones, E.W. et al., eds, Cold Spring Harbor Laboratory Press, Cold Spring Harbor, pp. 319–414.
2   Dever, T.E. et al. (1992) Cell 68, 585–596.
3   Ramirez, M. et al. (1992) Mol. Cell. Biol. 12, 5801–5815.
4   Hershey, J.W.B. (1991) Annu. Rev. Biochem. 60, 717–755.
5   Wek, R.C. et al. (1990) Mol. Cell. Biol. 10, 2820–2831.
6   Wek, R.C. et al. (1989) Proc. Natl Acad. Sci. USA 86, 4579–4583.
7   Roussou, I. et al. (1988) Mol. Cell. Biol. 8, 2132–2139.

[8] Ramirez, M. et al. (1991) Mol. Cell. Biol. 11, 3027–3036.
[9] Chen, J.-J. et al. (1991) Proc. Natl Acad. Sci. USA 88, 7729–7733.
[10] Chong, K.L. et al. (1992) EMBO J. 11, 1553–1562.
[11] Dever, T.E. et al. (1993) Proc. Natl Acad. Sci. USA 90, 4616–4620.
[12] Cigan, A.M. et al. (1989) Proc. Natl Acad. Sci. USA 86, 2784–2788.
[13] Marton, M.J. et al. (1993) Mol. Cell. Biol. 13, 3541–3556.
[14] Lanker, S. et al. (1992) Cell 70, 647–657.
[15] Rolfes, R.J. and Hinnebusch, A.G. (1993) Mol. Cell. Biol. 13, 5099–5111.
[16] Diallinas, G. and Thireos, G. (1994) Gene 143, 21–27.

Raf1 protein kinase (vertebrates)

Tom W. Beck and Ulf R. Rapp (National Cancer Institute, USA)

Raf1 is a serine/threonine-specific PK which functions in a signal transduction pathway(s) between the cell membrane and the nucleus[1]. On mitogen stimulation of resting cells, Raf1 becomes hyperphosphorylated, which correlates with an increase of Raf1 kinase activity. Raf1 then phosphorylates and activates MAP kinase kinase (**Mek**) which in turn activates MAP kinase (**Erk1/2**) leading to transactivation of transcription from promoters bearing AP-1 (Jun/Fos), Ets and other response elements. Raf1 is required for the activation of c-Jun via phosphorylation in the transactivation domain.

Subunit structure and isoforms

SUBUNIT	AMINO ACIDS	MOL. WT	SDS-PAGE
Raf1	648	73 023	72–74 kDa

Genetics

Chromosomal location (human): 3p25.

Domain structure

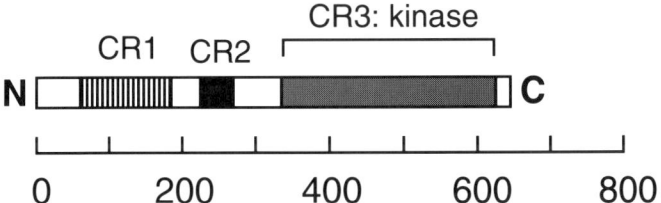

Since the early transforming versions of Raf1 were N-terminally truncated, constitutively active kinases, we proposed that Raf1 was composed of two functional domains; an N-terminal regulatory domain and a C-terminal catalytic domain[2]. Our model for the full-length protein predicted that it would be modulated by ligand binding to the N-terminal half, resulting in kinase activation. This model is supported by experiments demonstrating that the N-terminal regulatory half titrates out a direct Raf1/Ras interaction and/or a Ras-induced ligand required for Raf1 activation *in vivo*[3]. Based on homology with other Raf family members, Raf1 comprises three structural domains. CR1 contains a Cys finger domain (139–184) which is similar to that of **PKC**, *n*-chimaerin and Unc-13. CR2 contains both the major site of phosphorylation in response to serum *in vivo* (S259) and the major autophosphorylation site (T268)[4, 5]. CR3 is the catalytic domain. Additional phosphorylation *in vivo* occurs at S43, S499, and S621. Both S259 and S499 are regulatory phosphorylation sites for PKC *in vivo* and *in vitro*[6, 7].

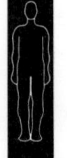

Database accession numbers

		PIR	SWISSPROT	EMBL/GENBANK	REF
Raf1	Human	A00637	P04049	L00212	2
	Rat	S00726	P11345	M15427	8
Mil	Chicken	S00644		X07017	9
Raf	*Xenopus*	S01930	P09560	S74063	10

Amino acid sequence of human Raf1

```
MEHIQGAWKT ISNGFGFKDA VFDGSSCISP TIVQQFGYQR RASDDGKLTD  50
PSKTSNTIRV FLPNKQRTVV NVRNGMSLHD CLMKALKVRG LQPECCAVFR 100
LLHEHKGKKA RLDWNTDAAS LIGEELQVDF LDHVPLTTHN FARKTFLKLA 150
FCDICQKFLL NGFRCQTCGY KFHEHCSTKV PTMCVDWSNI RQLLLFPNST 200
IGDSGVPALP SLTMRRMRES VSRMPVSSQH RYSTPHAFTF NTSSPSSEGS 250
LSQRQRSTST PNVHMVSTTL PVDSRMIEDA IRSHSESASP SALSSSPNNL 300
SPTGWSQPKT PVPAQRERAP VSGTQEKNKI RPRGQRDSSY YWEIEASEVM 350
LSTRIGSGSF GTVYKGKWHG DVAVKILKVV DPTPEQFQAF RNEVAVLRKT 400
RHVNILLFMG YMTKDNLAIV TQWCEGSSLY KHLHVQETKF QMFQLIDIAR 450
QTAQGMDYLH AKNIIHRDMK SNNIFLHEGL TVKIGDFGLA TVKSRWSGSQ 500
QVEQPTGSVL WMAPEVIRMQ DNNPFSFQSD VYSYGIVLYE LMTGELPYSH 550
INNRDQIIFM VGRGYASPDL SKLYKNCPKA MKRLVADCVK KVKEERPLFP 600
QILSSIELLQ HSLPKINRSA SEPSLHRAAH TEDINACTLT TSPRLPVF
```

Sites of interest: S43, S621, phosphorylation; S259, S499, phosphorylation by
PKC; T268, autophosphorylation

Homologues in other species
Raf1 has been sequenced from mouse (G. Heidecker and U.R. Rapp,
unpublished), rat, chicken (c-*mil*), *Xenopus laevis*, *D. melanogaster*
(**DmRaf**), and *C. elegans* (A. Golden and P.W. Sternberg, unpublished).

Physiological substrates and specificity determinants
The only known physiological substrate is MAP kinase kinase (**Mek**)[11].

Assay
The enzyme is usually assayed in immune complexes as a mitogen-dependent
transfer of radioactivity from $[\gamma\text{-}^{32}\text{P}]\text{ATP}$ to Raf1 (autophosphorylation),
histone type V-S, or any of several peptide substrates (e.g. Raf1 residues 32–
50)[12–15]. The best assay for Raf1 activity is incubation with dephosphorylated
MAP kinase kinase (**Mek**), which can then be assayed by phosphorylation of,
or reactivation of, MAP kinase (**Erk1/2**)[11]. The recent cloning of Mek should
allow direct assays of Raf1 phosphorylation.

Enzyme activators and inhibitors
Raf1 is activated by PKC α, β and γ isoforms[6, 7]. The CR1 domain from Raf1,
A-Raf or B-Raf is an inhibitor (J. Bruder and U.R. Rapp, unpublished)[3].

Pattern of expression

Raf1 mRNA is expressed in all human and mouse cell lines and all mouse tissues examined to date, as 3.1 and 3.4 kb RNAs in mouse and human, respectively[16]. The levels of expression vary by as much as 5–10-fold between different mouse tissues. In mouse testes, Raf1 is expressed in both the somatic and germ cell compartments[17].

References

1 Rapp, U.R. (1991) Oncogene 6, 495–500.
2 Bonner, T. et al. (1986) Nucl. Acids Res. 14, 1009–1015.
3 Bruder, J.T. et al. (1992) Genes Devel. 6, 545–556.
4 McGrew, et al. (1992) Oncogene 7, 33–42.
5 Morrison, D.K. et al. (1993) J. Biol. Chem. 268, 17309–17316.
6 Kolch, W. et al. (1993) Nature 364, 249–252.
7 Sozeri, O. et al. (1992) Oncogene 7, 2259–2262.
8 Ishikawa, F. et al. (1987) Oncogene Res. 243–253.
9 Koenen, M. et al. (1988) Oncogene 2, 179–185.
10 le Guellec, R. et al. (1988) Nucl. Acids Res. 16, 10357.
11 Kyriakis, J.M. et al. (1992) Nature 358, 417–421.
12 App, H. et al. (1991) Mol. Cell. Biol. 11, 913–919.
13 Moelling, K. et al. (1984) Nature 312, 558–561.
14 Morrison, D.K. et al. (1988) Proc. Natl Acad. Sci. USA 85, 8855–8859.
15 Schultz, A.M. (1988) Oncogene 2, 187–193.
16 Storm, U.R. et al. (1990) Oncogene 5, 345–351.
17 Wadewitz, A.G. et al. (1993) Oncogene 8, 1055–1062.

Tom W. Beck and Ulf R. Rapp (National Cancer Institute, USA)

A-Raf was originally identified by screening a mouse spleen cDNA library at reduced stringency with a v-*raf* probe[1]. It follows the same general functional rules and biochemical properties which have been established for **Raf-1** (S. Grugal and U.R. Rapp, unpublished)[2,3]. Mitogen stimulation of resting cells results in A-Raf hyperphosphorylation, an increase of Raf1 kinase activity, and activation of the MAP kinase cascade resulting in transactivation of transcription from promoters bearing AP-1 (Jun/Fos), Ets and other response elements.

Subunit structure and isoforms

SUBUNIT	AMINO ACIDS	MOL. WT	SDS-PAGE
A-Raf	606	67530	69 kDa

Genetics

Chromosomal location (human): Xp11.2.

Domain structure

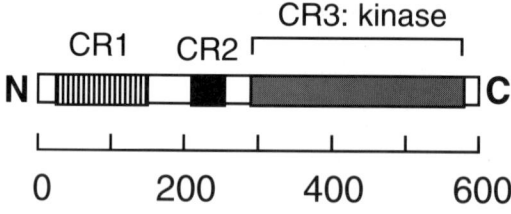

A-Raf has a shorter N-terminal region than Raf1, and lacks some of the residues which are phosphorylated in Raf1 *in vivo* including S43 and S499, but retains S259 and S621 equivalents.

Database accession numbers

	PIR	SWISSPROT	EMBL/GENBANK	REF
Human	A26436	P10398	X04790	2
Mouse	A25382	P04627	M13071	4
Rat	S00726	P14056	X06942	1

Amino acid sequence of human A-Raf

```
MEPPRGPPAN GAEPSRAVGT VKVYLPNKQR TVVTVRDGMS VYDSLDKALK  50
VRGLNQDCCV VYRLIKGRKT VTAWDTAIAP LDGEELIVEV LEDVPLTMHN 100
FVRKTFFSLA FCDFCLKFLF HGFRCQTCGY KFHQHCSSKV PTVCVDMSTN 150
RQQFYHSVQD LSGGSRQHEA PSNRPLNELL TPQGPSPRTQ HCDPEHFPFP 200
APANAPLQRI RSTSTPNVHM VSTTAPMDSN LIQLTGQSFS TDAAGSRGGS 250
DGTPRGSPSP ASVSSGRKSP HSKSPAEQRE RKSLADDKKK VKNLGYRDSG 300
YYWEVPPSEV QLLKRIGTGS FGTVFRGRWH GDVAVKVLKV SQPTAEQAQA 350
FKNEMQVLRK TRHVNILLFM GFMTRPGFAI ITQWCEGSSL YHHLHVADTR 400
```

Amino acid sequence of human A-Raf *continued*

```
FDMVQLIDVA RQTAQGMDYL HAKNIIHRDL KSNNIFLHEG LTVKIGDFGL 450
ATVKTRWSGA QPLEQPSGSV LWMAAEVIRM QDPNPYSFQS DVYAYGVVLY 500
ELMTGSLPYS HIGCRDQIIF MVGRGYLSPD LSKISSNCPK AMRRLLSDCL 550
KFQREERPLF PQILATIELL QRSLPKIERS ASEPSLHRTQ ADELPACLLS 600
AARLVP
```

Homologues in other species
A-Raf has been sequenced from human, mouse and rat.

Physiological substrates and specificity determinants
MAP kinase kinase (**Mek**) is the only known substrate.

Assay
The enzyme is assayed as an immune complex, as described for the Raf1 protein, using autophosphorylation, histone type V-S, several peptides (e.g. Raf1 32–50) or MAP kinase kinase as substrate (S. Grugal and U.R. Rapp, unpublished).

Enzyme activators and inhibitors
Protein kinase C α or γ isoforms are activators. The CR1 domains from Raf1, A-Raf or B-Raf are inhibitors (J. Bruder and U.R. Rapp, unpublished).

Pattern of expression
A-Raf is expressed as a 2.6 kb mRNA in some human and mouse cell lines. In the mouse, A-Raf is predominantly expressed in urogenital tissues, including epididymis, ovary, prostate, kidney and bladder. However, lower levels are detectable in other tissues[5]. In mouse testes, A-Raf is expressed in the somatic compartment, but not in germ cells[6].

References
1. Ishikawa, F. et al. (1987) Oncogene Res. 1, 243–253.
2. Beck, T.W. et al. (1987) Nucl. Acids Res. 15, 595–609.
3. Rapp, U.R. (1991) Oncogene 6, 495–500.
4. Huleihel, M. et al. (1986) Mol. Cell. Biol. 6, 2655–2662.
5. Storm, U.R. et al. (1990) Oncogene 5, 345–351.
6. Wadewitz, A.G. et al. (1993) Oncogene 8, 1055–1062.

B-Raf
B-Raf protein kinase (vertebrates)

Tom W. Beck and Ulf R. Rapp (National Cancer Institute, USA)

A B-Raf oncogene was originally identified in an NIH3T3 cell transformant which was transfected with Ewing sarcoma DNA and a full-length B-Raf cDNA, and protein has since been characterized[1-4]. It follows the same general functional rules and biochemical properties established for Raf1[2,4,5]. NGF, EGF or TPA stimulation of PC12 cells results in B-Raf becoming hyperphosphorylated, an increase of B-Raf kinase activity, and activation of the MAP kinase cascade (J. Troppmair and U.R. Rapp, unpublished)[2,4].

Subunit structure and isoforms

SUBUNIT	AMINO ACIDS	MOL. WT	SDS-PAGE
B-Raf	765	85 000	92–95 kDa

Genetics

Chromosomal location (human): 7q33–36.

Domain structure

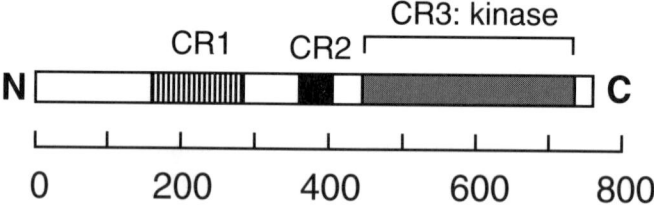

B-Raf is structurally similar to Raf1 and A-Raf but has a larger N-terminus. The CR2 region of B-Raf diverges slightly from this region in Raf1 and A-Raf. B-Raf has residues corresponding to all the major phosphorylation sites identified in Raf1.

Database accession numbers

		PIR	SWISSPROT	EMBL/GENBANK	REF
B-Raf	Human	S13798	P15056	X65187	3, 4
	Mouse	A40951	P28028		6
R-Mil	Chicken			X67052	7
	Quail			M80845	8

Amino acid sequence of human B-Raf

```
MAALSGGGGG GAEPGQALFN GDMEPEAGAG RPAASSAADP AIPEEVWNIK  50
QMIKLTQEHI EALLDKFGGE HNPPSIYLEA YEEYTSKLDA LQQREQQLLE 100
SLGNGTDFSV SSSASMDTVT SSSSSSLSVL PSSLSVFQNP TDVARSNPKS 150
PQKPIVRVFL PNKQRTVVPA RCGVTVRDSL KKALMMRGLI PECCAVYRIQ 200
DGEKKPIGWD TDISWLTGEE LHVEVLENVP LTTHNFVRKT FFTLAFCDFC 250
RKLLFQGFRC QTCGYKFHQR CSTEVPLMCV NYDQLDLLFV SKFFEHHPIP 300
```

Amino acid sequence of human B-Raf *continued*

```
QEEASLAETA LTSGSSPSAP ASDSIGPQIL TSPSPSKSIP IPQPFRPADE 350
DHRNQFGQRD RSSSAPNVHI NTIEPVNIDD LIRDQGFRGD GGSTTGLSAT 400
PPASLPGSLT NVKALQKSPG PQRERKSSSS SEDRNRMKTL GRRDSSDDWE 450
IPDGQITVGQ RIGSGSFGTV YKGKWHGDVA VKMLNVTAPT PQQLQAFKNE 500
VGVLRKTRHV NILLFMGYST KPQLAIVTQW CEGSSLYHHL HIIETKFEMI 550
KLIDIARQTA QGMDYLHAKS IIHRDLKSNN IFLHEDLTVK IGDFGLATVK 600
SRWSGSHQFE QLSGSILWMA PEVIRMQDKN PYSFQSDVYA FGIVLYELMT 650
GQLPYSNINN RDQIIFMVGR GYLSPDLSKV RSNCPKAMKR LMAECLKKKR 700
DERPLFPQIL ASIELLARSL PKIHRSASEP SLNRAGFQTE DFSLYACASP 750
KTPIQAGGYG AFPVH
```

Sites of interest: T372, autophosphorylation

Homologues in other species

B-Raf has been sequenced from human, mouse, chicken and quail.

Physiological substrates and specificity determinants

MAP kinase kinase (**Mek**) is believed to be a physiological substrate (J. Troppmair and U.R. Rapp, unpublished).

Assay

The enzyme (from PC12 cells) has been assayed in immune complexes as mitogen-dependent transfer of radioactivity from $[\gamma^{-32}P]ATP$ to either B-Raf (autophosphorylation), casein, phosvitin or a peptide substrate (B-Raf 359–386) as an immune complex[2, 4]. In contrast to Raf1, little or no mitogen-dependent stimulation of B-Raf kinase activity was observed with histone H1, histone H5 or syntide[8]. Nevertheless, B-Raf activity can be measured using the MAP kinase assay as described for Raf1 (J. Troppmair and U.R. Rapp, unpublished).

Enzyme activators and inhibitors

PKC is an activator[2]. The CR1 domain from Raf1, A-Raf or B-Raf is an inhibitor (J. Bruder and U.R. Rapp, unpublished). Autophosphorylation reaction can be competed by exogenous substrates[4].

Pattern of expression

B-Raf is expressed in the mouse as multiple mRNA species of 2.6, 4.5, 10 and 13 kb, with highest levels in cerebellum and testes[9]. In mouse testes, B-Raf is expressed in the germ cell compartment, but not in the somatic compartment[10]. In quail neuroretina, alternate splicing gives rise to B-Raf/c-Rmil protein products possessing and lacking residues 393–433[8].

References

1 Ikawa, S. et al. (1988) Mol. Cell. Biol. 8, 2651–2654.
2 Oshima, M. et al. (1991) J. Biol. Chem. 266, 23753–23760.
3 Sithanandam, G. et al. (1990) Oncogene 5, 1775–1780.
4 Stephens, R.M. et al. (1992) Mol. Cell. Biol. 12, 3733–3742.

[5] Rapp, U.R. (1991) Oncogene 6, 495–500.
[6] Miki, T. et al. (1991) Proc. Natl Acad. Sci. USA 88, 5167–5171.
[7] Calogeraki, I. et al. (unpublished).
[8] Eychene, A. et al. (1992) Oncogene 7,1315–1323.
[9] Storm, U.R. et al. (1990) Oncogene 5, 345–351.
[10] Wadewitz, A.G. et al. (1993) Oncogene 8, 1055–1062.

 **DmRaf** Raf homologue (*D. melanogaster*)

Michael B. Melnick and Norbert Perrimon
(Howard Hughes Medical Institute, Boston, USA)

DmRaf was originally cloned by homology with mammalian Raf1[1,2]. Mutations in *DmRaf* identify the *l(1)pole hole* locus which has been implicated in signalling from both the *torso*[3], *sevenless*[4,5] and EGF receptor[6] tyrosine kinases in *D. melanogaster*.

Subunit structure and isoforms

SUBUNIT	AMINO ACIDS	MOL. WT	SDS-PAGE
DmRaf	782	88 600	

Genetics

Chromosomal location: X, 2F6.
Mutations:
400B8 (G484R), *C2Z2* (E516V), *11-29* (truncation), null mutants
PB26 (W634STOP), *C110* (R217L), residual activity

Domain structure

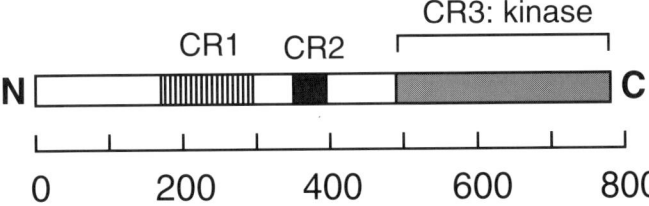

DmRaf contains the three highly conserved domains found in vertebrate Rafs, i.e. CR1, CR2 and the kinase domain (CR3). Domains CR1 and CR2 regulate activation of DmRaf. The C110 mutation in the CR1 domain disrupts the interaction with Ras1 but still shows partial activity (Hou et al., unpublished).

Database accession numbers

	PIR	SWISSPROT	EMBL/GENBANK	REF
DmRaf			L10626	5

 ## Amino acid sequence of *D. melanogaster* Raf

```
MSSESSTEGD SDLYDPLAEE LHNVQLVKHV TRENIDALNA KFANLQEPPA  50
MYLIGESSKA ELNTTWVLGN PTPTSKLFIK YPTYVYVCIS RLWFLPTEYQ 100
ELTSKLHELE AKEQELMERL NSQDQQEDSS LVERFKEQPH YQNQTQILQQ 150
QRQLARVHHG NDLTDSLGSQ PGSQCGTLTR QPKILLRAHL PNQQRTSVEV 200
ISGVRLCDAL MKALKLRQLT PDMCEVSTTH SGRHIIPWHT DIGTLHVEEI 250
FVRLLDKFPI RTHIKHQIIR KTFFSLVFCE GCRRLLFTGF YCSQCNFRFH 300
QRCANRVPML CQPFPMDSYY QLLLAENPDN GVGFPGRGTA VRFNMSSRSR 350
SRRCSSSGSS SSSKPPSSSS GNHRQGRPPR ISQDDRSNSA PNVCINNIRS 400
VTSEVQRSLI MQARPPLPHP CTDHSNSTQA SPTSTLKHNR PRARSADESN 450
```

Amino acid sequence of *D. melanogaster* Raf *continued*

```
KNLLLRDAKS SEENWNILAE EILIGPRIGS GSFGTVYRAH WHGPVAVKTL 500
NVKTPSPAQL QAFKNEVAML KKTRHCNILL FMGCVSKPSL AIVTQWCEGS 550
SLYKHVHVSE TKFKLNTLID IGRQVAQGMD YLHAKNIIHR DLKSNNIFLH 600
EDLSVKIGDF GLATAKTRWS GEKQANQPTG SILWMAPEVI RMQELNPYSF 650
QSDVYAFGIV MYELLAECLP YGHISNKDQI LFMVGRGLLR PDMSQVRSDA 700
PQALKRLAED CIKYTPKDRP LFRPLLNMLE NMLRTLPKIH RSASEPNLTQ 750
SQLQNDEFLY LPSPKTPVNF NNFQFFGSAG NI
```

Homologues in other species
Vertebrate homologues include **Raf1**, **A-Raf**, **B-Raf**, and the c-*mil* proto-oncogene in birds. The vertebrate functional homologue is **Raf1** (Brand and Perrimon, unpublished). Other similar vertebrate genes are **A-Raf**, **B-Raf**, and the c-mil proto-oncogene in birds.

Pattern of expression
DmRaf encodes a single 3.3 kb RNA which is ubiquitously expressed throughout development[3].

References
1 Mark, G.E. et al. (1987) Mol. Cell. Biol. 7, 2134–2140.
2 Nishida, Y. et al. (1988) EMBO J. 7, 775–781.
3 Ambrosio, L. et al. (1989). Nature 342, 288–291.
4 Dickson, B. et al. (1992). Nature 360, 600–602.
5 Melnick, M.B. et al. (1993) Development 118, 127–138.
6 Brand, A.H. and Perrimon, N. (1994) Genes and Devel. 8, 629–639.

 **LIN-45 Raf protein kinase (*C. elegans*)**

Andy Golden (California Institute of Technology, USA)

The *C. elegans raf* gene was cloned using degenerate oligonucleotides and PCR protocols. A genomic fragment was cloned and shown to rescue the mutant phenotypes of a mutation in the *lin-45* gene. This gene had been previously identified in a genetic screen for mutations that disrupt hermaphrodite vulval differentiation. Double mutant analysis with an activated *let-60 ras* mutation suggests that the *lin-45 raf* gene acts downstream of *let-60 ras* in the vulval induction pathway.

Subunit structure and isoforms

SUBUNIT	AMINO ACIDS	MOL. WT	SDS-PAGE
CeRaf	*813	*90 491	

*Assuming that translation initiates at the first ATG.

Domain structure

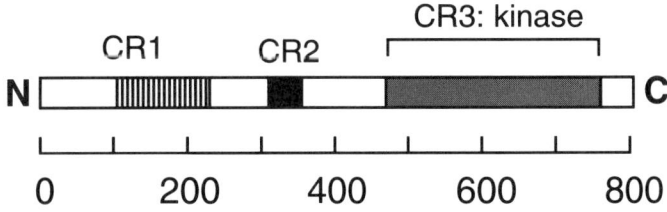

CeRaf contains the three conserved domains, CR1, CR2 and CR3, found in other members of the Raf family (see **Raf1**).

Genetics

Chromosomal location: IV, beteen *unc-44* and *deb-1*.
Physical map: YAC Y22D1.
In vivo mutation:
  sy96, hypomorph. A G-A transition in the splice acceptor site of intron 4. Yields a mixed population of mRNAs, most of which are predicted to encode truncated products (at amino acid 229) that would lack a kinase domain[1].
In vitro mutation:
  K507W, in the ATP binding site; 30% of transgenic animals expressing this variant are defective in vulval differentiation.

Database accession numbers

	PIR	SWISSPROT	EMBL/GENBANK	REF
CeRaf			L15347	*1*

Amino acid sequence of *C. elegans* Raf

```
MSRINFKKSS ASTTPTSPHC PSPRLISLPR CASSSIDRKD QASPMASPST  50
PLYPKHSDSL HSLSGHHSAG GAGTSDKEPP KFKYKMIMVH LPFDQHSRVE 100
VRPGETARDA ISKLLKKRNI TPQLCHVNAS SDPKQESIEL SLTMEEIASR 150
LPGNELWVHS EYLNTVSSIK HAIVRRTFIP PKSCDVCNNP IWMMGFRCEF 200
CQFKFHQRCS SFAPLYCDLL QSVPKNEDLV KELFGIASQV EGPDRSVAEI 250
VLANLAPTSG QSPAATPDSS HPDLTSIKRT GGVKRHPMAV SPQNETSQLS 300
PSGPYPRDRS SSAPNINAIN DEATVQHNQR ILDALEAQRL EEESRDKTGS 350
LLSTQARHRP HFQSGHILSG ARMNRLHPLV DCTPLGSNSP SSTCSSPPGG 400
LIGQPTLGQS PNVSGSTTSS LVAAHLHTLP LTPPQSAPPQ KISPGFFRNR 450
SRSPGERLDA QRPRPPQKPH HEDWEILPNE FIIQYKVGSG SFGTVYRGEF 500
FGTVAIKKLN VVDPTPSQMA AFKNEVAVLK KTRHLNVLLF MGWVREPEIA 550
IITQWCEGSS LYRHIHVQEP RVEFEMGAII DILKQVSLGM NYLHSKNIIH 600
RDLKTNNIFL MDDMSTVKIG DFGLATVKTK WTVNGGQQQQ QPTGSILWMA 650
PEVIRMQDDN PYTPQSDVYS FGICMYEILS SHLPYSNINN RDQILFMVGR 700
GYLRPDRSKI RHDTPKSMLK LYDNCIMFDR NERPVFGEVL ERLRDIILPK 750
LTRSQSAPNV LHLDSQYSVM DAVMRSQMLS WSYIPPATAK TPQSAAAAAA 800
RNKKAYYNVY GLI
```

Assay

The *lin-45 raf* gene has been assayed *in vivo* for its effects on development using chromosomal mutations. It has also been assayed *in vivo* using transgenic animals carrying *in vitro*-generated dominant-negative mutations. The *sy96* mutation affects vulval differentiation and fertility in the hermaphrodite, spicule formation in the male, and viability and P12 neuroectoblast specification in both sexes.

Reference

[1]   Han, M. et al. (1993) Nature 363, 133–140.

Lawrence S. Mathews (University of Michigan, Ann Arbor, USA)

Type II activin receptors were first identified as one of two affinity-labelled complexes generated by binding and chemical cross-linking of radioiodinated activin A to activin-responsive cells. A cDNA encoding ActRII was isolated by using an expression cloning strategy[1]. cDNAs encoding a second gene product, ActRIIB were isolated by using PCR with primers based on the sequence of ActRII[2]. ActRII and ActRIIB are receptor serine/threonine PKs. Manipulation of expression of these receptors in *Xenopus laevis* embryos disrupts activin signalling in early differentiative events[3, 4].

Subunit structure and isoforms

SUBUNIT	AMINO ACIDS	MOL. WT	SDS-PAGE
ActRII	494	56 000	80–90 kDa
ActRIIB-1	518	58 700	
ActRIIB-2	494	55 900	80–90 kDa
ActRIIB-3	510	57 800	
ActRIIB-4	486	55 000	

Two alternative splicing events in the ActRIIB gene, one in the extracellular and one in the intracellular domain, generate four potential protein products (ActRIIB-1 to -4). ActRII and ActRIIB are believed to exist as monomers, but can also form dimers with type I ActRs[5].

Genetics

ActRII

Chromosomal location:
  Human   2
  Mouse   2
Site-directed mutants: K219E (inactive)

ActRIIB

Chromosomal location (human): 3.
Site-directed mutants: K217R (inactive)

Domain structure

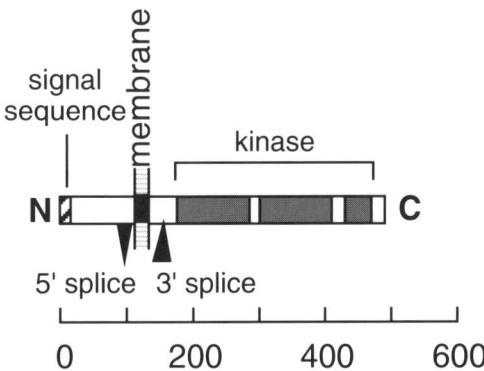

The diagram refers to ActRII and the IIB isoform which most closely resembles it, IIB-2. The latter contains the extra amino acids from the 5' alternative splice site, but not from the 3' alternative splice site. Deletion of amino acids V106–T113 in the extracellular domain by alternative splicing yields isoforms (B-3/B-4) which bind ligand with lower affinity[2]. The function of a second alternative splice which encodes 24 amino acids (V157–Q180) in the intracellular domain is unknown. The extracellular ligand-binding domains contain two sites of N-linked glycosylation. The intracellular kinase domains contain two inserts unique to the receptor protein serine kinase subfamily. The intracellular juxtamembrane regions are P-rich, and the C-terminal regions are S/T-rich.

Database accession numbers

		PIR	SWISSPROT	EMBL/GENBANK	REF
II	Mouse		P27038	M65827	1
	Human		P27037	X63128	6
	Rat			L10639	7
	Xenopus		P27039	S70930	8
	Chicken	S23089			9
	Bovine			L21717	
IIB	Mouse		P27040	M84120	2
	Rat			M87067	10
	Xenopus		P27041	M88594	3

Amino acid sequence of human ActRII

```
MGAAAKLAFA VFLISCSSGA ILGRSETQEC LFFNANWEKD RTNQTGVEPC  50
YGDKDKRRHC FATWKNISGS IEIVKQGCWL DDINCYDRTD CVEKKDSPEV 100
YFCCCEGNMC NEKFSYFPEM EVTQPTSNPV TPKPPYYNIL LYSLVPLMLI 150
AGIVICAFWV YRHHKMAYPP VLVPTQDPGP PPPSPLLGLK PLQLLEVKAR 200
GRFGCVWKAQ LLNEYVAVKI FPIQDKQSWQ NEYEVYSLPG MKHENILQFI 250
GAEKRGTSVD VDLWLITAFH EKGSLSDFLK ANVVSWNELC HIAETMARGL 300
AYLHEDIPGL KDGHKPAISH RDIKSKNVLL KNNLTACIAD FGLALKFEAG 350
KSAGDTHGQV GTRRYMAPEV LEGAINFQRD AFLRIDMYAM GLVLWELASR 400
CTAADGPVDE YMLPFEEEIG QHPSLEDMQE VVVHKKKRPV LRDYWQKHAG 450
MAMLCETIEE CWDHDAEARL SAGCVGERIT QMQRLTNIIT TEDIVTVVTM 500
VTNVDFPPKE SSL
```

Sites of interest: M1–G19, signal sequence; N43, N66 glycosylation; Y136–Y161, transmembrane domain

Amino acid sequence of mouse ActRIIB-1

```
MTAPWAALAL LWGSLCAGSG RGEAETRECI YYNANWELER TNQSGLERCE  50
GEQDKRLHCY ASWANSSGTI ELVKKGCWLD DFNCYDRQEC VATEENPQVY 100
FCCCEGNFCN ERFTHLPEPG GPEVTYEPPP TAPTLLTVLA YSLLPIGGLS 150
LIVLLAFWMY RHRKPPYGHV DIHEVRQCQR WAGRRDGCAD SFKPLPFQDP 200
GPPPPSPLVG LKPLQLLEIK ARGRFGCVWK AQLMNDFVAV KIFPLQDKQS 250
```

Amino acid sequence of mouse ActRIIB-1 *continued*

```
WQSEREIFST PGMKHENLLQ FIAAEKRGSN LEVELWLITA FHDKGSLTDY 300
LKGNIITWNE LCHVAETMSR GLSYLHEDVP WCRGEGHKPS IAHRDFKSKN 350
VLLKSDLTAV LADFGLAVRF EPGKPPGDTH GQVGTRRYMA PEVLEGAINF 400
QRDAFLRIDM YAMGLVLWEL VSRCKAADGP VDEYMLPFEE EIGQHPSLEE 450
LQEVVVHKKM RPTIKDHWLK HPGLAQLCVT IEECWDHDAE ARLSAGCVEE 500
RVSLIRRSVN GTTSDCLVSL VTSVTNVDLL PKESSI
```

Sites of interest: M1–G18, signal sequence; N42, N65 glycosylation; V124–T131, spliced out in B-3, B-4; L135–Y160, transmembrane domain; V175–Q198, spliced out in B-2, B-4

Homologues in other species

Sequences have been obtained for ActRII cDNAs from human, mouse, rat, cow, *Xenopus laevis* and chicken. The human and mouse sequences differ by 2 substitutions; the human and *Xenopus* sequences differ by 63 substitutions. Sequences have been obtained for ActRIIB cDNAs from mouse, rat and *Xenopus laevis*. The mouse and *X. laevis* sequences differ by 86 substitutions.

Assay

There is currently no assay for ActRII kinase activity. ActRIIB has been purified from the mouse embryonal carcinoma cell line, P19, and was found to both autophosphorylate and phosphorylate activin, histone and enolase on serine, threonine and tyrosine[11].

Pattern of expression

ActRII and ActRIIB are expressed very early in embryonic development, and are widely, but not ubiquitously, expressed in adult tissues.

References

1   Mathews, L.S. et al. (1991) Cell 65, 973–982.
2   Attisano, L. et al. (1992) Cell 68, 97–108.
3   Mathews, L.S. et al. (1992) Science 255, 1702–1705.
4   Hemmati-Brivanlou, A. et al. (1992) Nature 359, 609–614.
5   Mathews, L.S. et al. (1993) J. Biol. Chem. 263, 19013–19018.
6   Donaldson, C.J. et al. (1992) Biochem. Biophys. Res. Commun. 184, 310–331.
7   Shinozaki, H. et al. (1992) FEBS Lett. 312, 53–56.
8   Kondo, M. et al. (1991) Biochem. Biophys. Res. Commun. 181, 684–690.
9   Ohuchi, H. et al. (1992) FEBS Lett. 303, 185–189.
10 Legerski, R. et al. (1992) Biochem. Biophys. Res. Commun. 183, 672–679.
11 Nakamura, T. et al. (1992) J. Biol. Chem. 267, 18924–18928.

Herbert Y. Lin and Bruce Tidor (Department of Medicine, Harvard Medical School and Department of Biology, Massachusetts Institute of Technology, USA)

The transforming growth factor-β (TGFβ) type II receptor is a transmembrane serine/threonine kinase which binds TGFβ1 and TGFβ3 with greater affinity than TGFβ2[1]. It is thought to mediate the multiple effects of TGFβ, such as increased extracellular matrix production and inhibition of cell growth. The signal transduction pathway and the substrate specificities of this receptor kinase are not known.

Subunit structure and isoforms

SUBUNIT	AMINO ACIDS	MOL. WT	SDS-PAGE
TGFβRII	567	65 000	80–90 kDa

The higher molecular weight on SDS-PAGE is due to glycosylation.

Genetics

Chromosomal location (human): 3 (J. Lawrence, personal communication).
Site-directed mutants: K277E, K277R (inactive)

Domain structure

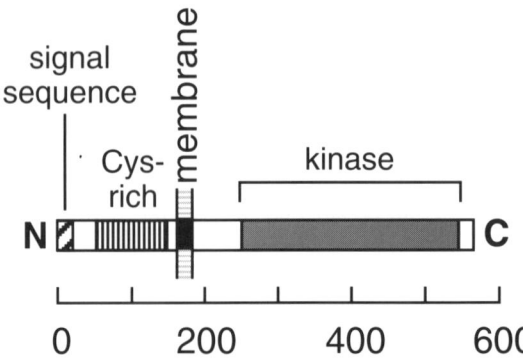

The kinase domain has four inserts when compared to mouse PKAα. It is unclear what the function of these inserts are, but they are in regions where they may serve to position loops that interact with substrate. It is not known whether dimerization is necessary for kinase function. There are several *in vitro* autophosphorylation sites[1].

Database accession numbers

	PIR	SWISSPROT	EMBL/GENBANK	REF
Human			M85079	1
Rat			L09653	2

Amino acid sequence of human TGFβRII

```
MGRGLLRGLW PLHIVLWTRI ASTIPPHVQK SVNNDMIVTD NNGAVKFPQL  50
CKFCDVRFST CDNQKSCMSN CSITSICEKP QEVCVAVWRK NDENITLETV 100
CHDPKLPYHD FILEDAASPK CIMKEKKKPG ETFFMCSCSS DECNDNIIFS 150
EEYNTSNPDL LLVIFQVTGI SLLPPLGVAI SVIIIFYCYR VNRQQKLSST 200
WETGKTRKLM EFSEHCAIIL EDDRSDISST CANNINHNTE LLPIELDTLV 250
GKGRFAEVYK AKLKQNTSEQ FETVAVKIFP YEEYASWKTE KDIFSDINLK 300
HENILQFLTA EERKTELGKQ YWLITAFHAK GNLQEYLTRH VISWEDLRKL 350
GSSLARGIAH LHSDHTPCGR PKMPIVHRDL KSSNILVKND LTCCLCDFGL 400
SLRLDPTLSV DDLANSGQVG TARYMAPEVL ESRMNLENAE SFKQTDVYSM 450
ALVLWEMTSR CNAVGEVKDY EPPFGSKVRE HPCVESMKDN VLRDRGRPEI 500
PSFWLNHQGI QMVCETLTEC WDHDPEARLT AQCVAERFSE LEHLDRLSGR 550
SCSEEKIPED GSLNTTK
```

Sites of interest: M1–T23, signal sequence; L160–Y189, transmembrane
domain; S352, T530, S548, potential autophosphorylation
sites determined by molecular modelling

Homologues in other species
The human (Hep G2 cells), mink (MVLu cells), and rat (pituitary) have been
sequenced in full. A partial porcine (LLC-PK1 cells) clone has been sequenced.

Physiological substrates and specificity determinants

TGFβ type II receptors have been shown to autophosphorylate on serines and
threonines[1]. No other substrate has been identified.

Assay
A GST-fusion protein has been shown to autophosphorylate upon addition of
[γ-^{32}P]ATP.

Enzyme activators and inhibitors
The truncated type II receptor acts as a dominant negative inhibitor[3, 4].

Pattern of expression
The type II receptor is widely expressed at the plasma membrane of most
cells. Notable exceptions include retinoblastoma cells, PC12 cells, haematopoietic precursors, and some chemically generated mutant cell lines.

References
[1] **Lin, H.Y. et al. (1992) Cell 68, 775–785; erratum 70, 1069.**
[2] Tsuchida, K. et al. (1993) Biochem. Biophys. Res. Commun. 191, 790–795.
[3] Chen, R.H. et al. (1993) Science 260, 1335–1338.
[4] Brand, T. et al. (1993) J. Biol. Chem. 268, 11500–11503.

DAF-1 — DAF-1 receptor PK (*C. elegans*)

Miguel Estevez and Donald L. Riddle
(University of Missouri-Columbia, USA)

The nematode *C. elegans* has a life cycle consisting of four larval stages (L1–L4) followed by an adult reproductive stage. If the animals experience overcrowding or starvation as L1 stage larvae, they can moult into an alternate L3 stage called the dauer larva. The dauer larva is a developmentally arrested dispersal stage that can survive for months without feeding while searching for food in which to resume development[1]. Mutations in the *daf-1* gene cause the nematodes to develop into dauer larvae at a restrictive temperature even when food is adundant[2,3]. The *daf-1* gene encodes a transmembrane receptor serine/threonine PK, and DAF-1 was the first receptor serine/threonine PK identified. Mammalian activin[4] and TGFβ[5] receptors were subsequently shown to be similar in structure. The DAF-1 receptor functions to promote growth (non-dauer development) in adundant food.

Subunit structure and isoforms

SUBUNIT	AMINO ACIDS	MOL. WT	SDS-PAGE
DAF-1	669	75 014	

Genetics
Chromosomal location: IV[2].
The position of three mutations in the *daf-1* gene (two transposon insertions, *m402* and *m412*, and one EMS-induced point mutation, *m213*)[6] have been identified, and all three are in the coding region for the kinase domain.

Domain structure

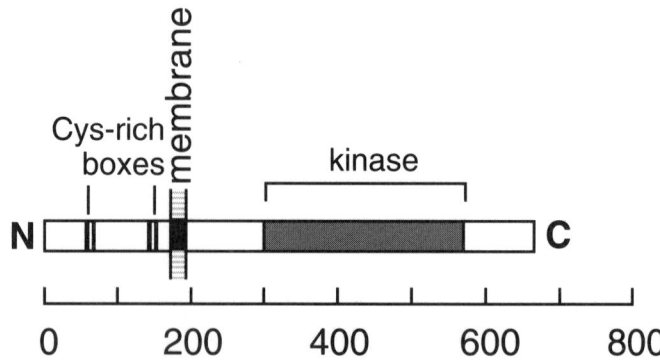

Database accession numbers

	PIR	SWISSPROT	EMBL/GENBANK	REF
DAF-1			M32877	6

340

Amino acid sequence of *C. elegans* DAF-1

```
MRIRHVVFCL LALVYGAETS DDDLDERTNI FIRDKLIPAL KLAEVTKVNF  50
TRLHLCHCSR EVGCNARTTG WVPGIEFLNE TDRSFYENTC YTDGSCYQSA 100
RPSPEISHFG CMDEKSVTDE TEFHDTAAKV CTNNTKDPHA TVWICCDKGN 150
FCANETIIHL APGPQQSSTW LILTILALLT FIVLLGIAIF LTRKSWEAKF 200
DWYIRFKPKP GDPLRETENN VPMVTMGDGA GSSVPEVAPI EQQGSTMSTS 250
AGNSFPPGIM PNNMKDMLDV LEETSGSGMG PTTLHKLTIG GQIRLTGRVG 300
SGRFGNVSRG DYRGEAVAVK VFNALDEPAF HKETEIFETR MLRHPNVLRY 350
IGSDRVDTGF VTELWLVTEY HPSGSLHDFL LENTVNIETY YNLMRSTASG 400
LAFLHNQIGG SKESNKPAMA HRDIKSKNIM VKNDLTCAIG DLGLSLSKPE 450
DAASDIIANE NYKCGTVRYL APEILNSTMQ FTVFESYQCA DVYSFSLVMW 500
ETLCRCEDGD VLPREAATVI PYIEWTDRDP QDAQMFDVVC TRRLRPTENP 550
LWKDHPEMKH IMEIIKTCWN GNPSARFTSY ICRKRMDERQ QLLLDKKAKA 600
VAQTAGVTVQ DRKILGPQKP KDESPANGAP RIVQKEIDRE DEQENWRETA 650
KTPNGHISSN DDSSRPLLG
```

Sites of interest: N49, N79, N133, N154, potential N-linked glycosylation;
L171–L191, transmembrane domain

Enzyme activators and inhibitors
No ligand for the DAF-1 protein has been demonstrated by direct binding. However, *daf-7*, another gene whose function is required to prevent dauer larva formation, encodes a growth factor belonging to the TGFβ superfamily that is potentially the ligand (unpublished).

Pattern of expression
The *daf-1* gene encodes a 2.5 kb transcript that is present at all stages of development. Transgenic nematodes carrying a fusion of the *daf-1* promoter with a β-galactosidase reporter gene suggest that expression is localized to nuerones in the head and tail.

References
[1] Cassada, R.C. and Russell, R.L. (1975) Devel. Biol. 46, 326–342.
[2] Riddle, D.L. et al. (1981) Nature 290, 668–671.
[3] Swanson, M.M. and Riddle, D.L. (1981) Devel. Biol. 84, 27–40.
[4] Matthews, L. and Vale, W. (1991) Cell 65, 973–982
[5] Lin, H.Y. et al. (1992) Cell 68, 775–785; erratum, Cell 70, 1069.
[6] Georgi, L.L. et al. (1990) Cell 61, 635–645.

 **ZmPK1** ZmPK1 (*Zea mays*)

John C. Walker (University of Columbia-Missouri, USA)

ZmPK1 is a protein kinase found in the plasma membranes of young maize seedlings. *In vitro* the enzyme will autophosphorylate on serine. ZmPK1 is a potential plant receptor serine/threonine PK[1] but its physiological function is not known.

Subunit structure and isoforms

SUBUNIT	AMINO ACIDS	MOL. WT	SDS-PAGE
ZmPK1	817	88 182	100 kDa

Genetics
Chromosomal location: 6, long arm; maps to position 112 on the maize RFLP map.

Domain structure

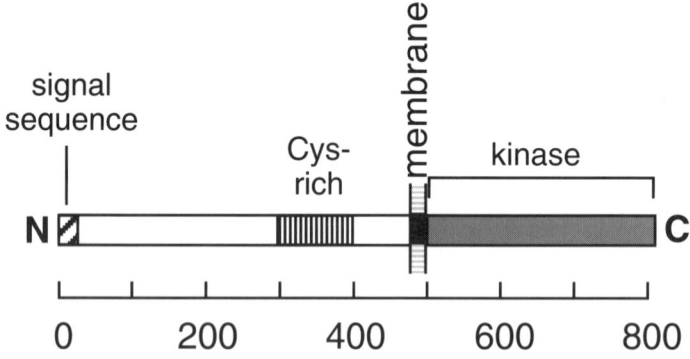

The extracellular region is related to the self-incompatibility locus (*S*-locus) glycoproteins of *Brassica* spp. The Cys-rich region contains 12 cysteines that are conserved between all *S*-locus related proteins. The protein kinase domain is related to the Raf family and other plant receptor-like protein kinases[2].

Database accession numbers

	PIR	SWISSPROT	EMBL/GENBANK	REF
ZmPK1	S10903	P17801	X52384	*1*

 ## Amino acid sequence of *Zea mays* ZmPK1

```
MPRPLAALLS TACILSFFIA LFPRAASSRD ILPLGSSLVV ESYESSTLQS  50
SDGTFSSGFY EVYTHAFTFS VWYSKTEAPA ANNKTIVWSA NPDRPVHARR 100
SALTLQKDGN MVLTDYDGAA VWRADGNNFT GVQRARLLDT GNLVIEDSGG 150
NTVWQSFDSP TDTFLPTQLI TAATRLVPTT QSRSPGNYIF RFSDLSVLSL 200
IYHVPQVSDI YWPDPDQNLY QDGRNQYNST RLGMLTDSGV LASSDFADGQ 250
ALVASDVGPG VKRRLTLDPD GNLRLYSMND SDGSWSVSMV AMTQPCNIHG 300
```

Amino acid sequence of *Zea mays* ZmPK1 *continued*

```
LCGPNGICHY SPTPTCSCPP GYATRNPGNW TEGCMAIVNT TCDRYDKRSM 350
RFVRLPNTDF WGSDQQHLLS VSLRTCRDIC ISDCTCKGFQ YQEGTGSCYP 400
KAYLFSGRTY PTSDVRTIYL KLPTGVSVSN ALIPRSDVFD SVPRRLDCDR 450
MNKSIREPFP DVHKTGGGES KWFYFYGFIA AFFVVEVSFI SFAWFFVLKR 500
ELRPSELWAS EKGYKAMTSN FRRYSYRELV KATRKFKVEL GRGESGTVYK 550
GVLEDDRHVA VKKLENVRQG KEVFQAELSV IGRINHMNLV RIWGFCSEGS 600
HRLLVSEYVE NGSLANILFS EGGNILLDWE GRFNIALGVA KGLAYLHHEC 650
LEWVIHCDVK PENILLDQAF EPKITDFGLV KLLNRGGSTQ NVSHVRGTLG 700z
YIAPEWVSSL PITAKVDVYS YGVVLLELLT GTRVSELVGG TDEVHSMLRK 750
LVRMLSAKLE GEEQSWIDGY LDSKLNRPVN YVQARTLIKL AVSCLEEDRS 800
KRPTMEHAVQ TLLSADD
```

Sites of interest: M1–S28, signal peptide; F473–L498, transmembrane domain

Pattern of expression

ZmPK1 is expressed in the root, mesocotyl and coleoptile of young maize seedlings. The protein is found in plasma membrane enriched fractions. There is also a low level of mRNA expression detected in the silks.

References

[1] Walker, J.C. and Zhang, R. (1990) Nature 345, 743–746.
[2] Walker, J.C. (1993) Plant J. 3, 451–456.

Srk

S-locus receptor PK (*Brassica* spp.)

Joshua C. Stein and June B. Nasrallah (Cornell University, USA)

SRK was identified as a gene derived from the self-incompatibility locus (*S* locus), a complex and polymorphic locus which is responsible for inhibiting self-pollination[1]. Srk has a receptor-like structure, possesses intrinsic serine/threonine PK activity, and is capable of autophosphorylation[2,3]. The pattern of expression of *SRK* and its high degree of sequence polymorphism among different self-incompatibility haplotypes suggest a role for Srk as a receptor in pollen/stigma recognition[1]. Plants bearing null alleles of *SRK* exhibit loss of the self-incompatibility response[4,5]. Potential ligands and substrates remain to be identified.

Subunit structure and isoforms

SUBUNIT	AMINO ACIDS	MOL. WT	SDS-PAGE
Srk_6	857	98 071	

Each *S*-locus haplotype (>50 are known) is believed to encode a distinct Srk allele. Srk_6, Srk_2 and Srk_{910} have been characterized. These show as much as 32% sequence divergence at the amino acid level. The *S*-locus glycoprotein (Slg) is highly similar to the extracellular domain of Srk, and is also encoded at the *S* locus. When derived from the same *S*-locus haplotype, Srk and Slg have as high as 90% amino acid identity[1]. *SRK* transcript variants potentially encode N-terminally truncated and C-terminally truncated forms[1].

Genetics

Chromosomal location: Physically linked to the *SLG* gene, within the *S* locus[6].
Site-directed mutants:
SRK_{910} K557A (inactive)[2]
SRK_6 K556R (inactive)[3]

Domain structure

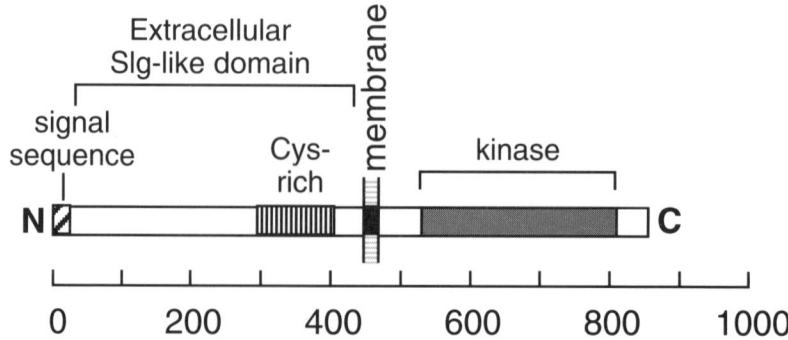

Database accession numbers

	PIR	*SWISSPROT*	*EMBL/GENBANK*	*REF*
SRK$_6$	A41369		M76647	1
SRK$_2$				1
SRK$_{910}$			M97667	2

Amino acid sequence of *B. oleracea* Srk$_6$

```
MKGARNIYHH SYMSFLLVFV VMILIHPALS IYINTLSSTE SLTISSNKTL  50
VSPGSIFEVG FFRTNSRWYL GMWYKKVSDR TYVWVANRDN PLSNAIGTLK 100
ISGNNLVLLD HSNKPVWWTN LTRGNERSPV VAELLANGNF VMRDSSNNDA 150
SEYLWQSFDY PTDTLLPEMK LGYNLKTGLN RFLTSWRSSD DPSSGNFSYK 200
LETQSLPEFY LSRENFPMHR SGPWNGIRFS GIPEDQKLSY MVYNFIENNE 250
EVAYTFRMTN NSFYSRLTLI SEGYFQRLTW YPSIRIWNRF WSSPVDPQCD 300
TYIMCGPYAY CDVNTSPVCN CIQGFNPRNI QQWDQRVWAG GCIRRTQLSC 350
SGDGFTRMKK MKLPETTMAT VDRSIGVKEC KKRCISDCNC TAFANADIRN 400
GGSGCVIWTE RLEDIRNYAT DAIDGQDLYV RLAAADIAKK RNASGKIISL 450
TVGVSVLLLL IMFCLWKRKQ KRAKASAISI ANTQRNQNLP MNEMVLSSKR 500
EFSGEYKFEE LELPLIEMET VVKATENFSS CNKLGQGGFG IVYKGRLLDG 550
KEIAVKRLSK TSVQGTDEFM NEVTLIARLQ HINLVQVLGC CIEGDEKMLI 600
YEYLENLSLD SYLFGKTRRS KLNWNERFDI TNGVARGLLY LHQDSRFRII 650
HRDLKVSNTL LDKNMTPKIS DFGMARIFER DETEANTMKV VGTYGYMSPE 700
YAMYGIFSEK SDVFSFGVIV LEIVSGKKNR GFYNLDYEND LLSYVWSRWK 750
EGRALEIVDP VIVDSLSSQP SIFQPQEVLK CIQIGLLCVQ ELAEHRPAMS 800
SVVWMFGSEA TEIPQPKPPG YCVRRSPYEL DPSSSWQCDE NESWTVNQYT 850
CSVIDAR
```

Sites of interest: M1–Y32, signal sequence; N47, N120, N196, N260, N314,
N389, N442, N-linked glycosylation; I447–W466, transmembrane domain

Homologues in other species

SRK$_6$ and *SRK$_2$* were sequenced from *B. oleracea*. *SRK$_{910}$* was sequenced from
a *B. campestris* haplotype which was introgressed into *B. napus*.

Assay

Assays were developed using the catalytic domain expressed in bacteria,
using autophosphorylation *in vivo* or *in vitro*[2,3].

Pattern of expression[1,2]

SRK expression is limited to reproductive tissues. RNA transcripts
accumulate maximally in stigmas at one day prior to flower bud opening.
Moderate levels of transcript occur in anther tissues at earlier stages.

References

1   Stein, J.C. et al. (1991) Proc. Natl Acad. Sci. USA 88, 8816–8820.
2   Goring, D.R. and Rothstein, S.J. (1992) Plant Cell 4, 1273–1281.

3 Stein, J.C. and Nasrallah, J.N. (1993) Plant Physiol. 101, 1103–1106.
4 Goring, D.R. et al. (1993) Plant Cell 5, 531–539.
5 Nasrallah, J.B. et al. (1995) Plant J. (in press).
6 Boyes, D.C. and Nasrallah, J.N. (1993) Mol. Gen. Genet. 236, 369–373.

Melanie H. Cobb (University of Texas Southwestern Medical Center, USA)

CK1 has a wide variety of substrates, but effects on their function have been difficult to document. No clearcut mechanism of regulation has been defined. Early evidence suggested modest regulation by insulin and more recent data suggest regulatory effects of phosphatidylinositol bisphosphate[1]. The existence of multiple (at least eight) isoforms[2-4] may have confounded efforts to understand regulation of CK1. Recognition of some of its substrates is controlled by hierarchical phosphorylation: only phosphorylated forms of these substrates are recognized. An example of possible physiological significance is the phosphorylation of S10 of glycogen synthase, which occurs only after phosphorylation of S7 by another protein kinase[5]. Phosphorylation of S10 inhibits glycogen synthase activity. Genetic evidence from yeast homologues **Hrr25**[6] and **Yck1/2**[7,8] suggests a role in DNA repair and cell division.

Subunit structure and isoforms

SUBUNIT	AMINO ACIDS	MOL. WT	SDS-PAGE
CK1α	325	*37 600–39 000	37–39 kDa
CK1β	336	38 700	
CK1γ	373, 397, 434	*41, 42, 49 kDa	40–50 kDa
CK1δ		49 100	55 kDa

*Splice variants.

The CK1 family represents a distant branch of the protein kinase family. The hallmarks of protein kinase subdomains VIII and XI are difficult to identify. All forms consist of a single catalytic subunit. CK1α can occur in at least three alternatively spliced forms (J. Brockman and R. Anderson, unpublished)[2]. Inserted exons have been found between subdomains VI and VII and between VIa and VIb. In addition to CK1β and CK1δ[3], three isoforms of CK1γ have recently been sequenced[4].

Domain structure

All forms of CK1 comprise a kinase domain with unusual short insertions, and variable C-terminal extensions.

Database accession numbers

	PIR	SWISSPROT	EMBL/GENBANK	REF
CK1α			M76543	2
CK1β			M76544	2
CK1γ			M76542	2, 4
CK1δ			M76545, L07578	2, 3

Amino acid sequences of bovine CK1

CK1α

```
MASSSGSKAE FIVGGKYKLV RKIGSGSFGD IYLAINITNG EEVAVKLESQ  50
KARHPQLLYE SKLYKILQGG VGIPHIRWYG QEKDYNVLVM DLLGPSLEDL 100
FNFCSRRFTM KTVLMLADQM ISRIEYVHTK NFIHRDIKPD NFLMGIGRHC 150
NKLFLIDFGL AKKYRDNRTR QHIPYREDKN LTGTARYASI NAHLGIEQSR 200
RDDMESLGYV LMYFNRTSLP WQGLKAATKK QKYEKISEKK MSTPVEVLCK 250
GFPAEFAMYL NYCRGLRFEE APDYMYLRQL FRILFRTLNH QYDYTFDWTM 300
LKQKAAQQAA SSSGQGQQAQ TPTGF
```

CK1β

```
MASSSRPKTD VLVGGRYKLV REIGFGSFGH VYLAIDLTNH EQVAVKLESE  50
NTRQPRLLHE KELYNFLQGG VGIPQIRWYG QETDYNVLVM DLLGPSLEDL 100
FNFCSRRFSM KTVLMLADQM ISRIEYVHSR NLIHRDIKPD NFLMGTGPQW 150
KKLFLVDFGL AKKYRDNRTG QHIPHRSGKS FIGTPFCASI SAHLGIEQSR 200
RDDMESIGYV LMYFNRGSLP WQGLKAATLK QKCEKISEMK MTTPVDVLCK 250
GFPIEFAMYL KYCLRLSFEE APDYRYLRQL FRLLFRKLSY QHDYAFDWIV 300
LKQKAEQQAS SSSGEGQQAQ TPTGKSDNTK SEMKHS
```

Homologues in other species

Partial or complete sequences for CK1α, β, γ and δ are known from bovine brain. CK1δ has been sequenced from rabbit muscle. CK1α, β and γ isoforms have been sequenced from rat.

Physiological substrates and specificity determinants

The consensus sequence for phosphorylation by CK1 is S or T preceded by a string of acidic residues (underlined, optimum of 4) with the most important at P-3[9]. A single phosphate group at P-3 is sufficient[10]. Sequences around sites believed to be phosphorylated by CK1 in intact cells are shown below. Double underlining indicates residues which must be phosphorylated to create a CK1 site, the latter shown in bold:

Glycogen synthase:	RTLSVASLPGL[11]
α$_{s2}$-Casein:	LSTSEENSKK[10]
p53:	MEESQSDISLELP[12]
V40 large T antigen:	ADSQHSTPP[13]
Inhibitor-2:	GDDDDAYSDTET[11]

Assay

CK1 is usually assayed as the transfer of radioactivity from $[\gamma\text{-}^{32}\text{P}]$ATP to α$_s$-casein. As casein is also phosphorylated by other kinases, the peptide DDDDVASLPGLRRR provides a more specific substrate[9].

Enzyme activators and inhibitors

INHIBITORS	SPECIFICITY	K_i (μM)	REF
CK1-7	Good	8.5	14
CK1-8	Poor	80	

Pattern of expression

One or more forms are ubiquitously distributed in mammalian tissues and cell lines. Isoforms may differ in subcellular distribution, although they are found in cytoplasm, in nuclei, and associated with membranes and the cytoskeleton.

References

1 Brockman, J.L. et al. (1991) J. Biol. Chem. 266, 2508–1512.
2 Rowles, J. et al. (1991) Proc. Natl. Acad. Sci. USA 88, 9548–9552.
3 Graves, P. et al. (1993) J. Biol. Chem. 268, 6394–6401.
4 Zhai et al. (1994) (unpublished data).
5 Cochet, C. et al. (1982) Biochem. Pharmacol. 32, 1357–1361.
6 Hoekstra, M.F. et al. (1991) Science 253, 1031–1034.
7 Robinson, L.C. et al. (1992) Proc. Natl Acad. Sci. USA 89, 28–32.
8 Wang, P.C. et al. (1992) Mol. Biol. Cell 3, 275–286.
9 Flotow, H. et al. (1991) J. Biol. Chem. 266, 3724–3727.
10 Meggio, F. et al. (1979) FEBS Lett. 106, 76–80.
11 Flotow, H. et al. (1990) J. Biol. Chem. 265, 14264–14269.
12 Milne, D.M. et al. (1992) Oncogene 7, 1361–1369.
13 Umphress, J.L. et al. (1992) Eur. J. Biochem. 203, 239–243.
14 Chijiwa, T. et al. (1989) J. Biol. Chem. 264, 4924–4927.

Yck1/2

Casein kinase I homologues (*S. cerevisiae*)
(YCK1/CKI2, YCK2/CKI1)

E. Jane Albert Hubbard and Marian Carlson (Columbia University, USA)

Casein kinases are ubiquitous Ca^{2+}- and cyclic nucleotide-independent protein kinases, originally characterized on the basis of their preference for acidic substrates such as casein[1]. Yck1 and Yck2 are a functionally redundant pair of casein kinase I homologues in *S. cerevisiae*. The genes were isolated by virtue of their effects in high dosage. *YCK1* was found as a multicopy suppressor of growth defects associated with reduced **Snf1** activity, and *YCK2* was identified by the protection it provides from the deleterious effects of sudden shifts to high salinity. Increased dosage of either gene has the same effect in either assay[2]. The same pair of genes were independently cloned using degenerate oligonucleotides based on peptide sequence obtained from purified protein[3]. Yck1 and Yck2 perform an essential function in yeast cells. Although loss of one of the two genes causes no adverse effect, the absence of both genes is lethal[2,3]. Spores that lack both Yck1 and Yck2 germinate and go through 3–10 rounds of division on rich medium, but then form projections that elongate over the next 1–2 days before growth ceases entirely[2].

Subunit structure and isoforms

SUBUNIT	AMINO ACIDS	MOL. WT	SDS-PAGE
Yck1	538	61 715	62 kDa
Yck2	546	62 079	62 kDa

Genetics

Chromosomal location[3]:
- *YCK1*   right arm of chromosome VIII
- *YCK2*   left arm of chromosome XIV

Domain structure

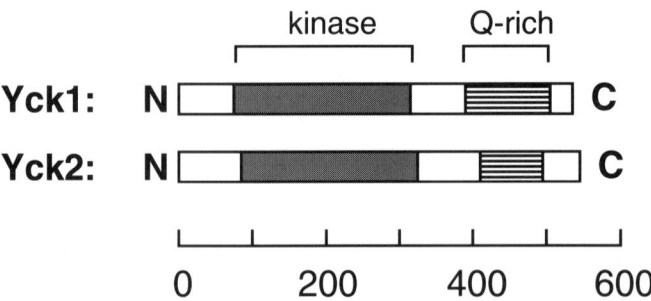

The kinase domains are 94% identical between the two homologues; overall, Yck1 and Yck2 are 77% identical. They are ≈40% identical to **Hrr25**[4]. As with other members of the CKI family, the kinase domain sequence differs from those of other kinases by the absence of the conserved APE motif in subdomain VIII and the conserved Arg residue in subdomain XI. The C-terminal portions of Yck1 and Yck2 contain regions that are 47% and 43% glutamine residues, respectively. Both proteins end in a pair of cysteine residues, a motif thought to be a site for prenylation, which would then mediate attachment to membranes[3, 5].

Database accession numbers

	PIR	SWISSPROT	EMBL/GENBANK	REF
Yck1		P23291	M74552	2
Cki2	B43764		X60327	3
Yck2		P23292	M74553	2
Cki1	A43764		X60326	3

Amino acid sequences of *S. cerevisiae* Yck1/2

Yck1

```
MSMPIASTTL AVNNLTNING NANFNVQANK QLHHQAVDSP ARSSMTATTA  50
ANSNSNSSRD DSTIVGLHYK IGKKIGEGSF GVLFEGTNMI NGVPVAIKFE 100
PRKTEAPQLR DEYKTYKILN GTPNIPYAYY FGQEGLHNIL VIDLLGPSLE 150
DLFDWCGRKF SVKTVVQVAV QMITLIEDLH AHDLIYRDIK PDNFLIGRPG 200
QPDANNIHLI DFGMAKQYRD PKTKQHIPYR EKKSLSGTAR YMSINTHLGR 250
EQSRRDDMEA LGHVFFYFLR GHLPWQGLKA PNNKQKYEKI GEKKRSTNVY 300
DLAQGLPVQF GRYLEIVRSL SFEECPDYEG YRKLLLSVLD DLGETADGQY 350
DWMKLNDGRG WDLNINKKPN LHGYGHPNPP NEKSRKHRNK QLQMQQLQMQ 400
QLQQQQQQQQ YAQKTEADMR NSQYKPKLDP TSYEAYQHQT QQKYLQEQQK 450
RQQQQKLQEQ QLQEQQLQQQ QQQQQQLRAT GQPPSQPQAQ TQSQQFGARY 500
QPQQQPSAAL RTPEQHPNDD NSSLAASHKG FFQKLGCC
```

Yck2

```
MSQVQSPLTA TNSGLAVNNN TMNSQMPNRS NVRLVNGTLP PSLHVSSNLN  50
HNTGNSSASY SGSQSRDDST IVGLHYKIGK KIGEGSFGVL FEGTNMINGL 100
PVAIKFEPRK TEAPQLKDEY RTYKILAGTP GIPQEYYFGQ EGLHNILVID 150
LLGPSLEDLF DWCGRRFSVK TVVQVAVQMI TLIEDLHAHD LIYRDIKPDN 200
FLIGRPGQPD ANKVHLIDFG MAKQYRDPKT KQHIPYREKK SLSGTARYMS 250
INTHLGREQS RRDDMEAMGH VFFYFLRGQL PWQGLKAPNN KQKYEKIGEK 300
KRLTNVYDLA QGLPIQFGRY LEIVRNLSFE ETPDYEGYRM LLLSVLDDLG 350
ETADGQYDWM KLNGGRGWDL SINKKPNLHG YGHPNPPNEK SKRHRSKNHQ 400
YSSPDHHHHY NQQQQQQQAQ AQAQAQAQAK VQQQQLQQAQ AQQQANRYQL 450
QPDDSHYDEE REASKLDPTS YEAYQQQTQQ KYAQQQQKQM QQKSKQFANT 500
GANGQTNKYP YNAQPTANDE QNAKNAAQDR NSNKSSKGFF SKLGCC
```

Pattern of expression

Subcellular fractionation suggests that Yck1 and Yck2 are membrane-bound[3].

References
[1] Tuazon, P.T. and Traugh, J.A. (1991) Adv. Second Messenger Phosphoprotein Res. 23, 123–164.
[2] Robinson, L.C. et al. (1992) Proc. Natl Acad. Sci. USA 89, 28–32.
[3] Wang, P.-C. et al. (1992) Mol. Biol. Cell 3, 275–286.
[4] Hoekstra, M.F. et al. (1991) Science 253, 1031–1034.
[5] Molenaar, C.M. et al. (1988) EMBO J. 7, 971–976.

Merl F. Hoekstra (ICOS Corporation, USA)

The *HRR25* gene of *S. cerevisiae* was identified through *hrr25-1* mutant defects in DNA strand-break repair[1]. Phenotypic analysis suggested that *HRR25* encodes a regulator of DNA strand-break repair. Hrr25 mutants are sensitive to enzymatic agents like the HO endonuclease, to chemical agents like methyl methanesulphonate, and are weakly sensitive to physical agents like X-rays, all of which cause DNA strand interruptions. Hrr25 mutants are also unable to proceed through meiosis I, consistent with the idea that *hrr25* mutants are unable to process the strand breaks and lesions that occur in meiosis. Deletion analysis revealed that Hrr25 is helpful but not essential for mitotic growth. Hrr25Δ mutants have a severely reduced doubling time (~12 h) and the resulting cells are delayed in the G2/M phase of the cell cycle. *HRR25* encodes a budding yeast isoform of casein kinase I[2-5] (CKI). The CKI family is a group of protein kinases which recognize serine and threonine residues flanked by acidic regions in the substrate[4, 5].

Subunit structure and isoforms

SUBUNIT	AMINO ACIDS	MOL. WT	SDS-PAGE
Hrr25	494	57 393	58 kDa

Members of the CKI family are normally active as monomers. Related enzymes in *S. cerevisiae* include Cki1 and Cki2 (**Yck1/2**) as well as a fourth, less-well characterized form called Cki3[4-6].

Genetics
Chromosomal location: XVIR.
Site-directed mutation *hrr25-R38* is catalytically inactive *in vitro* and unable to complement *hrr25-1* or *hrr25Δ* mutants *in vivo*. The *hrr25-1* point mutant is an EMS mutagenesis-induced allele that maps to protein kinase subdomain VII.

Domain structure

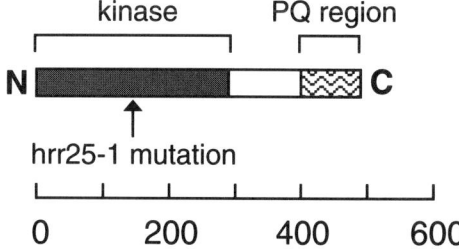

Along with other members of the CKI family, the catalytic domain of Hrr25 lacks the APE sequence in subdomain VIII, and the Arg residue in subdomain XI is missing or difficult to assign. Between subdomains X and XI is the sequence TKKQKY, which is also seen in the bovine and rat forms of CKIα and δ. This sequence has been suggested to act as a nuclear localization signal, but is also closely related to the PIP$_2$ binding motif seen in proteins like **CKIα**, Hrr25 has a C-terminal tail, with a region containing 50% P and Q.

Database accession numbers

	PIR	SWISSPROT	EMBL/GENBANK	REF
Hrr25	A40860	P29295	M68605	

Amino acid sequence of *S. cerevisiae* Hrr25

```
MDLRVGRKFR IGRKIGSGSF GDIYHGRTLI SGEEVAIKLE SIRSRHPQLD  50
YESRVYRYLS GGVGIPFIRW FGREGEYNAM VIDLLGPSLE DLFNYCHRRF 100
SFKTVIMLAL QMFCRIQYIH GRSFIHRDIK PDNFLMGVGR RGSTVHVIDF 150
GLSKKYRDFN THRHIPYREN KSLTGTARYA SVNTHLGIEQ SRRDDLESLG 200
YVLIYFCKGS LPWQGLKATT KKQKYDRIME KKLNVSVETL CSGLPLEFQE 250
YMAYCKNLKF DEKPDYLFLA RLFKDLSIKL EYHNDHLFDW TMLRYTKAMV 300
EKQRDLLIEK GDLNANSNAA SASNSTDNKS ETFNKIKLLA MKKFPTHFHY 350
YKNEDKHNPS PEEIKQQTIL NNNAASSLPE ELLNALDKGM ENLRQQQPQQ 400
QVQSSQPQPQ PQQLQQQPNG QRPNYYPEPL LQQQQRDSQE QQQQVPMATT 450
RATQYPPQIN SNNFNTNQAS VPPQMRSNPQ QPPQDKPAGQ SIWL
```

Homologues in other species

Activities with the biochemical properties of casein kinase I have been reported to be ubiquitously distributed in yeasts, plants and mammalian cells[7]. Sequenced kinases with significant homology to the Hrr25 include the essential gene pair **Yck1/2** from *S. cerevisiae*, and the various isoforms of **CKI** from mammals. Hrr25 is 68% identical to bovine CKIα and 53%/51% identical to Yck1/Yck2. Structural homologues of Hrr25 have been identified in the *C. elegans* sequencing project and four homologues called hhp1, hhp2, cki1 and cki2 have been identified in Schizosaccharmyces pombe[6, 8, 9].

Physiological substrates and specificity determinants

In vivo substrates for Hrr25 have not yet been identified. *In vitro* Hrr25 is capable of phosphorylating casein to high stoichiometry and the enzyme, when purified as a recombinant form from *E. coli*, has a wide specificity. Hrr25 can phosphorylate serine, threonine and tyrosine residues when assayed by autophosphorylation, or by phosphorylation of soluble proteins in a *E. coli* lysate.

Assay

Hrr25 is assayed as cyclic nucleotide-independent transfer of radioactivity from [γ-^{32}P]ATP to casein[2]. Synthetic peptide substrates for Hrr25p include DDDEESITR, DDDVASLPGLRRR, EEEEVASLPGLRRR and DDDDDDSLPGLRRR. These peptides are much poorer substrates than casein.

Enzymes activators and inhibitors

Both auto- and casein phosphorylation are inhibited by CKI-7 (IC_{50} $10\,\mu M$). Hrr25 and several homologues from fission yeast have dual-specificity activity[6].

Regulation of expression

HRR25 mRNA levels show little fluctuation during cell cycle progression or following mating factor stimulation of haploid yeast cells. Hrr25 shows no fluctuation in protein levels during a synchronous cell cycle.

References

1 Hoekstra, M.F. et al. (1991) Science 253, 1031–1034.
2 DeMaggio, A.J. (1992) Proc. Natl Acad. Sci. USA 89, 7008–7012.
3 Rowles, J. et al (1991) Proc. Natl Acad. Sci. USA 88, 9548–9552.
4 Wang, P.C. et al. (1992) Mol. Biol. Cell. 3, 275–286.
5 Robinson, L.C. et al. (1992) Proc. Natl Acad. Sci. USA 89, 28–32.
6 Hoekstra, M.F. et al. (1994) Mol. Biol. Cell 5, 877–886.
7 Tuazon, P.T. and Traugh, J.A. (1991) Adv. Sec. Mess. Phosphoprot. Res. 23, 123–164.
8 Ohillion, N. and Hoekstra, M.F. (1994) EMBO J. 13, 2777–2788.
9 Wang, P.C. et al. (1994) J. Biol. Chem. 269, 12014–12023.

 **Pkn1** Pkn1 (*Myxococcus xanthus*)

Jose Munoz-Dorado, Sumiko Inouye and Masayori Inouye
(Robert Wood Johnson Medical School, USA)

Pkn1 is a protein serine/threonine kinase of the eukaryotic like protein kinase family, the first to be reported in a prokaryote. Genetic evidence suggests that it is involved in the maintenance of the stability of fruiting bodies. In addition to Pkn1, more than 20 putative protein kinase genes were identified by Southern blot hybridization using the oligonucleotide corresponding to subdomain VII of the eukaryotic protein kinases[2]. DNA sequences of eight have been determined, and the deduced amino acid sequences show that all of them contain catalytic domains with significant similarities to eukaryotic protein kinases.

Subunit structure and isoforms

SUBUNIT	AMINO ACIDS	MOL. WT	SDS-PAGE
Pkn1	693	74 168	78–81 kDa

Domain structure

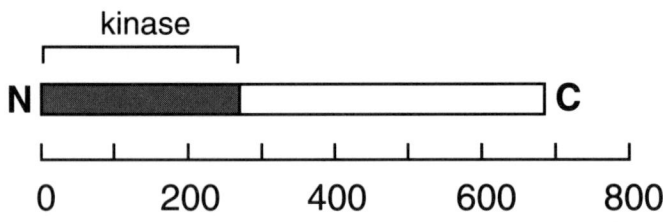

The N-terminal region shows a significant similarity with the catalytic domain of eukaryotic serine/threonine PKs.

Database accession numbers

	PIR	SWISSPROT	EMBL/GENBANK	REF
Pkn1			M73498	[1]

Amino acid sequence of *M. xanthus* Pkn1

```
MPEVSSGGGC GACGRRHGAD ASCPTLVRAD VRAGGTAHPR CAPVVEAQDP  50
LVGVRCGSFR LVRRLGRGGM GAVYLGEHVS IGSRVAVKVL HAHLTMYPEL 100
VQRFHAEARA VNLIGHENIV SIFDMDATPP RPYLIMEFLD GAPLSAWVGT 150
PLAAGAVVSV LSQVCDALQA AHARGIVHRD LKPDNIFLVR RNGNAPFVKV 200
LDFGIAKLAD AHMPQTHAGI IVGTPEYMAP EQSLGRGVDG RADLYALGVI 250
AYQLLTGRLP FNDEGLAAQL VAHQLRPPPP PSSVYPAVSA ALEHVILRAL 300
AKKPEDRYAS IAAFRNALQV ALAEHVRVSA RKTRPGGLAV LERAPVAPDM 350
PTEGQSRGRL GVDARAGHVP SSLASTSQRR LAPAAPAVPR ASLVEVPVQV 400
VLRPGESPVR LRGSGLSRGG LFLHGGRVLP PLCSRLPVVL ELASGPLSVM 450
CEVVRVVPPA QARVWGMPTG FGVQFVEATA VLKAAVDALL QGEPVRAVPQ 500
VPLTEDPAVA RLLEAWRQRS AGDAYAVLAL EPDSDMGTVR LRTREAWRSL 550
ESLEQHSLTP PQRAQVDALR VRVREAAEAL GATVQRALYD AWRGNHRGVA 600
KCLEAGLTAE QLESLRREFL ARRPQAMGTA RSHFQSGGAL ERDGQLSQAL 650
DQYERGLKLA PLEVDMLQRY RRLRRVLGGR ATAPTGHDRA RSP
```

References
[1] Munoz-Dorado, J. et al. (1991) Cell 67, 995–1006.
[2] Zhang, W. et al. (1992) J. Bacteriol. 174, 5450–5453.

George F. Vande Woude (ABL–Basic Research Program, Frederick, MD, USA)

The *MOS* protooncogene (c-*MOS*) encodes a PK that possesses serine/threonine autophosphorylation activity[1]. Mos is required for, and regulates, oocyte maturation (meiosis)[2]. In response to hormonal stimulation and the resultant decrease in **PKA**, Mos initiates progression through meiosis I by activating maturation promoting factor (MPF) from pre-MPF, leading to germinal vesicle (nuclear) breakdown, chromosome condensation and spindle formation. Mos is also a component of cytostatic factor (CSF), an activity of unfertilized eggs that stabilizes MPF and arrests maturing oocytes at meiosis II[3]. During *Xenopus laevis* oocyte maturation, increased Mos activity is associated with the progressive phosphorylation of the protein on serine residues. While the major phosphorylation site of *Xenopus* Mos is the conserved S3, mutation of S105 totally abolishes biological activity[4].

Subunit structure and isoforms

SUBUNIT	AMINO ACIDS	MOL. WT	SDS-PAGE
Mos	346	37 819	39 kDa

Mos is a monomer.

Genetics
Chromosomal location:

Human	8q11–q12[5]
Mouse	4

Domain structure

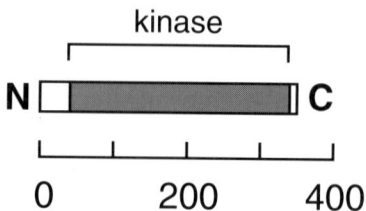

The region from residue 49 to 153 of human Mos is highly conserved among Mos proteins from different species and is part of the consensus ATP-binding site shared by the catalytic domains of protein kinases[6]. Within this region, the conserved Lys87 has been shown to be necessary for the transforming activity of the viral *MOS* (v-*MOS*) oncogene[7]. A second conserved region spanning residues 181–297, interrupted by a non-conserved region from residues 227–239, shares homology with the kinase domain of the **Src** family of protein kinases[6]. Other segments interspersed in these regions are *MOS*-specific.

Database accession numbers

	PIR	SWISSPROT	EMBL/GENBANK	REF
Human			J00119	8, 9
Monkey			X12449	8
Mouse			J00372	8, 9
Rat			X00422	8
Chicken			M19412	8, 9
Xenopus			M13311	9
v-mos			M16719	10

Amino acid sequence of human Mos

```
MPSPLALRPY LRSEFSPSVD ARPCSSPSEL PAKLLLGATL PRAPRLPRRL  50
AWCSIDWEQV CLLQRLGAGG FGSVYKATYR GVPVAIKQVN KCTKNRLASR 100
RSFWAELNVA RLRHDNIVRV VAASTRTPAG SNSLGTIIME FGGNVTLHQV 150
IYGAAGHPEG DAGEPHCRTG GQLSLGKCLK YSLDVVNGLL FLHSQSIVHL 200
DLKPANILIS EQDVCKISDF GCSEKLEDLL CFQTPSYPLG GTYTHRAPEL 250
LKGEGVTPKA DIYSFAITLW QMTTKQAPYS GERQHILYAV VAYDLRPSLS 300
AAVFEDSLPG QRLGDVIQRC WRPSAAQRPS ARLLLVDLTS LKAELG
```

Homologues in other species

v-*MOS* was identified as the transforming gene of the acute transforming retrovirus, Moloney murine sarcoma virus. The v-*MOS* oncogene encodes a 37 kDa Env–Mos fusion protein, the amino acid sequence of which is colinear with the murine c-*MOS* with the exception of 11 N-terminal residues contributed by the viral *env* gene[10]. The c-*MOS* gene has been sequenced from the human, African green monkey, mouse, rat, chicken and *Xenopus laevis* (Mos[xe]).

Physiological substrates and specificity determinants

On gel filtration of cytosolic extracts from matured *Xenopus* oocytes, the Mos elution profile is dependent on the state of tubulin polymerization. Mos elutes at an apparent molecular weight of approximately 500 kDa when tubulin is in the dimer form. Tubulin is found in stoichiometric amounts in Mos immuno-precipitates prepared from *Xenopus* oocytes or Mos[xe]-transformed murine NIH3T3 cell extracts, and Mos specifically phosphorylates tubulin *in vitro*[11]. Double immunofluorescence studies of Mos[xe]-transformed cells showed a similar staining pattern for Mos and tubulin antibodies. After injection into *Xenopus* oocytes, Mos rapidly activates MAP kinase (**Erk1/2**)[12] and directly phosphorylates MAP kinase kinase (MAPKK, MEK, MKK) *in vivo* and *in vitro*[15]. *In vitro* vimentin, the major component of intermediate filaments, can be phosphorylated by Mos[13]. Cyclin B2 was proposed as an *in vitro* Mos substrate; however, other studies have argued against this possibility[2].

Assays

The maturation initiating activity of Mos is assayed by injection of Mos protein into fully-grown *Xenopus* oocytes, while the cytostatic factor activity of Mos is tested by the mitotic cleavage arrest that results from injection of Mos protein into cells of the *Xenopus* early embryo. The transformation activity is measured by transfection of murine NIH3T3 cell with *MOS* gene constructs and quantification of the resulting transformed cell foci. The Mos autophosphorylation activity is assayed in immune complex kinase assays.

Enzyme activators and inhibitors

The biological activities of Mos can be inhibited with antisense oligonucleotides or antisera directed against, respectively, the *MOS* mRNA or protein[2].

Pattern of expression

The highest expression of *MOS* mRNA is in germ cells (oocytes, spermatocytes and haploid round spermatids) of ovaries and testes[2]. Low levels of *MOS* RNA have been detected in brain, kidney, cardiac and skeletal muscle, mammary glands and epididymis. In cells transfected with v-*MOS*, low-level *MOS* expression leads to transformation, while high-level expression is cytotoxic[14]. In all species examined, the *MOS* gene is encoded by a single, unspliced exon.

References

1 Maxwell, S.A. and Arlinghaus, R.B. (1985) Virology 143, 321–333.
2 Yew, N. et al. (1993) Curr. Opin. Genet. Devel. 3, 19–25.
3 Sagata, N. et al. (1989) Nature 342, 512–518.
4 Freeman, R.S. et al. (1992) J. Cell Biol. 116, 725–735.
5 Testa, J. et al. (1988) Genomics 3, 44–47.
6 Hunter, T. and Cooper, J.A. (1986) In The Enzymes, Boyer, P.D. and Krebs, E.G. eds, Academic Press, Orlando, pp. 192–246.
7 Hannick, M. and Donoghue, D.J. (1985) Proc. Natl Acad. Sci. USA 82, 7894–7898.
8 Sagata, N. et al. (1988) Nature 335, 519–525.
9 Paules, R.S. et al. (1988) Oncogene 3, 59–68.
10 Seth, A. and Vande Woude, G.F. (1985) J. Virol. 173, 144–152.
11 Zhou, R. et al. (1991) Science 251, 671–675.
12 Posada, J. et al. (1993) Mol. Cell Biol. 13, 2546–2553.
13 Singh, B. and Arlinghaus, R.B. (1991) Virology 173, 144–156.
14 Papkoff, J. et al. (1982) Cell 29, 417–426.
15 Resing, K.A. et al. (1995) Biochem. in press

Research sponsored by the National Cancer Institute, DHHS, under contract NO1–CO–46000 with ABL.

Pim1

Pim1 protein kinase (vertebrates)

Jos Domen and Anton Berns (Netherlands Cancer Institute, Netherlands)

Expression of the proto-oncogene *PIM1*, which encodes a protein-serine/threonine kinase, has been found to be elevated in various murine haematopoietic malignancies due to proviral insertion near the gene. During transformation, Pim1 cooperates with c-Myc. Overexpression in mice enhances susceptibility to retrovirally and chemically induced carcinogenesis[1,2]. Variable, and sometimes high, expression levels, but thus far no molecular activation, have been reported in human haematopoietic malignancies. Recent analysis of *PIM1*-null mutant mice has revealed impaired responses to IL-3[3] and IL-7[4] in haematopoietic cells.

Subunit structure and isoforms

SUBUNIT	AMINO ACIDS	MOL. WT	SDS-PAGE
Pim1 p34	312	35 536	34 kDa
Pim1 p44	397	44 451	44 kDa

Two isoforms are known in the mouse, through usage of different translation initiation codons[5]. Initiation at the canonical AUG produces a 34 kDa protein, while initiation at an upstream CUG leads to production of a 44 kDa protein. The proteins are produced in approximately equimolar amounts. Usage of an upstream CUG codon has not been detected in human cells. Both forms are enzymatically active. It is not known whether they fulfil distinct functions in mouse cells. The 44 kDa form is present in cells as a larger, 100 kDa complex, which is relatively stable with a half-life of ≈45 min. The 34 kDa form seems to be monomeric and very unstable, half-life 5–10 min.

Domain structure

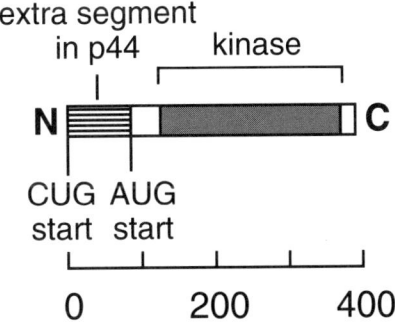

Pim1 can autophosphorylate at unidentified sites. The N-terminal extension, only present in mouse Pim1, is a preferred substrate for autophosphorylation[5].

Database accession numbers

	PIR	SWISSPROT	EMBL/GENBANK	REF
Mouse	A21169	P06803	M13945	6
Rat		P26794	X63675	7
Human	A27476	P11309	M27903	8

Amino acid sequence of mouse Pim1 p44

```
MGPAAPLALP PPALPDPAGE PARGQPRQRP QSSSDSPSAL RASRSQSRNA  50
TRSLSPGRRL SPSSLRRRCC SSRHRRRTDT LEVGMLLSKI NSLAHLRARP 100
CNDLHATKLA PGKEKEPLES QYQVGPLLGS GGFGSVYSGI RVADNLPVAI 150
KHVEKDRISD WGELPNGTRV PMEVVLLKKV SSDFSGVIRL LDWFERPDSF 200
VLILERPEPV QDLFDFITER GALQEDLARG FFWQVLEAVR HCHNCGVLHR 250
DIKDENILID LSRGEIKLID FGSGALLKDT VYTDFDGTRV YSPPEWIRYH 300
RYHGRSAAVW SLGILLYDMV CGDIPFEHDE EIIKGQVFFR QTVSSECQHL 350
IKWCLSLRPS DRPSFEEIRN HPWMQGDLLP QAASEIHLHS LSPGSSK
```

Sites of interest: M85, initiating methionine in p34 and in human Pim1

Homologues in other species

Both mouse and human Pim1 have been sequenced[6, 8]. In addition a partial, PCR-derived, rat cDNA sequence has been published[7]. All sequences are highly homologous (at least 94% identical at the amino acid level).

Physiological substrates and specificity determinants

While no Pim1 phosphorylation sites have been mapped directly, comparison of the sequences of good substrates shows that Pim1 has a preference for serines surrounded by basic residues (arginines) and possibly a nearby proline. No physiological substrate for Pim1 has been identified *in vivo*. Although the protein had originally been described as a PTK[9], more recent analysis has shown conclusively that, in line with the primary sequence, both mouse[5] and human[5, 10, 11] Pim1 exclusively exhibit serine/threonine PK activity.

Assay

Pim1 can be assayed in immune complexes[5]. Both autophosphorylation and phosphorylation of exogenous substrates (preferred substrates are histone H2B and the protamine, salmine) occur efficiently.

Enzyme activators and inhibitors

Pim1 kinase is almost completely inhibited by addition of 1 mM $ZnCl_2$ to the kinase reaction, an inhibition not commonly observed for other protein kinases[5].

Pattern of expression

Biochemical fractionation has shown that Pim1 is mainly located in the cytoplasm[5]. *PIM1* mRNA is expressed mainly in haematopoietic tissues and gonads, throughout development. Relatively high expression has further been noted in embryonic stem cells, while low expression levels are seen in cultured fibroblasts. The mRNA is unstable and can be induced by a variety of growth factors and lymphokines, including IL-2 and IL-3.

References
1 van Lohuizen, M. et al. (1989) Cell 56, 673–682.
2 Breuer, M. et al. (1989) Nature 340, 61–63.
3 Domen, J. et al. (1993) Blood 82, 1445–1452.
4 Domen, J. et al. (1993) J. Exp. Med. 178, 1665–1673.
5 Saris, C.J.M. et al. (1991). EMBO J. 10, 655–664.
6 Selten, G. et al. (1986) Cell 46, 603–611.
7 Wingett, D. et al. (1992) Nucl. Acids Res. 20, 3183–3189.
8 Reeves, R. et al. (1990) Gene 90, 303–307.
9 Telerman, A. et al. (1988) Mol. Cell. Biol. 8, 1498–1503.
10 Hoover, D. et al. (1991) J. Biol. Chem. 266, 14018–14023.
11 Padma, R. and Nagarajan, L. (1991) Cancer Res. 51, 2486–2489.

Jun Miyoshi (Osaka University, Japan)

SHOK cells have been established from Syrian hamster embryo as an alternative to NIH3T3 cells. Two activated forms of *COT* oncogenes have been detected by transfection of SHOK cells with tumour DNAs extracted from human thyroid and colon carcinoma. Both genes were oncogenically activated during transfection process because of truncation of the C-terminal domain. Cot proto-oncogene product is a cytosolic serine/threonine PK and is implicated in signal transduction by growth factors[1-6].

Subunit structure and isoforms

SUBUNIT	AMINO ACIDS	MOL. WT	SDS-PAGE
Cot (M1)	467	52 939	58 kDa
Cot (M30)	438	49 626	52 kDa

Two forms of Cot proteins are generated by alternative initiation of translation at codon M1 and M30.

Genetics

Seven coding exons were identified in the human and mouse *cot* locus. The *COT* and CT12*COT* oncogenes are rearranged in exon 6 and the intron between exon 6 and 7, respectively.
Site-directed mutants: K167F (inactive).

Domain structure

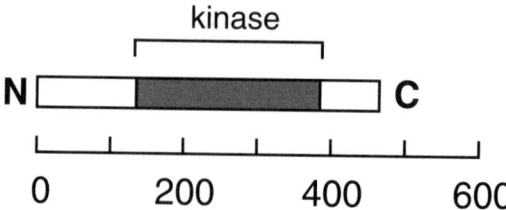

The N-terminal region may be involved in stability of Cot protein. Since both the oncogenically activated versions truncated in the C-terminal region have shown constitutive kinase activities, it seems likely that the C-terminal region is involved in negative regulation of the protein kinase activity.

Database accession numbers

	PIR	SWISSPROT	EMBL/GENBANK	REF
Human			S43914	2, 6

Amino acid sequences of human Cot (M1)

```
MEYMSTGSDN KEEIDLLIKH LNVSDVIDIM ENLYASEEPA VYEPSLMTMC  50
QDSNQNDERS KSLLLSGQEV PWLSSVRYGT VEDLLAFANH ISNTAKHFYG 100
QRPQESGILL NMVITPQNGR YQIDSDVLLI PWKLTYRNIG SDFIPRGAFG 150
KVYLAQDIKT KKRMACKLIP VDQFKPSDVE IQACFRHENI AELYGAVLWG 200
ETVHLFMEAG EGGSVLEKLE SCGPMREFEI IWVTKHVLKG LDFLHSKKVI 250
HHDIKPSNIV FMSTKAVLVD FGLSVQMTED VYFPKDLRGT EIYMSPEVIL 300
CRGHSTKADI YSLGATLIHM QTGTPPWVKR YPRSAYPSYL YIIHKQAPPL 350
EDIADDCSPG MRELIEASLE RNPNHRPRAA DLLKHEALNP PREDQPRCQS 400
LDSALLERKR LLSRKELELP ENIADSSCTG STEESEMLKR QRSLYIDLGA 450
LAGYFNLVRG PPTLEYG
```

Sites of interest: M30, initiating methioine in M30 form; R397, in the Cot oncogene product residues C-terminal to here are replaced by 18 unrelated residues; D425, in the CT12-Cot oncogene product residues C-terminal to here are replaced by a single G

Homologues in other species

The *COT* and CT12*COT* oncogene cDNAs have been sequenced from SHOK cells transformed by human tumour DNAs. Coding exons of the *COT* proto-oncogene have been sequenced from the human (placenta) and mouse (Balb/c liver). The *COT* proto-oncogene cDNAs have been sequenced from the human (PLC/PRF/5 cells) and mouse (C57BL/6 thymus).

Assay

The protein kinase activity by autophosphorylation in immune complexes. Although physiological substrates are not known, mixed histone can serve as a substrate in the immune complex kinase assay.

Pattern of expression

The mouse 3 kb *COT* mRNA was detected in a variety of tissues, but levels of expression were relatively high in the spleen, thymus and thyroid gland. RNA PCR analysis showed that the bone marrow was one of the major tissues expressing *COT* mRNA in young adult mice. Human *COT* transcripts were detected in spleen, but not readily detected in brain, liver, lung, muscle, or heart. Thus, expression of the *COT* proto-oncogene appears to be relatively limited to haematopoietic (possibly lymphoid) cells. Cot proteins have a cytosolic distribution.

References

1 Higashi, T. et al. (1990) Proc. Natl Acad. Sci. USA 87, 2409–2413.
2 Miyoshi, J. et al. (1991) Mol. Cell. Biol. 11, 4088–4096.
3 Aoki, M. et al. (1991) Oncogene 6, 1515–1519.
4 Sasai, H. et al. (1993) Br. J. Cancer 67, 262–267.
5 Ohara, R. et al. (1993) Jpn J. Cancer Res. 84, 518–525.
6 Aoki, M. et al. (1993) J. Biol. Chem. 268, 22723–22732.

Elizabeth M.J. Douville and John C. Bell (University of Ottawa, Canada)

When expressed in bacteria, this enzyme phosphorylates all three hydroxy amino acids and has therefore been included in the family of dual-specificity kinases[1]. The tyrosine phosphorylating activity of this enzyme has not been detected in mammalian cells.

Subunit structure and isoforms

SUBUNIT	AMINO ACIDS	MOL. WT	SDS-PAGE
Esk-1	856	96 000	98 kDa
Esk-2	830	94 000	96 kDa

There are at least two isoforms generated by alternative splicing. One form, Esk-1, contains an exon encoding a hydrophobic segment which has the characteristics of a transmembrane domain[2]. The Esk proteins are most homologous to the serine/threonine PKs **Snf1**, **Pim1** and the Ca^{2+}/calmodulin dependent PKs[2].

Genetics

Chromosomal location (mouse): 9, close to the *dilute* locus (Nancy Jenkins, unpublished results).

Domain structure

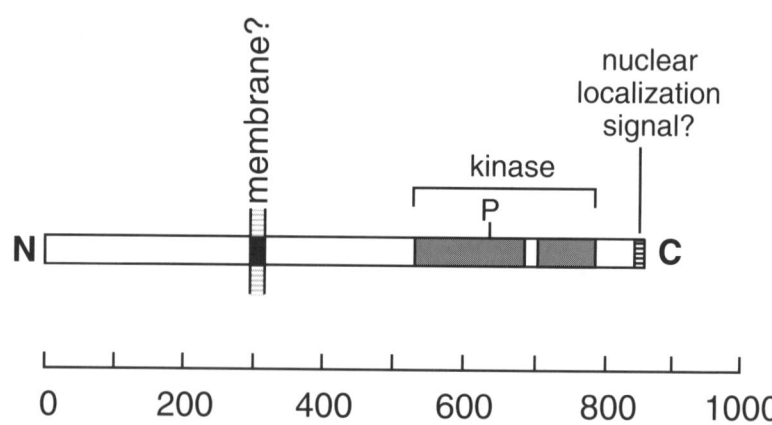

The transmembrane domain in Esk-1 is putative only. The phosphorylation site shown (T675) has been mapped[5] in bacteria.

Database accession numbers

	PIR	SWISSPROT	EMBL/GENBANK	REF
Mouse			M86377	2
Human			M86699	4

Amino acid sequence of mouse Esk-1

```
MEAEELIGSS VTIDSIMSKM RDIKNKINED CTDELSLSKI CADHTETVNQ  50
IMRVGNTPEN WLNFLLKLEK NSSPLNDDLL NKLIGRYSQA IEVLPPDKYG 100
QNESFARIQV RLAELKAIQE PDDARDYFQM ARENCKKFAF VHVSFAQFEL 150
SQGNLKKSEQ LLHKAVETGA VPLQMLETAM RNLHLQKKQL LPEEDKKSVS 200
ASTVLSAQEP FSSSLGNVQN RSISCESRGQ AGAARVLYGE NLPPQDAEVR 250
HQNPFKQTHA AKRSCPFGRV PVNLLNSPDF YVKTDSSAVT QLTTRLALSS 300
VPLPYVTCLL HLQLLALAGL AKGSGPDRDA ILPGSRPRGS DSYELRGLKP 350
IQTIYLKDSL VSNEKSSELM SDLIALKSKT DSSLTKLEET KPEIAERRPM 400
QWQSTRKPEC VFQNPAAFAP LRHVPDVTPK ADKESPPISV PKWLDPKSAC 450
ETPSSSSLDD YMKCFKTPVV KNDFPPACPS STPYSQLARL QQQQQQGLST 500
PLQSLQISGS SSINECISVN GRIYSILKQI GSGGSSKVFQ VLNEKKQINA 550
IKYVNLEDAD SQTIESYRNE IAFLNKLQQH SDKIIRLYDY EITEQYIYMV 600
MECGNIDLNS WLKKKKSINP WERKSYWKNM LEAVHIIHQH GIVHSDLKPA 650
NFVIVDGMLK LIDFGIANQM QPDTTSIVKD SQVGTVNYMA PEAIRDMSSS 700
RENSKIRTKV SPRSDVWSLG CILYYMTYGR TPFQHIINQV SKLHAIINPA 750
HEIEFPEISE KDLRDVLKCC LVRNPKERIS IPELLTHPYV QIQPHPGSQM 800
ARGATDEMKY VLGQLVGLNS PNSILKTAKT LYERYNCGEG QDSSSSKTFD 850
KKRERK
```

Sites of interest: L296–A321, putative transmembrane domain: these residues are missing in Esk-2; Y596–E602, putative PI-3-kinase P-Tyr binding site; T675, phosphorylation; Y810, putative autophosphorylation; S845–K856, putative nuclear localization signal

Homologues in other species
Human homologues, PYT[3] and **Ttk**[4], have been cloned from human fibroblast and tumour cell lines respectively.

Assay
Esk is assayed by autophosphorylation in immunoprecipitates[2-4], but can phosphorylate myelin basic protein and enolase, predominantly on T residues[2].

Pattern of expression
Esk gene expression is detected in proliferating cells and in adult tissues with a significant stem cell population. Highest expression levels are found in testes, thymus, spleen and bone marrow. Esk kinase autophosphorylation activity is found in both cytosolic and membrane fractions.

References
1   Lindberg, R.A. et al. (1992) Trends Biochem. Sci. 17, 114–119.
2   Douville, E.M.J. et al. (1992) Mol. Cell. Biol. 12, 2681–2689.
3   Lindberg, R.A. et al. (1993) Oncogene 8, 351–359.
4   Mills, G.B. et al. (1992) J. Biol. Chem. 267, 16000–16006.

Rosemarie E. Schmandt, David Hogg and Gordon B. Mills
(University of Toronto, Canada)

Ttk is a member of the family of dual-specificity PKs which can phosphorylate serine, threonine and tyrosine residues[1]. Ttk is expressed only in cells which are rapidly proliferating *in vivo*, including testes, thymus and bone marrow, as well as malignant tissues and cell lines undergoing log phase growth. Ttk is not detectable in quiescent cells. Ttk is not highly expressed in the G1 phase of the cell cycle, however, levels increase through S phase and peak in G2/M. This pattern of Ttk expression suggests a role for Ttk in cell cycle progression.

Subunit structure and isoforms

SUBUNIT	AMINO ACIDS	MOL. WT	SDS-PAGE
Ttk	857	96946	97 kDa

Comparison of the sequences of human Ttk cDNA and the murine homologue, **Esk**[2], suggest the possibility of isoforms, but this has not been verified.

Genetics
Chromosomal location (human): 6q14.

Domain structure

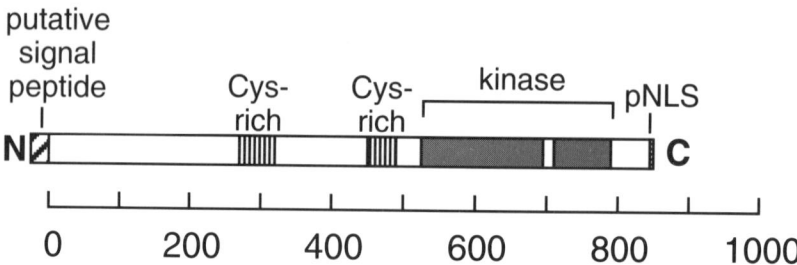

The sequence predicts two potential translation initiation start sites. If the CTG start site is selected rather than the ATG site, the sequence for Ttk would encode a putative signal peptide. The N-terminal domain possesses six of eight potential N-linked glycosylation sites, and numerous potential O-glycosylation sites. The N-terminal domain contains Cys residues in two clusters, one of 5 and one of 4 cysteines, each spaced 10–15 amino acids apart. The kinase domain contains a predominant autophosphorylation site consisting of a threonine doublet followed by a serine residue. The kinase domain also encodes a highly charged kinase insert between kinase domains VIII and IX, with potential sites for phosphorylation by **PKC**, **PKG**, **PhK**, **CKI** and **CKII**. The C-terminal six residues are highly positively charged and form a putative nuclear localization signal (pNLS). A potentially alternatively spliced form of Ttk was identified in a murine cDNA embryonal carcinoma cell line[2]. This spliced domain is compatible with a transmembrane domain. Such a splice variant remains to be identified in human cells.

Database accession numbers

	PIR	SWISSPROT	EMBL/GENBANK	REF
Human			M86699	[1]

Amino acid sequences of human Ttk

```
MNKVRDIKNK FKNEDLTDEL SLNKISADTT DNSGTVNQIM MMANNPEDWL  50
SLLLKLEKNS VPLSDALLNK LIGRYSQAIE ALPPDKYGQN ESFARIQVRF 100
AELKAIQEPD DARDYFQMAR ANCKKFAFVH ISFAQFELSQ GNVKKSKQLL 150
QKAVERGAVP LEMLEIALRN LNLQKKQLLS EEEKKNLSAS TVLTAQESFS 200
GSLGHLQNRN NSCDSRGQTT KARFLYGENM PPQDAEIGYR NSLRQTNKTK 250
QSCPFGRVPV NLLNSPDCDV KTDDSVVPCF MKRQTSRSEC RDLVVPGSKP 300
SGNDSCELRN LKSVQNSHFK EPLVSDEKSS ELIITDSITL KNKTESSLLA 350
KLEETKEYQE PEVPESNQKQ WQAKRKSECI NQNPAASSNH WQIPELARKV 400
NTEQKHTTFE QPVFSVSKQS PPISTSKWFD PKSICKTPSS NTLDDYMSCF 450
RTPVVKNDFP PACQLSTPYG QPACFQQQQH QILATPLQNL QVLASSSANE 500
CISVKGRIYS ILKQIGSGGS SKVFQVLNEK KQIYAIKYVN LEEADNQTLD 550
SYRNEIAYLN KLQQHSDKII RLYDYEITDQ YIYMVMECGN IDLNSWLKKK 600
KSIDPWERKS YWKNMLEAVH TIHQHGIVHS DLKPANFLIV DGMLKLIDFG 650
IANQMQPDTT SVVKDSQVGT VNYMPPEAIK DMSSSRENGK SKSKISPKSD 700
VWSLGCILYY MTYGKTPFQQ IINQISKLHA IIDPNHEIEF PDIPEKDLQD 750
VLKCCLKRDP KQRISIPELL AHPYVQIQTH PVNQMAKGTT EEMKYVLGQL 800
VGLNSPNSIL KAAKTLYEHY SGGESHNSSS SKTFEKKRGK K
```

Sites of interest: M1, N-terminus assuming use of ATG initiation (CTG site would encode the extension MESEDLSGRE LTIDSI); N90, N186, N210, N247, N303, N342, putative glycosylation; T659, T660, autophosphorylation

Homologues in other species

Ttk was initially sequenced from an expression library from a human NK-like cell line, YT2C2. Additional sequence was obtained from thymus and ovarian cancer cell libraries. A murine homologue of Ttk, designated **Esk**[2] was cloned from an embryonal carcinoma cell line, P19 EC.

Assay

The enzyme is assayed by incubation with [γ-^{32}P]ATP in the presence or absence of exogenous substrate such as enolase or myelin basic protein[1,2].

Pattern of expression

Ttk is expressed in all rapidly dividing normal tissues, including testes, thymus and bone marrow, in malignant tumours, and in all cell lines undergoing log phase growth.

References
[1] Mills, G.B. et al. (1992) J. Biol. Chem. 267, 16000–16006.
[2] Douville, E. M. J. et al. (1992) Mol. Cell. Biol. 12, 2681–2689.

NinaC gene product (*D. melanogaster*)

Craig Montell (Johns Hopkins University School of Medicine, Baltimore, USA)

Null mutations in *ninaC*[1] result in phototransduction defects and light-induced retinal degeneration[2,3]. Elimination of p174 results in a null phenotype; however, disruption of p132 has little effect[2]. The roles of the putative kinase and myosin domains appear to be distinct. Mutation of the kinase domain results in an electrophysiological phenotype but no retinal degeneration[4]. Deletion of the myosin domain causes both electrophysiological and retinal degeneration phenotypes[4]. Two *ts* mutations in the myosin domain result in retinal degeneration, but no ERG phenotype[4]. Thus, the electrophysiological and retinal degeneration phenotypes are not strictly coupled. It has been proposed that the role of the kinase domain is to regulate other rhabdomeric proteins important in phototransduction, and that the myosin domain has three distinct roles: the first is to transport the linked kinase domain into the rhabdomeres so that it can access its rhabdomeric substrates; a second is to maintain the structure of the rhabdomere during periods of illumination; and a third role is to carry another unlinked protein important in phototransduction into the rhabdomeres[4]. The unlinked protein appears to be calmodulin[5]. Most of the CaM in wildtype photoreceptor cells is in the rhabdomeres[5]; however, some CaM is present in subrhabdomeral bodies. Elimination of the rhabdomeric isoform p174 causes most of the CaM to be present in the cell body, and mutation of p132, localized to the cell body, results in CaM localization exclusively in the rhabdomeres[5].

Subunit structure and isoforms

SUBUNIT	AMINO ACIDS	MOL. WT	SDS-PAGE
p132	1135	131 834	132 kDa
p174	1501	174 268	174 kDa

p132 and p174 are isoforms produced by alternative splicing, which differ only at the C-terminus. Both bind calmodulin[5].

Genetics
Chromosomal location: 28A1–3[1].

Domain structure

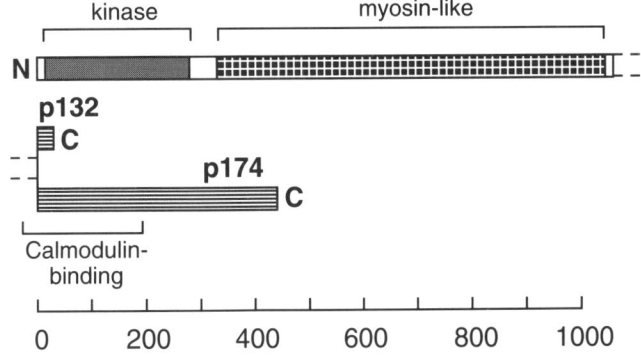

The kinase domain is followed by a domain homologous to the head region of the myosin heavy chain[1]. The myosin head domains of p132 and p174 are linked to tails of 82 and 448 amino acids that are identical for the first 28 amino acids and differ by the C-terminal 54 and 420 amino acids.

Database accession numbers

	PIR	SWISSPROT	EMBL/GENBANK	REF
p174	B29813	P10676	J03131	1
p132	A29813	P10677	J03131	1

Amino acid sequences of *D. melanogaster* NinaC

p174

```
MMYLPYAQLP DPTDKFEIYE EIAQGVNAKV FRAKELDNDR IVALKIQHYD   50
EEHQVSIEEE YRTLRDYCDH PNLPEFYGVY KLSKPNGPDE IWFVMEYCAG  100
GTAVDMVNKL LKLDRRMREE HIAYIIRETC RAAIELNRNH VLHRDIRGDN  150
ILLTKNGRVK LCDFGLSRQV DSTLGKRGTC IGSPCWMAPE VVSAMESREP  200
DITVRADVWA LGITTIELAD GKPPFADMHP TRAMFQIIRN PPPTLMRPTN  250
WSKQINDFIS ESLEKNAENR PMMVEMVEHP FLTELIENED EMRSDIAEML  300
ELSRDVKTLY KEPELFVDRG YVKRFDEKPE KMYPEDLAAL ENPVDENIIE  350
SLRHRILMGE SYSFIGDILL SLNSNEIKQE FPQEFHAKYR FKSRSENQPH  400
IFSVADIAYQ DMLHHKEPQH IVLSGESYSG KSTNARLLIK HLCYLGDGNR  450
GATGRVESSI KAILMLVNAG TPVNNDSTRC VLQYCLTFGK TGKMSGAVFN  500
MYMLEKLRVA TTDGTQHNFH IFYYFYDFIN QQNQLKEYNL KADRNYRYLR  550
VPPEVPPSKL KYRRDDPEGN VERYREFENI LRDIDFNHKQ LETVRKVLAA  600
ILNIGNIRFR QNGKYAEVEN TDIVSRIAEL LRVDEKKFMW SLTNFIMVKG  650
GIAERRQYTT EEARDARDAV ASTLYSRLVD FIINRINMNM SFPRAVFGDT  700
NAIIIHDMFG FECFNRNGLE QLMINTLNEQ MQYHYNQRIF ISEMLEMEAE  750
DIDTINLNFY DNKTALDNLL TKPDGLFYII DDASRSCQDQ NLIMDRVSEK  800
HSQFVKKHTA TEISVAHYTG RIIYDTRAFT DINRDFVPPE MIETFRSSLD  850
ESIMLMFTNQ LTKAGNLTMP FEAVQHKDES ERKSYALNTL SAGCISQVNN  900
LRTLAANFRF TCLTLLKMLS QNANLGVHFV RCIRADLEYK PRSFHSDVVQ  950
QQMKALGVLD TVIARQKGFS SRLPFDEFLR RYQFLAFDFD EPVEMTKDNC 1000
RLLFLRLKME GWALGKTKVF LRYYNDEFLA RLYELQVKKV IKVQSMMRAL 1050
LARKRVKGGK VFKLGKKGPE HHDVAASKIQ KAFRGFRDPV RLPPLVNEKS 1100
GQLNENTADF IRPFAKKWRE KSIFQVLLHY RAARFQDFVN LSQQVHIYNQ 1150
RMVAGLNKCT RAVPFERINM REVNSSQLGP LPVPIKKMPF RLDQIPFYDT 1200
QYMVDPANSI SRQAFPNQLL TQHMEDDEPW DSPLQRNPSM TSCALTYNAY 1250
KKEQACQTNW DRMGESDNIY NQGYFRDPQQ LRRNQMQMNM NAYNNAYNSY 1300
NSNYNNQNWG EHRSGSRRNS LKGYAAPPPP PPPMPSSNYY RNNPNQQQRN 1350
YQQRSSYPPS DPVRELQNMA RNEGDNSEDP PFNFKAMLRK TNYPRGSETN 1400
TYDFNNRRGS DSGDQHTFQP PKLRSTGRRY QDDEGYNSSS GNYGVSRKFG 1450
QQQRAPTLRQ SPASVGRSFE DSNARSFEEA GSYVEEEIAP GITLSGYAVD 1500
I
```

Sites of interest: K1081, end of common region, residues after this being replaced in p132 by the sequence below

p132 (unique region)

                                    GKKTQVDRL REYDEEHIDI 1100
SETPSEAEEM FLEARMDEAL AAVRIAKIEQ ASAEE

Pattern of expression

Both p132 and p174 are expressed specifically in *D. melanogaster* photo-receptor cells[1]. The p132 protein is restricted primarily to the subrhabdomeral cytoplasm and p174 to the rhabdomeres, the microvillar structure where phototransduction occurs[2].

References

[1] Montell, C. and Rubin, G. M. (1988) Cell 52, 757–772.
[2] Porter, J. A. et al. (1992) J. Cell Biol. 116, 683–693.
[3] Matsumoto, H. et al. (1987) Proc. Natl Acad. Sci. USA 84, 985–989.
[4] Porter, J. A. and Montell, C. (1993) J. Cell Biol. 122, 601–612.
[5] Porter, J.A. et al. (1993) Science 262, 1038–1042.

SplA

Spore lysis A PK (*D. discoideum*)
(Previously DPyk1)

Glen H. Nuckolls and James A. Spudich (Stanford University, USA)

The *splA* gene is necessary for the formation of spores in *D. discoideum*. Strains in which this gene has been disrupted undergo relatively normal morphological development, but spore cells lyse prior to maturation.

Subunit structure and isoforms

Only a partial sequence available.

Genetics

Strains in which the *splA* gene was disrupted produced inviable spores that failed to assemble a spore coat.

Domain structure

Only a partial sequence available. The kinase domain was found in a 1.1 kb partial cDNA to end ≈30 amino acids from the C-terminus.

Database accession numbers

	PIR	SWISSPROT	EMBL/GENBANK	REF
SplA	A35670	P18160	M33785	*1*

Partial amino acid sequence of *D. discoideum* SplA

```
RPFGGWETQS SLSHPPSRPP PPPPPPPQLP VRSEYEIDFN ELEFGQTIGK  50
GFFGEVKRGY WRETDVAIKI IYRDQFKTKS SLVMFQNEVG ILSKLRHPNV 100
VQFLGACTAG GEDHHCIVTE WMGGGSLRQF LTDHFNLLEQ NPHIRLKLAL 150
DIAKGMNYLH GWTPPILHRD LSSRNILLDH NIDPKNPLVS SRQDIKCKIS 200
DFGLSRLKKE QASQMTQSVG CIPYMAPEVF KGDSNSEKSD VYSYGMVLFE 250
LLTSDEPQQD MKPMKMAHLA AYESYRPPIP LTTSSKWKEI LTQCWDSNPD 300
SRPTFKQIIV HLKEMEDQGV SSFASVPVQT IDTGVYA
```

Sites of interest: The consensus sequence for almost all tyrosine kinases would predict a P instead of C221[2]

Physiological substrates and specificity determinants

Physiological substrates have not yet been identified. Features of the sequence of the catalytic domain might suggest that this kinase has specificity for serine or threonine as well as tyrosine[2].

Assay

When expressed in *E. coli* the *DPYK1* partial cDNA caused the phosphorylation of a variety of polypeptides as assayed by antiphosphotyrosine staining of Western blots[1].

Pattern of expression

Northern blots revealed an increase in expression early in development[1].

References

1. J.L. Tan and J.A. Spudich (1990) Mol. Cell. Biol. 10, 3578–3583.
2. Lindberg, R.A. et al. (1992) Trends Biochem. Sci. 17, 114–119.

Glen H. Nuckolls and James A. Spudich (Stanford University, USA)

The expression of DPyk2 is developmentally regulated and so it is suspected that this kinase plays some role in signal transduction during development.

Subunit structure and isoforms

The entire gene has not yet been cloned.

Genetics

The *DPYK2* locus was targeted by homologous recombination and disruption of the gene was confirmed by genomic Southern analysis. The resulting strains exhibited normal development and produced viable spores.

Domain structure

The entire gene has not yet been cloned. A 1.3 kb partial cDNA clone shows the kinase domain to end ≈50 amino acids from the C-terminus.

Database accession numbers

	PIR	SWISSPROT	EMBL/GENBANK	REF
DPyk2	B35670	P18161	M33784	[1]

Partial amino acid sequence of *D. discoideum* DPyk2

```
RFYNTTNSTK DITFLVCDNP DSTKEKSNVS NTSSIISASN LNRHITPNSH  50
MRPRGRSISE SLIMSPINKE SLNDIQRAIE SEKIKKTKFE ELKSILGERE 100
YIIDINDIQF IQKVGEGAFS EVWEGWWKGI HVAIKKLKII GDEEQFKERF 150
IREVQNLKKG NHQNIVMFIG ACYKPACIIT EYMAGGSLYN ILHNPNSSTP 200
KVKYSFPLVL KMATDMALGL LHLHSITIVH RDLTSQNILL DELGNIKISD 250
FGLSAEKSRE GSMTMTNGGI CNPRWRPPEL TKNLGHYSEK VDVYCFSLVV 300
WEILTGEIPF SDLDGSQRSA QVAYAGLRPP IPEYCDPELK LLLTQCWEAD 350
PNDRPPFTYI VNKLKEISWN NPIGFVSDQF YQYSEPSTPR LALSNQSSNS 400
SSISLSPTKL
```

Sites of interest: The consensus sequence of almost all tyrosine kinases would predict a P instead of N272[2]

Physiological substrates and specific determinants

Physiological substrates have not been identified. Features of the sequence of the catalytic domain might suggest that this kinase has specificity for serine or threonine as well as tyrosine[2].

Assay

When expressed in *E. coli* the *DPYK2* partial cDNA caused the phosphorylation of a variety of polypeptides as assayed by antiphosphotyrosine staining of Western blots[1].

Pattern of expression

Northern blots revealed an increase in the level of message early in development[1].

References
1   **Tan, J.L. and Spudich, J.A. (1990) Mol. Cell. Biol. 10, 3578–3583.**
2   Lindberg, R.A. et al. (1992) Trends Biochem. Sci. 17, 114–119.

John Rosamond (University of Manchester, UK)

The Cdc7 protein kinase plays a key role in the mitotic cell cycle of *S. cerevisiae*, with its activity being required at the G1/S phase boundary for the initiation of DNA synthesis. The enzyme is also needed during meiosis, although the requirement here appears to be for commitment to recombination rather than for the initiation of pre-meiotic S phase. In addition to this involvement in growth and development, Cdc7 has been implicated in an error-prone DNA repair pathway, although it is unclear whether this represents an independent lesion or is simply another manifestation of a failure to initiate DNA replication.

Subunit structure and isoforms

SUBUNIT	AMINO ACIDS	MOL. WT	SDS-PAGE
Cdc7	507	58 330	56 kDa

Cdc7 protein kinase activity appears to require interaction with the Dbf4 protein[1]. However, the subunit composition of the active form of the kinase is not known.

Genetics[2, 3]

Chromosomal location: IV, centromere linked.
Mutations:
  K76M, D182N, E288Q, T281E (inactive)
  G384D, G314D, G139E, E390K (temperature sensitive)

Domain structure

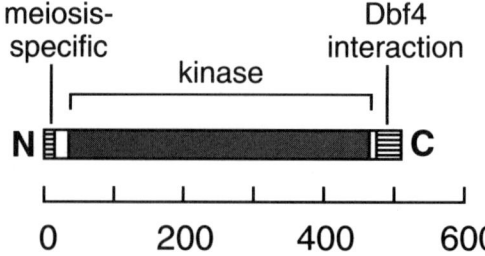

The Cdc7 protein kinase consists of an unusual catalytic domain with short N- and C-terminal extensions. Mutants lacking the N-terminal region retain normal mitotic function but are unable to undergo meiosis. The kinase domain is one of the largest known amongst the protein kinase family and contains significantly larger inserts relative to other kinases between subdomains I–II, VII–VIII and X–XI. The significance of this is not known. The C-terminal domain is a potential regulatory domain that is involved directly in the interaction with Dbf4[1].

Database accession numbers

	PIR	SWISSPROT	EMBL/GENBANK	REF
Cdc7	A25228	P06243	M12624	4
			X15632	5

Amino acid sequence of *S. cerevisiae* Cdc7

```
MTSKTKNIDD IPPEIKEEMI QLYHDLPGIE NEYKLIDKIG EGTFSSVYKA  50
KDITGKITKK FASHFWNYGS NYVALKKIYV TSSPQRIYNE LNLLYIMTGS 100
SRVAPLCDAK RVRDQVIAVL PYYPHEEFRT FYRDLPIKGI KKYIWELLRA 150
LKFVHSKGII HRDIKPTNFL FNLELGRGVL VDFGLAEAQM DYKSMISSQN 200
DYDNYANTNH DGGYSMRNHE QFCPCIMRNQ YSPNSHNQTP PMVTIQNGKV 250
VHLNNVNGVD LTKGYPKNET RRIKRANRAG TRGFRAPEVL MKCGAQSTKI 300
DIWSVGVILL SLLGRRFPMF QSLDDADSLL ELCTIFGWKE LRKCAALHGL 350
GFEASGLIWD KPNGYSNGLK EFVYDLLNKE CTIGTFPEYS VAFETFGFLQ 400
QELHDRMSIE PQLPDPKTNM DAVDAYELKK YQEEIWSDHY WCFQVLEQCF 450
EMDPQKRSSA EDLLKTPFFN ELNENTYLLD GESTDEDDVV SSSEADLLDK 500
DVLLISE
```

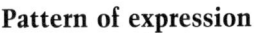

Sites of interest: T281, putative phosphorylation

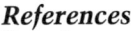

Assay
The enzyme is assayed by the transfer of radioactivity from $[\gamma\text{-}^{32}\text{P}]$ATP to histone, usually H1 or H2A.

Pattern of expression
The protein has a nuclear localization. Transcript and protein levels are constant throughout mitosis, when the protein is present in excess of the amount needed for one round of cell division[6]. The transcript is periodically regulated during meiosis and is maximal at about the time of pre-meiotic recombination.

References
1 Jackson, A.L. et al. (1993) Mol. Cell. Biol. 13, 2899–2908.
2 Buck, V. et al. (1991) Mol. Gen. Genet. 227, 452–457.
3 Hollingsworth, R.E et al. (1992) Genetics 132, 53–62.
4 **Patterson, M. et al. (1986) Mol. Cell. Biol. 6, 1590–1598.**
5 Ham, J. et al. (1989) Nucl. Acids Res. 17, 5781–5792.
6 Sclafani, R.A. et al. (1988) Mol. Cell. Biol. 8, 293–300.

Cdc15 Cell division cycle protein 15 (*S. cerevisiae*)

Achim Wach, Bert Schweitzer, Thomas Hinz and Peter Philippsen (University of Basel, Switzerland, and University of Giessen, Germany)

The *CDC15* gene of the yeast *S. cerevisiae* was originally characterized by a mutation which arrests the cell cycle late in mitosis (post anaphase) at 36°C but not at 23°C[1]. When arrested, pole to pole microtubules are still present, but apparently exit from mitosis (e.g. nuclear envelope reorganization or cytokinesis) is blocked. Upon return to permissive temperature, cells recover well from the mitotic block[2] and resume normal growth passing through the next 1 to 2 cell cycles as a synchronized culture. However, at permissive temperature *CDC15* mutants have a slightly increased level of chromosome loss[3]. The *CDC15* gene was isolated in an effort to clone all fragments of chromosome I[4]. The DNA sequence of the *CDC15* allele[5] then revealed that *CDC15* encodes a 974 residue protein. Its first 270 N-terminal amino acids contain all 11 conserved motifs typically found in PK domains. The Cdc15 protein could be precipitated with antibodies directed against a C-terminal peptide. The protein kinase activity of these immuno-precipitated molecules showed that the Cdc15 protein is a serine/threonine PK, since casein was phosphorylated *in vitro* at S/T residues (T. Hinz, PhD Thesis, University of Giessen, 1992).

Subunit structure and isoforms

SUBUNIT	AMINO ACIDS	MOL. WT	SDS-PAGE
Cdc15	974	110284	110–115 kDa

Genetics

Chromosome location: I, right arm, 20 kb distant from the centromere.
Mutations: The allele *cdc15-1* encodes a heat-sensitive Cdc15 isoform.

Domain structure

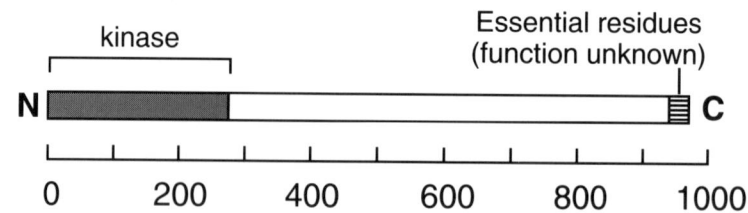

Database accession numbers

	PIR	SWISSPROT	EMBL/GENBANK	REF
Cdc15	S15038	P27636	X52683	5

Amino acid sequence of *S. cerevisiae* Cdc15

```
MNSMADTDRV NLTPIQRASE KSVQYHLKQV IGRGSYGVVY KAINKHTDQV  50
VAIKEVVYEN DEELNDIMAE ISLLKNLNHN NIVKYHGFIR KSYELYILLE 100
YCANGSLRRL ISRSSTGLSE NESKTYVTQT LLGLKYLHGE GVIHRDIKAA 150
NILLSADNTV KLADFGVSTI VNSSALTLAG TLNWMAPEIL GNRGASTLSD 200
IWSLGATVVE MLTKNPPYHN LTDANIYYAV ENDTYYPPSS FSEPLKDFLS 250
KCFVKNMYKR PTADQLLKHV WINSTENVKV DKLNKFKEDF TDADYHWDAD 300
FQEEKLNISP SKFSLRAAPA PWAENNQELD LMPPTESQLL SQLKSSSKPL 350
TDLHVLFSVC SLENIADTII ECLSRTTVDK RLITAFGSIF VYDTQHNHSR 400
LRLKFIAMGG IPLIIKFEHL AKEFVIDYPQ TLIECGIMYP PNFASLKTPK 450
YILELVYRFY DLTSTAFWCR WCFKHLDISL LLNNIHERRA QSILLKLSSY 500
APWSFEKILP SLIDSKLKKK ILISPQITYV VFKSINYMIT TNDDKIHKSA 550
IPSSSSLPLS SSPTRNSPVN SVQSPSRSPV HSLMATRPSS PMRHKSISNF 600
PHLTISSKSR LLIELPEGFF TWLTSFFVDM AQIKDLSVLK YFTKLCYLTV 650
HINSTFLNDL LDNDAFFAFI RNIDTIIPFI DDAKTAAFIW KQITAICVEM 700
SLDMDQMSAS LFSTAMNFIR KKNNTSISGL EIILNCLHFT LRNVNDDVAP 750
TVGSSESHSV FLIKVNNDAA IELPIDQLVD LFYALNDDDV NLSKLISIFT 800
KICSLPGFEN LTINIIFHPN FYEKIVSFFD TYFNSLLIQI DLLKFIKLIF 850
SKSLLKLYDY TGQPDPIKQT EPNRRNKATV FKLRAILVQI TEFLNNNWNK 900
DVPKRNSNQV GGDSVLICQL CEDIRSLSKK GSLQKVSSVT AAIGSSPTKD 950
ERSNLRSSKD KSDGFSVPIT TFQT
```

Sites of interest: M1, it is not clear whether this or M4 is the starting residue;
K900–V902, this sequence was originally reported as NGC[5]

Physiological substrates and specificity determinants

Physiological substrates are presently unknown. However, it was proposed that Cdc15 is necessary (directly or indirectly) for Clb2/Cdc28 protein kinase destruction[6]. In addition, Cdc15 itself is autophosphorylated during *in vitro* assays (T. Hinz, A. Wach and P. Philippsen, unpublished data).

Assay

Cdc15 activity is assayed with immunoprecipitated Cdc15 by transfer of radioactivity from [γ-^{32}P]ATP to the artificial substrate casein.

Pattern of expression

There is a low level of apparently constitutive *CDC15* mRNA production (B. Schweitzer and P. Philippsen, unpublished data). It is not known whether the Cdc15 protein kinase activity is regulated during the cell cycle.

References

1 Pringle, J.R. and Hartwell, L.H. (1981) In The Molecular Biology of the Yeast *Saccharomyces cerevisiae*, Life Cycle and Inheritance, Strathern, J.N., Jones, E.W. and Broach, J.R., eds, Cold Spring Harbor, New York, pp. 97–142.
2 Weinert, T.A. and Hartwell, L.H. (1993) Genetics 134, 63–80.

[3] Hartwell, L.H. and Smith, D. (1985) Genetics 110, 381–395.
[4] Steensma, H.Y. et al. (1987) Mol. Cell. Biol. 7, 410–419.
[5] Schweitzer, B. and Philippsen, P. (1991) Yeast 7, 265–273.
[6] Surana, U. et al. (1993) EMBO J. 12, 1969–1978.

Vps15

Jeffrey H. Stack and Scott D. Emr
(University of California, San Diego, USA)

The Vps15 protein kinase was identified in a selection for genes required for the efficient sorting of proteins to the yeast vacuole[1]. Vps15 functions specifically in the diversion of soluble vacuolar hydrolases from the Golgi complex to the vacuole[1,2]. Vps15 forms a membrane-associated hetero-oligomeric complex with another protein required for vacuolar protein sorting, the Vps34 phosphatidylinositol-3-kinase[3,4], and activates the Vps34 PI-3-kinase[3].

Subunit structure and isoforms

SUBUNIT	AMINO ACIDS	MOL. WT	SDS-PAGE
Vps15	1455	170 000	170 kDa

Genetics
Chromosomal location: 2.
Extensive mutational analyses have found a correlation between the *in vivo* phosphorylation level of Vps15 and vacuolar protein sorting function. The phenotypes of *vps15* kinase domain point mutants generated by site-directed mutagenesis range from silent to completely defective for protein sorting[2,5].

Domain structure

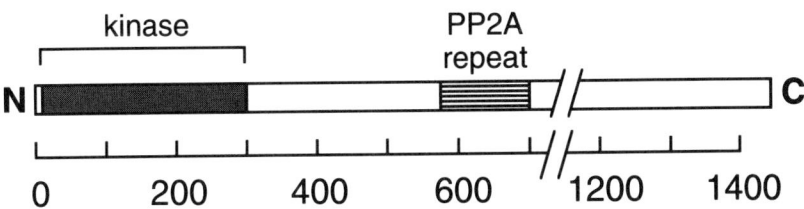

Vps15 contains a region with significant sequence similarity to the repeat motif which comprises the regulatory subunit A of protein phosphatase 2A[1]. The large C-terminal region contains no homology to any known protein sequence; however, truncation of as little as 30 amino acids from the C-terminus of Vps15 leads to a *ts* vacuolar protein sorting defect[5]. The kinase domain of Vps15 does not contain a canonical glycine-fold (domain I), but does contain a stretch of small uncharged residues that may fold into a similar structure. As Vps15 possesses kinase activity, it appears that this motif is conserved at the structural, if not the sequence, level.

Database accession numbers

	PIR	SWISSPROT	EMBL/GENBANK	REF
Vps15	S13343	P22219	M59835	2

Amino acid sequence of *S. cerevisiae* Vps15

```
MGAQLSLVVQ ASPSIAIFSY IDVLEEVHYV SQLNSSRFLK TCKALDPNGE   50
IVIKVFIKPK DQYSLRPFLQ RIRAQSFKLG QLPHVLNYSK LIETNRAGYM  100
IRQHLKNNLY DRLSLRPYLQ DIELKFIAFQ LLNALKDIHN LNIVHGDIKT  150
ENILVTSWNW CILTDFAAFI KPVYLPEDNP GEFLFYFDTS KRRTCYLAPE  200
RFNSKLYQDG KSNNGRLTKE MDIFSLGCVI AEIFAEGRPI FNLSQLFKYK  250
SNSYDVNREF LMEEMNSTDL RNLVLDMIQL DPSKRLSCDE LLNKYRGIFF  300
PDYFYTFIYD YFRNLVTMTT STPISDNTCT NSTLEDNVKL LDETTEKIYR  350
DFSQICHCLD FPLIKDGGEI GSDPPILESY KIEIEISRFL NTNLYFPQNY  400
HLVLQQFTKV SEKIKSVKEE CALLFISYLS HSIRSIVSTA TKPVKNLELL  450
AVFAQFVSDE NKIDRVVPYF VCCFEDSDQD VQALSLLTLI QVLTSVRKLN  500
QLNENIFVDY LLPRLKRLLI SNRQNTNYLR IVFANCLSDL AIIINRFQEF  550
TFAQHCNDNS MDNNTEIMES STKYSAKLIQ SVEDLTVSFL TDNDTYVKMA  600
LLQNILPLCK SFGRERTNDI ILSHLITYLN DKDPALRVSL IQTISGISIL  650
LGTVTLEQYI LPLLIQTITD SEELVVISVL QSLKSLFKTG LIRKKYYIDI  700
SKTTSPLLLH PNNWIRQFTL MIIIEIINKL SKAEVYCILY PIIRPFFEFD  750
VEFNFKSMIS CCKQPVSRSV YNLLCSWSVR ASKSLFWKKI ITNHVDSFGN  800
NRIEFITKNY SSKNYGFNKR DTKSSSSLKG IKTSSTVYSH DNKEIPLTAE  850
DINWIDKFHI IGLTEKDIWK IVALRGYVIR TARVMAANPD FPYNNSNYRP  900
LVQNSPPNLN LTNIMPRNIF FDVEFAEEST SEGQDSNLEN QQIYKYDESE  950
KDSNKLNING SKQLSTVMDI NGSLIFKNKS IATTTSNLKN VFVQLEPTSY 1000
HMHSPNHGLK DNANVKPERK VVVSNSYEGD VESIEKFLST FKILPPLRDY 1050
KEFGPIQEIV RSPNMGNLRG KLIATLMENE PNSITSSAVS PGETPYLITG 1100
SDQGVIKIWN LKEIIVGEVY SSSLTYDCSS TVTQITMIPN FDAFAVSSKD 1150
GQIIVLKVNH YQQESEVKFL NCECIRKINL KNFGKNEYAV RMRAFVNEEK 1200
SLLVALTNLS RVIIFDIRTL ERLQIIENSP RHGAVSSICI DEECCVLILG 1250
TTRGIIDIWD IRFNVLIRSW SFGDHAPITH VEVCQFYGKN SVIVVGGSSK 1300
TFLTIWNFVK GHCQYAFINS DEQPSMEHFL PIEKGLEELN FCGIRSLNAL 1350
STISVSNDKI LLTDEATSSI VMFSLNELSS SKAVISPSRF SDVFIPTQVT 1400
ANLTMLLRKM KRTSTHSVDD SLYHHDIINS ISTCEVDETP LLVACDNSGL 1450
IGIFQ
```

Sites of interest: G2, myristoylation

Assay

Vps15 can be assayed by autophosphorylation in immune complexes isolated under non-denaturing conditions, or after renaturation (unpublished results). This activity requires divalent cation, and is greatest in Mn^{2+} ($Mn^{2+}>Mg^{2+}$). The kinase activity of mutant forms of Vps15 correlates very well with the severity of the vacuolar protein sorting defect.

Pattern of expression

Differential centrifugation and sucrose gradients have demonstrated that the majority of Vps15 is associated with the cytoplasmic face of an intracellular membrane fraction, possibly corresponding to a late Golgi compartment[2, 3]. The Vps34 PI-3-kinase has been shown by native immunoprecipitation and chemical cross-linking experiments to form a membrane-associated complex with the Vps15 protein kinase[3]. Vps34 PI-3-kinase activity is severely defective in *vps15* kinase domain point mutants, suggesting that a Vps15-mediated protein phosphorylation event is required for the activation of the Vps34 PI-3-kinase[3].

References
[1] Herman, P.K. et al. (1992) Trends Cell Biol. 2, 363–368.
[2] Herman, P.K. et al. (1991) Cell 64, 425–437.
[3] Stack, J.H. et al. (1993) EMBO J. 12, 2195–2204.
[4] Schu, P.V. et al. (1993) Science, 260, 88–91.
[5] Herman, P.K. et al. (1991) EMBO J. 10, 1049–1060.

Ran1
Ran1 protein kinase (*S. pombe*)
(p52^{ran1}, p52^{pat1})

Maureen McLeod (SUNY, Brooklyn, USA)

Ran1 functions in *S. pombe* as a negative regulator of conjugation and meiosis[1]. It prevents the transcription of genes required for sexual differentiation[2].

Subunit structure and isoforms

SUBUNIT	AMINO ACIDS	MOL. WT	SDS-PAGE
Ran1	470	52 169	52 kDa

A protein, p21^{mei3}, has been identified that inhibits Ran1 kinase activity *in vitro* and is found associated with the kinase *in vivo*. These data indicate that p21^{mei3} is a regulatory subunit.

Genetics
Chromosomal location: II.
Site directed mutants: K47A (inactive).

Domain structure

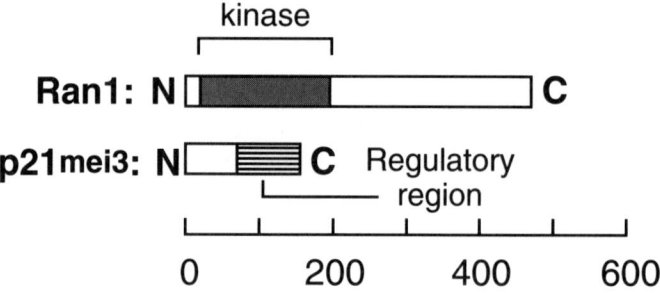

A truncated version of p21^{mei3}, containing amino acid residues 76–148 is active as an inhibitor of Ran1 protein kinase.

Database accession numbers

	PIR	SWISSPROT	EMBL/GENBANK	REF
ran1			X04728	3
mei3			X05142	4

Amino acid sequence of *S. pombe* Ran1

```
MMRENPELLL GQVLGDSLRF VSIIGAGAYG VVYKAEDIYD GTLYAVKALC  50
KDGLNEKQKK LQARELALHA RVSSHPYIIT LHRVLETEDA IYVVLQYCPN 100
GDLFTYITEK KVYQGNSHLI KTVFLQLISA VEHCSVGIY HRDLKPENIM 150
VGNDGNTVYL ADFGLATTEP YSSDFGCGSL FYMSPECQRE VKKLSSLSDM 200
LPVTPEPIES QSSSFATAPN DVWALGIILI NLCCKRNPWK RACSQTDGTY 250
```

Amino acid sequence of *S. pombe* Ran1 *continued*

```
RSYVHNPSTL LSILPISREL NSLLNRIFDR NPKTRITLPE LSTLVSNCKN 300
LTRRLRPAPL VSSRYLAYQQ QQQQQQMNLQ QGIXGYPHQG YMPTQNIGFP 350
WPPTPQFVSN WNHCATPTIP VSLQVLTPNS SLKVDPTTPL TAPIHATESF 400
WPSAAAAAAA VHNNANSYMP ITPTPYPNNA KIFGYPNQPP LTPIPFTGFV 450
LHPAPVGRAA DAVDPSRKSL
```

Amino acid sequence of *S. pombe* p21^{mei3}

```
MSSQNTSNSR HPASSASALP NRTNTARRST SPRTSTGSSS TNTNTKTVDH  50
AGTTFISSTS KRGRSTKAGS VHSVPMKRTK RVRRTPAQRI EHENKENIQT 100
EKVYRIKPVQ RVLSPSDLTN ELTILDLFHV PRPNETFDIT DRVNNTSR
```

Assay
The enzyme is assayed by autophosphorylation. No exogenous substrates have been identified.

Pattern of expression
The steady state level of *ran1*$^{+}$ transcript and Ran1 protein is constant throughout the *S. pombe* life cycle. The *mei3*$^{+}$ transcript and the p21^{mei3} protein, are developmentally regulated and are detected only during meiosis.

References
1 McLeod, M. and Beach, D. (1988) Nature 332, 509–514.
2 Nielsen, O. and Egel, R. (1990) EMBO J. 9, 1401–1406.
3 McLeod, M. and Beach, D. (1987) EMBO J. 6, 729–736.
4 McLeod, M. and Beach, D. (1986) EMBO J. 3645–3671.

David P. Leader (University of Glasgow, Scotland, UK)

The genomes of most herpesviruses contain two unique regions, the unique short (U$_S$) region and the unique long (U$_L$) region, flanked by repeated regions. The three subclasses of herpesviruses, alpha-, beta- and gamma-, show similarities in their U$_L$ regions, but differ in their U$_S$ regions. This entry concerns a protein kinase encoded in the unique short (U$_S$) region of the alphaherpesviruses (αHv), which is absent from the other subclasses[1]. The two αHv U$_S$ PKs that have been characterized – those of the human and porcine αHvs, herpes simplex virus type-1 (HSV-1) and pseudorabies virus (PRV), respectively – will be considered here. These PKs were initially referred to as "virus-induced protein kinase", and later designated HSV-PK and PRV-PK, respectively, when it was shown that they were encoded in the viral genome. However a protein kinase encoded in the U$_L$ region of all three subclasses of herpesvirus was subsequently discovered (**HvU$_L$**), which is why the present nomenclature is to be preferred. The αHvU$_S$ PKs are sometimes referred to as US3 PKs after the designation of the gene in HSV-1, but, because of the different organization of the U$_S$ region of the genomes of different αHvs, this nomenclature is at variance with the actual gene designations of other αHvs (see below). The function of the enzyme in the viral life cycle is unknown. It is not essential for growth of virus in tissue culture, but mutant viruses lacking the gene are severely attenuated when injected into mice[2].

Subunit structure and isoforms

SUBUNIT	AMINO ACIDS	MOL. WT	SDS-PAGE
HSV-1 U$_S$	481	53 000	68 kDa

The enzyme appears to be a homodimer[3].

Genetics

As already stated, this PK is encoded in the U$_S$ region of alphaherpesviruses. It is either designated by its position numbered from the conventional "left" of this region, by its position from the conventional "left" of the whole virus genome, or both. These are given below for the different alphaherpesvirus genes available at the moment:

HSV-1: gene US3[4]; PRV: gene US1[5]; VZV (varicella zoster virus): gene US2 or gene 66[6]; EHV-1 (equine herpes virus type 1) gene US2 or gene 69[7]; MDV (Marek's disease virus): neither numbering is available, but the gene is described by Ross *et al.*[8].

Domain structure

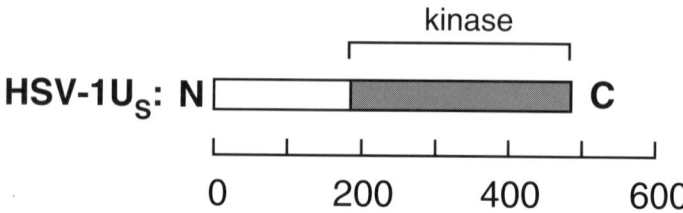

The protein kinase has two domains: a C-terminal kinase domain with strong similarity between different αHvU_S PKs, and homology with the eukaryotic family of serine/threonine PKs; and an N-terminal domain of variable length and little conservation except for the presence of an acidic "patch".

Database accession numbers

	PIR	SWISSPROT	EMBL/GENBANK	REF
HSV-1	A00656	P04413	X14112	[4]
HSV-2	B43674	P13287	X04798	[9]
PRV	A36655	P17613	D00676, D00545	[5]
VZV	E27345	P09251	X04370	[6]
EHV-1	F36802	P28926	M86664	[7]

Amino acid sequence of HSV-1 U_S

```
MACRKFCRVY GGQGRRKEEA VPPETKPSRV FPHGPFYTPA EDACLDSPPP  50
ETPKPSHTTP PSEAERLCHL QEILAQMYGN QDYPIEDDPS ADAADDVDED 100
APDDVAYPEE YAEELFLPGD ATGPLIGAND HIPPPCGASP PGIRRRSRDE 150
IGATGFTAEE LDAMDREAAR AISRGGKPPS TMAKLVTGMG FTIHGALTPG 200
SEGCVFDSSH PDYPQRVIVK AGWYTSTSHE ARLLRRLDHP AILPLLDLHV 250
VSGVTCLVLP KYQADLYTYL SRRLNPLGRP QIAAVSRQLL SAVDYIHRQG 300
IIHRDIKTEN IFINTPEDIC LGDFGAACFV QGSRSSPFPY GIAGTIDTNA 350
PEVLAGDPYT TTVDIWSAGL VIFETAVHNA SLFSAPRGPK RGPCDSQITR 400
IIRQAQVHVD EFSPHPESRL TSRYRSRAAG NNRPPYTRPA WTRYYKMDID 450
VEYLVCKALT FDGALRPSAA ELLCLPLFQQ K
```

Homologues in other species
Other αHvs which contain a U_S PK gene are described above. In addition, the αHvU_S from a different strain of PRV has been described (the NIA-3 strain, rather than the Ka strain), which shows a small number of differences in amino acid sequence[10]. No wildtype αHv yet described lacks such a gene.

Physiological substrates and specificity determinants

The consensus sequence for αHvU_S has been determined using synthetic peptides using pure enzyme from PRV, and a somewhat less pure preparation from HSV-1. Multiple N-terminal Arg residues are required. RRRRXS/TX is the best substrate, although peptides with only three Arg, or one more or less "spacer" residue, were also good substrates[11]. The site of phosphorylation in one physiological substrate for HSV-1 U_S PK has been determined, albeit indirectly by deletion analysis[12]. This is consistent with the prediction from model peptides:

HSV-1 protein UL34:   RRRRTRRSRE

Assay
The enzyme is usually assayed by the transfer of radioactivity from $[\gamma\text{-}^{32}P]$ATP to protamine[13]. Alternatively the peptide RRRRASVA can be used[11].

Enzyme activators and inhibitors

Spermine can stimulate the activity of the enzyme from PRV approximately two-fold[13]. Polyarginine appears to be a potent but non-specific inhibitor[11].

Pattern of expression

The enzyme appears to be constitutively active, and therefore control is exerted by the time of its synthesis during the replicative cycle of the virus. There are three main temporal classes of proteins in herpesviruses: immediate early, early and late. HvU_S is an early protein in both HSV-1 and PRV.

References

1 Leader, D.P. (1993) Pharmacol. Ther. 59, 343–387.
2 Meignier, B. et al. (1988) Virology 162, 251–254.
3 Purves, F.C. et al. (1987) Eur. J. Biochem. 167, 507–512.
4 McGeoch, D.J. et al. (1985) J. Mol. Biol. 181, 1–13.
5 Zhang, G. et al. (1990) J. Gen. Virol. 71, 1757–1765.
6 Davison, A.J. et al. (1986) J. Gen. Virol. 67, 1759–1816.
7 Telford, E.A.R. et al. (1992) Virology 189, 304–316.
8 Ross, L. et al. (1991) J. Gen. Virol. 72, 949–954.
9 McGeoch, D.J. et al. (1997) J. Gen. Virol. 68, 19–38.
10 van Zijl, M. et al. (1990) J. Gen. Virol. 71, 1747–1755.
11 Leader, D.P. et al. (1991) Biochim. Biophys. Acta 1091, 426–431.
12 Purves, F.C. et al. (1992) J. Virol. 66, 4295–4303.
13 Katan, M. et al. (1985) Eur. J. Biochem. 152, 57–65.

Duncan J. McGeoch, Lesley J. Coulter and Helen W. M. Moss (Institute of Virology and MRC Virology Unit, University of Glasgow, Scotland, UK)

Gene UL13 of herpes simplex virus type 1 (HSV-1) was characterized first by transcript mapping and DNA sequencing[1–3] and its product was subsequently reported to possess motifs diagnostic of Ser/Thr protein kinases[4,5]. The UL13 protein has now been identified as a component of the virus particle[6–8]. It is known to be responsible for phosphorylation of one virus-specified protein in infected cells[9], and it can phosphorylate others in cell extracts and in disrupted virions[7], but its function in the virus cycle remains obscure.

Subunit structure and isoforms

SUBUNIT	AMINO ACIDS	MOL. WT	SDS-PAGE
HSV-1 UL13	518	57 193	57 kDa

When UL13 was expressed in the absence of other virus components no activity was detectable, suggesting that association with at least one other viral protein may be required for activity[6,7].

Genetics

The UL13 gene of HSV-1 is leftward oriented in the prototype representation of the HSV-1 genome, with its ORF located between genomic residues 28504 and 26951[3,10]. The gene partially overlaps its neighbours: the promoter, transcript start site and first 82 codons of UL13 lie within the reading frame of its right neighbour, gene UL14, while the transcribed 5' non-coding region of the left neighbour, UL12, overlaps the final 34 codons of the UL13 reading frame[1,2]. Mutants of HSV-1 in which the UL13 gene is inactivated by insertion or deletion have been constructed[7,9]. These are capable of growth in tissue culture, although in some cell lines growth is reduced. Gene UL13 thus belongs to the numerous subset of HSV-1 genes non-essential for growth in cultured cells[10].

Domain structure

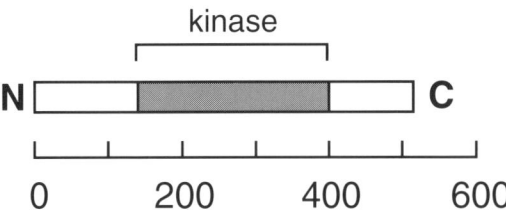

Comparisons among herpesvirus homologues indicate little conservation of sequences outside the kinase domain, and the N-terminal portions show considerable heterogeneity in length.

Database accession numbers

	PIR	SWISSPROT	EMBL/GENBANK	REF
U_L HSV-1	A03738	P04290	X14112	2, 3

Amino acid sequence of HSV-1 UL13

```
MDESRRQRPA GHVAANLSPQ GARQRSFKDW LASYVHSNPH GASGRPSGPS  50
LQDAAVSRSS HGSRHRSGLR ERLRAGLSRW RMSRSSHRRA SPETPGTAAK 100
LNRPPLRRSQ AALTAPPSSP SHILTLTRIR KLCSPVFAIN PALHYTTLEI 150
PGARSFGGSG GYGDVQLIRE HKLAVKTIKE KEWFAVELIA TLLVGECVLR 200
AGRTHNIRGF IAPLGFSLQQ RQIVFPAYDM DLGKYIGQLA SLRTTNPSVS 250
TALHQCFTEL ARAVVFLNTT CGISHLDIKC ANILVMLRSD AVSLRRAVLA 300
DFSLVTLNSN STIARGQFCL QEPDLKSPRM FGMPTALTTA NFHTLVGHGY 350
NQPPELLVKY LNNERAEFTN HRLKHDVGLA VDLYALGQTL LELVVSVYVA 400
PSLGVPVTRF PGYQYFNNQL SPDFALALLA YRCVLHPALF VNSAETNTHG 450
LAYDVPEGIR RHLRNPKIRR AFTDRCINYQ HTHKAILSSV ALPPELKPLL 500
VLVSRLCHTN PCARHALS
```

Homologues in other species

Homologues of HSV-1 UL13 are apparently ubiquitous among mammalian herpesviruses, in contrast to the HvU_S family. In the alphaherpesvirinae, UL13 homologues are known in varicella-zoster virus[11], equine herpesvirus 1[12] and pseudorabies virus[13]. In the betaherpesvirinae, sequences are known for human cytomegalovirus (HCMV) and human herpesvirus 6, and in the gammaherpesvirinae for Epstein–Barr virus and herpesvirus saimiri[5, 14]. The HCMV homologue, UL97, is known to phosphorylate the nucleoside analogue ganciclovir, thus enabling the antiviral action of this compound; the relationship of this necessarily adventitious activity to the *in vivo* role of the protein is not known[15, 16]. The sequences are highly diverged among the families, and not all the kinase motifs seen in the alphaherpesvirus sequences are retained in the others[4, 5]. The HvU_L subfamily is distant from the HvU_S subfamily, and is not closely related to any other group within the PK family.

Physiological substrates and specificity determinants

A study with UL13 mutants has shown that the viral protein ICP22 (encoded by gene US1 and thought to have a regulatory function) is a substrate of UL13 PK in infected cells[9]. When disrupted virions or nuclear extracts of infected cells are incubated with [γ-^{32}P]ATP, UL13-mediated phosphorylation of the abundant tegument protein VP22 (encoded by gene UL49) and of other proteins is observed[7]. In these *in vitro* situations, phosphorylation of the UL13 protein is also prominent, and may represent autophosphorylation[6, 7]. The relevance of any of these observations to function in natural infection is unresolved. There is no specific information on characteristics of amino acid sequences which are substrates for the UL13 PK: the sequences of observed substrates are consistent with the enzyme phosphorylating S/T residues in acidic surroundings.

Assay

The UL13 PK was first assayed in nuclear extracts of infected cells by incorporation of radioactivity from [γ-^{32}P]ATP or GTP into acid-precipitable material[6]. Casein was active as a substrate, but protamine was not. Notably, the enzyme was optimally active in the presence of 1.5 M NaCl. Attempts to purify UL13 PK resulted in loss of activity. The enzyme can also be detected in nuclear extracts and in disrupted virions by incubation in the presence of [γ-^{32}P]ATP followed by SDS-PAGE and autoradiography to detect labelled protein species[7].

Pattern of expression

The UL13 protein is found late in infection of cultured cells as a nuclear phosphoprotein[6, 8]. It is a component of virions, occurring at a readily detectable level in the tegument (the amorphous, proteinaceous layer surrounding the icosahedral capsid), and is probably identical to a previously listed virion protein, VP18.8[6, 7].

References

1   Costa, R.H. et al. (1983) J. Virol. 48, 591–603.
2   McGeoch, D.J. et al. (1986) Nucl. Acids Res. 14, 3435–3448.
3   McGeoch, D.J. et al. (1988) J. Gen. Virol. 69, 1531–1574.
4   Smith, R.F. and Smith, T.F. (1989) J. Virol. 63, 450–455.
5   Chee, M.S. et al. (1989) J. Gen. Virol. 70, 1151–1160.
6   Cunningham, C. et al. (1992) J. Gen. Virol. 73, 303–311.
7   Coulter et al. (1993) J. Gen. Virol. 74, 387–395.
8   Overton, H.A. et al. (1992) Virology 190, 184–192.
9   Purves, F.C. and Roizman, B. (1992) Proc. Natl Acad. Sci. USA 89, 7310–7314.
10 McGeoch, D.J. (1989) Annu. Rev. Microbiol. 43, 235–265.
11 Davison, A.J. and Scott, J.E. (1986) J. Gen. Virol. 67, 2279–2286.
12 Telford, E.A.R. et al. (1992) Virology 189, 304–316.
13 de Wind, N. et al. (1992) J. Virol. 66, 5200–5209.
14 Albrecht, J.-C. et al. (1992) J. Virol. 66, 5047–5058.
15 Littler, E. et al. (1992) Nature 358, 160–162.
16 Sullivan, V. et al. (1992) Nature 358, 162–164.

Geoffrey L. Smith (Oxford University, UK)

Vaccinia virus gene *B1R* encodes a PK that is expressed early during infection and which is essential for virus replication[1,2]. Conditional lethal *ts* mutants that map to B1R show a DNA negative phenotype at the non-permissive temperature, and DNA synthesis ceases after shift up[2-4]. The protein is located in cytoplasmic factories and is packaged into virus particles[5,6]. The enzyme has S/T specificity[5,6] and can phosphorylate 40S ribosomal proteins S2 and Sa both *in vitro*[7], and during vaccinia virus infection[8,9]. Another vaccinia gene product (B12R) is 36% identical to B1R[1], but does not show PK activity under conditions in which the B1R PK is active[10], and due to considerable divergence from the PK family is unlikely to have PK activity.

Subunit structure and isoforms

SUBUNIT	AMINO ACIDS	MOL. WT	SDS-PAGE
B1R	283	34 200	34 kDa

Genetics

The kinase is encoded by gene *B1R* (the first gene in the virus *Hind*III B fragment, transcribed rightwards). Two *ts* mutants (*ts2* and *ts25*) have amino acid substitutions (G79D and G227S) conferring the *ts* phenotype[2].

Domain structure

The protein contains little more than a canonical kinase domain.

Database accession numbers

		PIR	SWISSPROT	EMBL/GENBANK	REF
B1R	*WR	A34152	P16913	D00628	1
B1	*WR	A34152	P16913	J05178	2
B1R	**C	I42525	P20505	M35027	11

*Strain Western reserve; **strain Copenhagen.

Amino acid sequence of vaccinia virus B1R (strain WR)

```
MNFQGLVLTD NCKNQWVVGP LIGKGGFGSI YTTNDNNYVV KIEPKANGSL  50
FTEQAFYTRV LKPSVIEEWK KSHNIKHVGL ITCKAFGLYK SINVEYRFLV 100
INRLGADLDA VIRANNNRLP KRSVMLIGIE ILNTIQFMHE QGYSHGDIKA 150
SNIVLDQIDK NKLYLVDYGL VSKFMSNGEH VPFIRNPNKM DNGTLEFTPI 200
DSHKGYVVSR RGDLETLGYC MIRWLGGILP WTKISETKNC ALVSATKQKY 250
VNNTATLLMT SLQYAPRELL QYITMVNSLT YFEEPNYDEF RHILMQGVYY
```

Sites of interest: A106, replaced by V in strain Copenhagen

Physiological substrates and specificity determinants

S/T residues of ribosomal proteins Sa and S2 are phosphorylated *in vivo*[7-9].

Assay

See refs[4-6].

Pattern of expression

The gene is transcribed only early during infection (before DNA replication). The protein localizes to the cytoplasmic factories (sites of virus replication) within infected cells and is packaged into progeny virus particles.

References

1   Howard, S.T. and Smith, G.L. (1989) J. Gen. Virol. 70, 3187–3201.
2   Traktman, P. et al. (1989) J. Biol. Chem. 264, 21458–21461.
3   Rempel, R.E. et al. (1990) J. Virol. 64, 574–583.
4   Rempel, R.E. and Traktman, P. J. (1992) Virology 66, 4413–4426.
5   Banham, A.H. and Smith, G.L. (1992) Virology 191, 803–812.
6   Lin, S. et al. (1992) J. Virol. 66, 2717–2723.
7   Banham, A.H. et al. (1993) FEBS Lett. 321, 27–31.
8   Kaerlin, M. and Horak, I. (1976) Nature 259, 150–151.
9   Kaerlin, M. and Horak, I. (1978) Eur. J. Biochem. 90, 463–469.
10 Banham, A.H. and Smith, G.L. (1993) J. Gen. Virol. 74, 2807–2812.
11 Goebel et al. (1990) Virology 179, 247–266.

Geoffrey L. Smith (Oxford University, UK)

African swine fever virus (ASFV) strain Malawi (LIL20/1) gene *J9L* encodes a serine/threonine protein kinase which is packaged into virus particles[1].

Subunit structure and isoforms

SUBUNIT	AMINO ACIDS	MOL. WT	SDS-PAGE
J9L	299	35 100	

Genetics

The protein is encoded by gene *J9L* (the 9th gene in the virus *Sal*I j fragment, and is transcribed leftwards.

Domain structure

The protein contains little more than a canonical kinase domain.

Database accession numbers

	PIR	SWISSPROT	EMBL/GENBANK	REF
J9L	ASFV		X72954	[1]

Amino acid sequence of African swine fever virus J9L

```
MSRPEQQFKK VLKNPQAQYA VYPTIKVERI STTEHMYFIA TKPMFEGGRR  50
NNVFLGHQVG QPVVFKYVSK KEIPGNEVVV MKALQDTPGV IKLIEYTENA 100
MYHILIIEYI PNSIDLLHYH YFKKLEENEA KKIIFQMILI IQNIYEKGFI 150
HGDIKDENLI IDIDQKIIKV IDFGSAVRLN ETHPQYNMFG TWEYVCPEFY 200
YYGYYYQLPL TVWTIGMVAV NLFRFRAENF YLNDILKGEN YIPDNISETG 250
KQFITDCLTI NENKRLSFKG LVSHPWFKGL KKEIQPISEL GVDYKNVIT
```

Physiological substrates and specificity determinants

Phosphorylates calf thymus histones on serine residues *in vitro*[1].

Assay

Active under conditions used for the vaccinia virus B1R protein kinase[2].

References

[1] Baylis, S.A. et al. (1993) J. Virol. 67, 4549–4556.
[2] Banham, A.H. and Smith, G.L. (1992) Virology 191, 803–812.

Yoshiro Maru and Owen N. Witte (University of Tokyo, Japan, and University of California Los Angeles, USA)

BCR (breakpoint cluster region) was identified as the partner gene for *ABL* in the formation of the fused *BCR–ABL* oncogene found in Philadelphia chromosome-positive human leukaemias[1-3]. Bcr was subsequently shown to have intrinsic protein kinase activity (both auto- and trans-phosphorylation)[4], although it has no significant sequence similarity with members of the conventional PK family.

Subunit structure and isoforms

SUBUNIT	AMINO ACIDS	MOL. WT	SDS-PAGE
Bcr	1271	142 645	160 kDa

BCR cDNAs which lack amino acid residues 960–1003 have been reported. The protein overproduced via baculovirus vector forms a homotetramer[4].

Domain structure

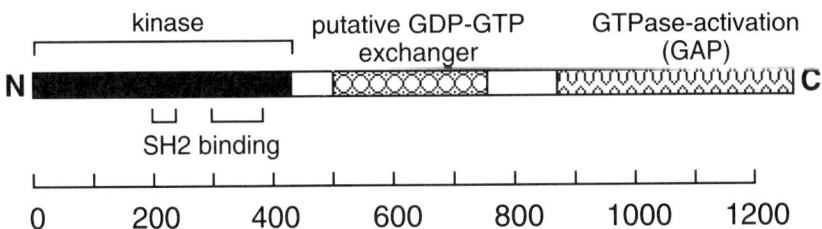

The kinase domain of Bcr has been defined by deletion analysis but is not related to those of other protein kinases[4]. Bcr also contains putative GDP–GTP exchanger and GAP domains[5,6]. The SH2 binding domain[7] overlaps with the kinase domain.

Database accession numbers

	PIR	SWISSPROT	EMBL/GENBANK	REF
Human			X02596	2,3

Amino acid sequence of human Bcr

```
MVDPVGFAEA WKAQFPDSEP PRMELRSVGD IEQELERCKA SIRRLEQEVN   50
QERFRMIYLQ TLLAKEKKSY DRQRWGFRRA AQAPDGASEP RASASRPQPA  100
PADGADPPPA EEPEARPDGE GSPGKARPGT ARRPGAAASG ERDDRGPPAS  150
VAALRSNFER IRKGHGQPGA DAEKPFYVNV EFHHERGLVK VNDKEVSDRI  200
SSLGSQAMQM ERKKSQHGAG SSVGDASRPP YRGRSSESSC GVDGDYEDAE  250
LNPRFLKDNL IDANGGSRPP WPPLEYQPYQ SIYVGGMMEG EGKGPLLRSQ  300
STSEQEKRLT WPRRSYSPRS FEDCGGGYTP DCSSNENLTS SEEDFSSGQS  350
```

Amino acid sequence of human Bcr *continued*

```
SRVSPSPTTY RMFRDKSRSP SQNSQQSFDS SSPPTPQCHK RHRHCPVVVS 400
EATIVGVRKT GQIWPNDGEG AFHGDADGSF GTPPGYGCAA DRAEEQRRHQ 450
DGLPYIDDSP SSSPHLSSKG RGSRDALVSG ALESTKASEL DLEKGLEMRK 500
WVLSGILASE ETYLSHLEAL LLPMKPLKAA ATTSQPVLTS QQIETIFFKV 550
PELYEIHKEF YDGLFPRVQQ WSHQQRVGDL FQKLASQLGV YRAFVDNYGV 600
AMEMAEKCCQ ANAQFAEISE NLRARSNKDA KDPTTKNSLE TLLYKPVDRV 650
TRSTLVLHDL LKHTPASHPD HPLLQDALRI SQNFLSSINE EITPRRQSMT 700
VKKGEHRQLL KDSFMVELVE GARKLRHVFL FTDLLLCTKL KKQSGGKTQQ 750
YDCKWYIPLT DLSFQMVDEL EAVPNIPLVP DEELNALKIK ISQIKSDIQR 800
EKRANKGSKA TERLKKKLSD QESLLLLMSP SMAFRVHSRN GKSYTFLISS 850
DYERAEWREN IREQQKSCFR SFSLASVELQ MLTNSCVKLQ TVHSIPLTIN 900
KEDDESPGLY GFLNVIVHSA TGFKQSSNLY CTLEVDSFGY FVNKAKTRVY 950
RDTAEPNWNE EFEIELEGSQ TLRILCYEKC YNKTKIPKED GESTDRLMGK 1000
GQVQLDPQAL QDRDWQRTVI AMNGIEVKLS VKFNSREFSL KRMPSRKQTG 1050
VFGVKIAVVT KRERSKVPYI VRQCVEEIER RGMEEVGIYR VSGVATDIQA 1100
LKAAFDVNNK DVSVMMSEMD VNAIAGTLKL YFRELPEPLF TDEFYPNFAE 1150
GIALSDPVAK ESCMLNLLLS LPEANLLTFL FLLDHLKRVA EKEAVNKMSL 1200
HNLATVFGPT LLRPSEKESK LPANPSQPIT MTDSWSLEVM SQVQVLLYFL 1250
QLEAIPAPDS KRQSILFSTE V
```

Sites of interest: C332, Cys essential for kinase activity; Y177, P-Tyr Grb2 binding site

Homologues in other species

Chicken Bcr has been cloned and found to have amino acid sequence identity of 80% in the kinase, 92% in the putative GDP–GTP exchanger, and 93% in the GAP regions (Y. Maru and O.N. Witte, unpublished).

Physiological substrates and specificity determinants

Physiological substrates are unknown, but phosphorylates histones and caseins *in vitro*.

Assay

See ref. 4. Mn^{2+} is preferred to Mg^{2+}.

Patterns of expression

BCR mRNA is ubiquitously expressed, but is most abundant in brain[8]. Immunohistochemical studies suggest a cytoplasmic localization[9].

References

1 Groffen, J. et al. (1984) Cell 36, 93–99.
2 Hariharan, I.K. et al. (1988) Oncogene 2, 113–117.
3 Lifshitz et al. (1988) Oncogene 2, 113–117.
4 Maru,Y. and Witte, O.N. (1991) Cell 67, 459–468.
5 Diekmann, D. et al. (1991) Nature 351, 400–402.

[6] Hart, M.J. et al. (1992) Science 258, 812–815.
[7] Pendergast, A.M. et al. (1991) Cell 66, 161–171.
[8] Collins, S. et al. (1987) Mol. Cell. Biol. 7, 2870–2876.
[9] Dhut, S. et al. (1988) Oncogene 3, 561–566.

BKDK | **Branched-chain α-ketoacid dehydrogenase kinase (vertebrates)**
(Branched-chain 2-oxo acid dehydrogenase kinase, [3 methyl-2-oxobutanoate dehydrogenase (lipoamide)] kinase)

Kirill M. Popov and Robert A. Harris (Indiana University, USA)

BKDK specifically phosphorylates and inactivates the E1 subunit (branched-chain α-ketoacid dehydrogenase) of the branched-chain α-ketoacid dehydrogenase complex[1]. This complex catalyses the committed step in the oxidative disposal of branched chain amino acids (leucine, isoleucine, valine).

Subunit structure and isoforms

SUBUNIT	AMINO ACIDS	MOL. WT	SDS-PAGE
BKDK	382	43 280	44 kDa

Domain structure

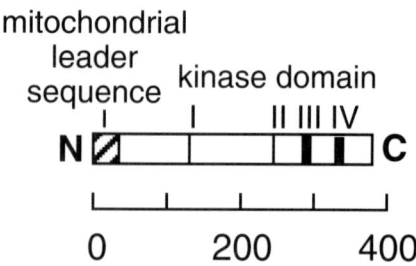

Although BKDK is a serine PK according to the sites phosphorylated (S293 and S303 in the E1α subunit[2]), it lacks significant sequence similarity with known eukaryotic Ser/Thr PKs[3]. A greater degree of similarity exists with members of the prokaryotic histidine PK family. Four subdomains (I, II, III, IV) of these kinases are highly conserved in all members of this family[4]. Considerable sequence similarity exists in BKDK within all four of these subdomains.

Database accession numbers

	PIR	SWISSPROT	EMBL/GENBANK	REF
Rat		Q00972	M93271	3

Amino acid sequence of rat BKDK

```
MILTSVLGSG PRSGSSLWPL LGSSLSLRVR STSATDTHHV ELARERSKTV  50
TSFYNQSAID VVAEKPSVRL TPTMMLYSGR SQDGSHLLKS GRYLQQELPV 100
RIAHRIKGFV VFLSSLVATL PYCTVHELYI RAFQKLTDFP PIKDQADEAQ 150
YCQLVRQLLD DHKDVVTLLA EGLRESRKHI EDEKLVRYFL DKTLTSRLGI 200
RMLATHHLAL HEDKPDFVGI ISTRLSPKKI IEKWVDFARR LCEHKYGNAP 250
RVRINGHVAA RFPFIPMPLD YILPELLKNA MRATMESHLD TPYNVPDVVI 300
TIANNDVDLI IRISDRGGGI AHKDLDRVMD YHFTTAEAST QDPRISPLFG 350
HLDMHSGGQS GPMHGFGFGL PTSRAYAEYL GGSLQLQSLQ GIGTDVYLRL 400
RHIDGREESF RI
```

Sites of interest: M1–R30, mitochondrial leader sequence; H162, N279, D315–G319, G365–G369, conserved regions I–IV

Physiological substrates and specificity determinants

BKDK will not phosphorylate the closely related E1α subunit of the pyruvate dehydrogenase complex[5].

Assay

The enzyme has been assayed by the rate of ATP-dependent inactivation of the branched-chain α-ketoacid dehydrogenase complex, and by the rate of incorporation of ^{32}P from $[\gamma\text{-}^{32}P]$ATP into its E1α subunit[6, 7].

Enzyme activators and inhibitors

INHIBITORS	SPECIFICITY	IC_{50}	CELL PERMEABILITY
α-Chloroiso-caproate	Excellent	7.5 μM	Excellent
Thiamine pyro-phosphate	Good	6 μM	None
α-Ketoiso-caproate	Excellent	70 μM	Excellent

Activator: Dihydrolipoyl transacylase (E2), the core of the branched-chain α-ketoacid dehydrogenase complex that binds both E1 and the kinase, stimulates branched-chain α-ketoacid dehydrogenase kinase activity several fold.

Pattern of expression

Branched-chain α-ketoacid dehydrogenase kinase is located in the mitochondrial matrix space in tight association with the branched-chain α-ketoacid dehydrogenase complex. It is expressed in all mammalian cells that contain mitochondria. Activity of the kinase is increased in the liver of rats fed diets deficient in protein[8].

References

1 Harris, R.A. et al. (1992) Adv. Enzyme Regul. 32, 267–284.
2 Cook, K.G. et al. (1983) FEBS Lett. 164, 47–50.
3 Popov, K.M. et al. (1992) J. Biol. Chem. 267, 13127–13130.
4 Stock, J.B. et al. (1989) Microbiol. Rev. 53, 450–490.
5 Popov, K.M. et al. (1991) Prot. Express. Purif. 2, 278–286.
6 Espinal, J. et al. (1988) Methods Enzymol. 166, 166–175.
7 Paxton, R. (1988) Methods Enzymol. 166, 313–320.
8 Espinal, J. et al. (1986) Biochem. J. 237, 285–288.

Stephen J. Yeaman (University of Newcastle upon Tyne, UK)

The pyruvate dehydrogenase multienzyme complex occupies a key position in intermediary metabolism, catalysing the oxidative decarboxylation of pyruvate to acetyl CoA. The activity of the complex is regulated by phosphorylation of the E1α component by PDK, causing complete inactivation of the complex[1]. The activity of PDK is under both acute and long-term control.

Subunit structure and isoforms

SUBUNIT	AMINO ACIDS	MOL. WT	SDS-PAGE
α	434	46270	48 kDa
β	?	?	45 kDa

The kinase from bovine kidney has an αβ structure. Limited proteolysis studies[2] and expression in *E. coli*[3] indicate that α is the catalytic subunit. The β subunit may exert a regulatory role[2]. There are no known isoforms.

Domain structure

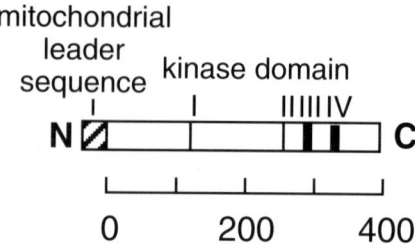

Although PDK phosphorylates serine residues, like **BKDK** it is more closely related to the histidine PK family than to the serine/threonine PK family. Four motifs (I, II, III, IV) are highly conserved in all members of this family.

Database accession numbers

		PIR	SWISSPROT	EMBL/GENBANK	REF
α	Rat			L22294	3

Amino acid sequence of rat PDK

(residues are numbered from the N-terminus of the mature protein)

```
                        MRLARL LRGGTSVRPL CAVPCASRSH    0
ASDSASGSGP ASESGVPGQV DFYARFSPSP LSMKQFLDFG SVNACEKTSF   50
MFLRQELPVR LANIMKEISL LPDNLLRTPS VQLVQSWYIQ SLQELLDFKD  100
KSAEDAKTIY EFTDTVIRIR NRHNDVIPTM AQGVTEYKES FGVDPVTSQN  150
VQYFLDRFYM SRISIRMLLN QHSLLFGGKG SPSHRKHIGS INPNCDVVEV  200
IKDGYENARR LCDLYYVNSP ELELEELNAK SPGQPIQVVY VPSHLYHMVF  250
ELFKNAMRAT MEHHADKGVY PPIQVHVTLG EEDLTVKMSD RGGGVPLRKI  300
DRLFNYMYST APRPRVETSR AVPLAGFGYG LPISRLYAQY FQGDLKLYSL  350
EGYGTDAVIY IKALSTESIE RLPVYNKAAW KHYRTNHEAD DWCVPSREPK  400
DMTTFRSS
```

Sites of interest: M(-26)–H(-1), mitochondrial leader sequence; H123, N255, D290–G294, G326–G330, conserved regions I–IV

Homologues in other species

The sequence of rat PDK shows significant similarity with two predicted protein products in the databases[3], i.e. ZK 370.5 from *C. elegans* and hypothetical phosphoprotein 3 from *Trypanosoma brucei*. It is also related to the other known mitochondrial PK, **BKDK**.

Physiological substrates and specificity determinants

The only known physiological substrate is the α subunit of E1 of the pyruvate dehydrogenase complex (PDH). No significant kinase activity has been detected against any other protein substrate. Three phosphorylation sites have been identified on PDH. The kinase shows a marked preference for site 1 with slower phosphorylation of sites 2 and 3. Sites 1 and 2 are within seven residues of each other on the E1α polypeptide. Phosphorylation of site 1 correlates with inactivation of the complex[3].

PDH site 1:   YHGHSMSDP
PDH site 2:   DPGVSYRTR
PDH site 3:   GMGTSVERA

Assay

PDH kinase is usually assayed by the transfer of radioactivity from $[\gamma\text{-}^{32}\text{P}]\text{ATP}$ into substrate (E1α polypeptide, or synthetic peptide corresponding to site 1)[4]. Alternatively, endogenous kinase activity in the PDH complex can be estimated from the first order rate constant of the ATP-dependent loss of complex activity[5].

Enzyme activators and inhibitors

Activity is modulated by a large number of metabolites and other compounds[6]:

EFFECTOR (μM)	ACTIVITY (% CONTROL)
NADH (250)	128
Acetyl CoA (250)	165
NADH, Acetyl CoA (250)	241
NAD (250)	87
Pyruvate (10)	226
Pyruvate (1000)	64
ADP (500)	22
Dichloroacetate (400)	60
Thiamine thiazolone pyrophosphate (40)	20

Pattern of expression

PDK is apparently located exclusively within the mitochondrial matrix, either tightly associated with the PDH complex or free in solution[7]. The PDH complex is loosely associated with the inner face of the inner mitochondrial membrane. Although PDK is a mitochondrial enzyme, both subunits are nuclear encoded. PDK appears to be expressed in the mitochondria of all mammalian cells. In rat heart its activity is increased in starvation or alloxan-induced diabetes[8].

References

[1] Reed, L.J. and Yeaman, S.J. (1987) The Enzymes XVIII, 77–95.
[2] Stepp, L.R. et.al. (1983) J. Biol. Chem. 258, 9454–9458.
[3] Popov, K.M. et al. (1993) J. Biol. Chem. 268, 26602–26606.
[4] Yeaman, S.J. et al. (1978) Biochemistry 17, 2364–2370.
[5] Denyer, G.S. et al. (1986) Biochem. J. 239, 347–354.
[6] Rahmatullah, M. and Roche, T.E. (1987) J. Biol. Chem. 262, 10265–10271.
[7] Jones, B.S. and Yeaman, S.J. (1991) Biochem. J. 275, 781–784.
[8] Kerbey, A.L. and Randle, P.J. (1981) FEBS Lett. 127, 188–192.

Index